*National Academy Press*

The National Academy Press was created by the National Academy of
Sciences to publish the reports issued by the Academy and by the
National Academy of Engineering, the Institute of Medicine, and the
National Research Council, all operating under the charter granted to
the National Academy of Sciences by the Congress of the United States.

STUDIES IN GEOPHYSICS

Explosive Volcanism: Inception, Evolution, and Hazards

Geophysics Study Committee
Geophysics Research Forum
Commission on Physical Sciences,
    Mathematics, and Resources
National Research Council

NATIONAL ACADEMY PRESS
Washington, D.C.   1984

NATIONAL ACADEMY PRESS   2101 Constitution Avenue, NW   Washington, DC 20418

NOTICE: The project that is the subject of this report was approved by the Governing Board of the National Research Council, whose members are drawn from the Councils of the National Academy of Sciences, the National Academy of Engineering, and the Institute of Medicine. The members of the committee responsible for this report were chosen for their special competences and with regard for appropriate balance.

This report has been reviewed by a group other than the authors according to procedures approved by a Report Review Committee consisting of members of the National Academy of Sciences, the National Academy of Engineering, and the Institute of Medicine.

The National Research Council was established by the National Academy of Sciences in 1916 to associate the broad community of science and technology with the Academy's purposes of furthering knowledge and of advising the federal government. The Council operates in accordance with general policies determined by the Academy under the authority of its congressional charter of 1863, which establishes the Academy as a private, nonprofit, self-governing membership corporation. The Council has become the principal operating agency of both the National Academy of Sciences and the National Academy of Engineering in the conduct of their services to the government, the public, and the scientific and engineering communities. It is administered jointly by both Academies and the Institute of Medicine. The National Academy of Engineering and the Institute of Medicine were established in 1964 and 1970, respectively, under the charter of the National Academy of Sciences.

The Geophysics Study Committee is pleased to acknowledge the support of the National Science Foundation, the Defense Advanced Research Projects Agency, the National Aeronautics and Space Administration, the National Oceanic and Atmospheric Administration, the U.S. Geological Survey, and the U.S. Department of Energy (Grant # DE-FGO1-82ER12018) for the conduct of this study.

Library of Congress Cataloging in Publication Data

Main entry under title:

Explosive volcanism.

(Studies in geophysics)
  1. Volcanism—Addresses, essays, lectures.
I. National Research Council (U.S.). Geophysics Study
Committee. II. Series.
QE522.E97 1984   551.2′1   83-23610
ISBN 0-309-03393-4

Printed in the United States of America

Panel on Explosive Volcanism

FRANCIS R. BOYD, JR., Carnegie Institution of Washington, *Chairman*
ARTHUR L. BOETTCHER, University of California, Los Angeles
LAWRENCE W. BRAILE, Purdue University
RICHARD W. CARLSON, Carnegie Institution of Washington
ROBERT L. CHRISTIANSEN, U.S. Geological Survey, Menlo Park
ROBERT W. DECKER, U.S. Geological Survey, Hawaii National Park
RICHARD S. FISKE, Smithsonian Institution
SUSAN WERNER KIEFFER, U.S. Geological Survey, Flagstaff
ROBERT G. LUEDKE, U.S. Geological Survey, Reston
BRUCE D. MARSH, The Johns Hopkins University
ROBERT G. McQUEEN, Los Alamos National Laboratory
JAMES G. MOORE, U.S. Geological Survey, Menlo Park
CARL J. RICE, The Aerospace Corporation
I. SELWYN SACKS, Carnegie Institution of Washington
LEE SIEBERT, Smithsonian Institution
TOM SIMKIN, Smithsonian Institution
ROBERT B. SMITH, University of Utah
ROBERT L. SMITH, U.S. Geological Survey, Reston
KENNETH H. WOHLETZ, Los Alamos National Laboratory

iii

Geophysics
Study Committee*

iv

Geophysics
Research Forum

DON L. ANDERSON, California Institute of Technology, *Chairman*
CHARLES R. BENTLEY, University of Wisconsin
JAMES H. COULSON, Tennessee Valley Authority
WILLIAM R. DICKINSON, University of Arizona
THOMAS DONAHUE, University of Michigan
JOHN V. EVANS, Communications Satellite Corporation
HOWARD R. GOULD, Exxon Production Research Company
DEVRIE S. INTRILGATOR, Carmel Research Center
KEITH A. KVENVOLDEN, U.S. Geological Survey, Menlo Park
THOMAS F. MALONE, West Hartford, Connecticut
ARTHUR E. MAXWELL, University of Texas at Austin
JOHN C. MAXWELL, University of Texas at Austin
PAUL W. POMEROY, Rondout Associates, Inc.
HUGH ODISHAW, University of Arizona
ALAN H. SHAPLEY, National Oceanic and Atmospheric Administration
JOHN SLAUGHTER, University of Maryland
FRANCIS G. STEHLI, University of Oklahoma
MURRAY STRASBERG, U.S. Navy
VERNER E. SUOMI, University of Wisconsin, Madison
EINAR A. TANDBERG-HANSSEN, National Aeronautics and Space Administration
BYRON D. TAPLEY, University of Texas at Austin
CHARLES A. WHITTEN, Silver Spring, Maryland

Ex Officio

LOUIS J. BATTAN, University of Arizona
OWEN GINGERICH, Smithsonian/Harvard Center for Astrophysics
ROBERT HOFSTADTER, Stanford University
THOMAS A. SENIOR, University of Michigan

Staff

PEMBROKE J. HART

v

Commission on Physical Sciences, Mathematics, and Resources

vi

Studies in Geophysics*

*Published to date.

Preface

In 1974 the Geophysics Research Board completed a plan, subsequently approved by the Committee on Science and Public Policy of the National Academy of Sciences, for a series of studies to be carried out on various subjects related to geophysics. The Geophysics Study Committee was established to provide guidance in the conduct of the studies.

One purpose of the studies is to provide assessments from the scientific community to aid policymakers in decisions on societal problems that involve geophysics. An important part of such an assessment is an evaluation of the adequacy of present geophysical knowledge and the appropriateness of present research programs to provide information required for those decisions. Some of the studies place more emphasis on assessing the present status of a field of geophysics and identifying the most promising directions for future research.

This study on explosive volcanism was begun soon after the cataclysmic eruptions of Mount St. Helens. It readily became apparent to the committee that an assessment of the explosive nature of volcanoes must cover all types of volcanic activity; any volcano can be explosive. Consideration of explosive volcanism must start with the generation of magma. Magma-forming processes are intimately connected with tectonics, and there are pronounced differences in the nature of volcanism between regions of compression and subduction and regions of rifting or broader extension. The cyclic and episodic aspects of volcanic activity form a basis for predicting eruptions and can be sources of information on the rates at which magma and energy are introduced to volcanic systems. Improved understanding of the physics of volcanic eruptions is an exciting goal that is vital to progress in hazard evaluation. Finally, the study of explosive volcanism must include an appreciation of the severe social problems that are caused by erupting volcanoes. None is of greater urgency than planning for a crisis. This report considers the progress in research on these aspects of explosive volcanism and the need for additional research efforts.

The study was developed through meetings of the Geophysics Study Committee and the Panel on Explosive Volcanism. The preliminary scientific findings of the panel were presented at an American Geophysical Union meeting that took place in San Francisco in December 1981. These presentations and the essays contained in this volume provide examples of current basic knowledge of explosive volcanism. They also pose many of the fundamental questions and uncertainties that require additional research. In completing their papers, the authors had the benefit of discussion at this symposium as well as comments of several scientific referees. Responsibility for the individual essays rests with the corresponding authors.

The Overview of the study summarizes the highlights of the essays and formulates conclusions and recommendations. In preparing it, the panel chairman and the committee had the benefit of meetings that took place at the symposium, the comments of the panel of authors, and selected referees. Responsibility for the Overview rests with the Geophysics Study Committee and the chairman of the panel.

Contents

Contents

Explosive Volcanism: Inception, Evolution, and Hazards

Overview
and
Recommendations

A burst of earthquake tremors with associated ground deformation and fumarolic activity at Mammoth Lakes in the Long Valley caldera, California, prompted the U.S. Geological Survey (USGS) to issue a notice of a potential volcanic hazard on May 25, 1982. Mammoth Lakes is a major outdoor recreation area that at times accommodates 30,000 or more vacationers. The focus of the earthquake activity is about 3 km from the center of population and at the time of the alert was about 3 km below the surface. The USGS notice (1982) states that "neither the probability nor the nature, scale, or timing of a possible eruption can be determined as yet." Field studies in the Long Valley area suggest that such an eruption might take a variety of forms but would most likely include explosions of groundwater superheated by contact with magma, ash falls, and minor pyroclastic flows, terminating with the extrusion of a rhyolite dome. More harrowing is the possibility of eruption of a major pyroclastic flow of the magnitude of the Bishop Tuff, erupted from Long Valley 700,000 years ago. Such an eruption could devastate an area of hundreds of square miles, cause great loss of life, and possibly interrupt the water supply from the Owens Valley aqueduct to Los Angeles.

This situation is analogous to the scene at Mount St. Helens immediately following the onset of seismic activity on March 20, 1980. That the eruption of Mount St. Helens could be forecast in considerable detail is a remarkable achievement. That the prediction did not include more than the *possibility* of a lateral blast at Mount St. Helens, however, and the fact that considerable uncertainty surrounds an expectation of events at Mammoth Lakes, is a symptom of our still imperfect understanding of volcanic processes.

The eruption at Mount St. Helens has profoundly heightened public awareness of volcanism. Such awareness has long existed among the residents of Hawaii and Alaska. Today those who live in the conterminous United States no longer think of volcanic eruptions as being remote in time or space. Curiosity about volcanic events and a desire to minimize loss of life in future eruptions have stimulated research. The intellectual and social climate for fostering advances in volcanology is excellent. Some avenues for research that appear especially promising are discussed in this volume.

Although explosive eruptions are usually associated with andesitic volcanoes, they can occur in any volcanic system. The energy for explosive eruptions is commonly supplied by expansion of volatiles exsolved from magma; the eruption of magmas that are relatively poor in volatiles is normally not accompanied by violent explosions. Nevertheless, interactions between magmas and groundwater, lakewater, or seawater also can produce explosive eruptions. About 1 percent of the eruptions at Kilauea Volcano in Hawaii have been explosive, with attendant loss of life, although this volcano is noted for relatively gentle, effusive outpouring of lava (see Chapter 9). The problem of explosive volcanism is thus neither separate nor distinct from the more general problem of volcanism, its causes, manifestations, and effects on society.

A discussion of the process of volcanism must begin with the generation of magma. Magma-forming processes are intimately connected with tectonics since pronounced differences exist in the nature of volcanism between regions of compression and subduction and regions of rifting or broader extension. The causes of these differences are incompletely understood. The cyclic or episodic nature of volcanic activity forms a basis for predicting eruptions and can be a source of information on the rates at which magma and energy are introduced to volcanic systems. Improved understanding of the physics of volcanic eruptions is an exciting goal and is vital to progress in hazard evaluation. Development of the physics of planetary volcanism with reference to the Moon, Mars, Venus, and especially Jupiter's moon, Io, has added a new dimension to research in this field.

The study of explosive volcanism also must include an appreciation of the severe social problems that are caused by erupting volcanoes. None is of greater urgency than planning for a crisis. Progress in research on these aspects of explosive volcanism is sketched in this Overview and discussed in detail in the chapters that follow.

TECTONISM AND VOLCANISM

A better understanding of the interrelationships of tectonism and volcanism is fundamental to progress in understanding the chemical aspects of magmatic processes. These interrelations are especially complex and varied in western North America. During the Tertiary a compressional subduction system was partly replaced by extensional tectonism. Broad changes in the compositions of erupted magmas from those that include andesites to a combination of basalt with associated large rhyolitic magma bodies occurred, with change in the style of extension. Remnants of the compressional system include a subducted plate, perhaps containing a gigantic hole, that underlies the Basin and Range and an active subduction system in the Pacific Northwest that underlies the Cascades. The portion of the Juan de Fuca plate beneath the Cascades has recently been shown to be a source of earthquakes whose foci delineate a subplanar zone that dips about 60° between the depths of 40 and 70 km (Crosson, 1983).

Studies of the Andean subduction system have shown major variations in the subduction angle both down dip and along strike (see Chapter 3). Beneath central Peru the initial dip of the Nasca plate is a normal 30°, but at a depth of about 100 km it flattens for 300 km beneath the continent before subducting again at a steeper angle. There is no volcanism in the area where there is no asthenosphere between the subducted plate and the overlying, rigid continental lithosphere. Equivalent complexities may well be present beneath western North America.

Tectonism influences magma generation through convective flow in the mantle and may influence the eruption of magma through fracturing in the lithosphere and crust. Magma erupts from vents that form a well-defined arc in the subduction system of the Cascades. In contrast, volcanic loci form a complex network of rectilinear zones in the Basin and Range, where extensional tectonics prevail (see Chapter 4). Many of these loci have migrated with time over the past 15 million years (m.y.), perhaps in response to differential movements of the crust and mantle. The migrations are to the east and northeast in the area that lies east of central Nevada and Idaho and to the west in California and western Oregon.

The tectonic setting of a magma's ascent from a mantle source to the surface influences whether it rises directly or is stored temporarily at intermediate depth. The nature of magmatic movements will in turn control the opportunities for chemical interaction between melts and rocks forming conduit walls and the possible development of secondary, more silicic melts. Some alkali basalts contain abundant inclusions of mantle rocks, indicating rapid transport from mantle sources with little opportunity for interaction with wall rocks. The mantle sources for alkali basalts may, however, have been metasomatically altered during or prior to partial fusion (see Chapter 1). Some basalts, such as those in the Columbia Plateau, have chemical and isotopic compositions that are commensurate with assimilation of crustal rocks (see Chapter 2). Major eruptions of rhyolite in areas such as Yellowstone are believed to be due to melting of deep crustal rocks by basaltic magma, and in those circumstances the basalts may have acted more as thermal than chemical parents.

Formation of calderas after major eruptions of rhyolite is clear evidence that the magma chambers were at a relatively shallow depth. Seismic data can be interpreted to show that magma is present beneath some parts of Yellowstone and that the top of a low-velocity body, 2 to 5 km deep, may represent a zone of partial melt or a large vapor-dominated hydrothermal system (see Chapter 7). Yellowstone rhyolites are more likely to have been formed by partial fusion of high-grade metamorphic rocks of the lower crust than of once-sedimentary, upper crustal rocks (see Chapter 6). The formation of rhyolite in large volume probably involves a combination of deep and shallow magma chambers. The absence of well-defined collapse features associated with flood basalt eruptions suggests that storage of these lavas, possibly accompanied by assimilation of wall rocks, was deeper than is characteristic for near-surface rhyolite chambers.

Eruptive configurations that maximize opportunities for chemical interaction between magma and wall rocks or that maximize opportunity for generation of anatectic magmas also maximize heat loss. Such configurations might include blocks broken from conduit walls and interlacing networks of dikes. Freezing or block choking in such systems could be inhibited by episodic injection of fresh magma. Basaltic magma may produce up to an equal volume of rhyolite melt in the course of crystallization of the basalt (see Chapter 5). But how do basaltic magmas contaminated by interaction with crustal rocks (e.g., Columbia Plateau basalts) erupt as phenocryst-free lava? Active seismic studies in the Yellowstone region outline another puzzle involving magma ascent. The lower crust beneath Yellowstone appears to be homogeneous and similar to thermally undisturbed, surrounding crust (see Chapter 7), whereas one might expect a complex eruptive plumbing system at high temperature. Yellowstone is the only major rhyolite caldera system in North America that has been examined in detail by active seismic methods.

VOLCANIC PERIODICITY

The periodic nature of some volcanic activity is an important characteristic for both theoretical studies and hazard assessments. The interval since the last eruption clearly is important in estimating the imminence of hazardous activity for volcanoes with a characteristic period. Moreover, data on periodicity will reveal information on the rates of energy and mass input to volcanic systems.

Some large volcanic systems, particularly those with rhyolitic magma chambers, undergo episodes of periodic, cyclic eruptivity separated by intervals of quiescence. The cyclic nature of the activity can involve both progressive changes in magmatic composition and variations in eruptive mode. Eruption of the Yellowstone rhyolites, for example, began 2 m.y. ago and has occurred in three episodes, each with a similar eruptive cycle of ash flows and caldera collapse.

The Smithsonian Institution maintains historical records of eruptive activity in a computer data bank (see Chapter 8). The sequence and age of prehistoric eruptions can be determined by stratigraphic studies coupled with age determinations using such methods as potassium-argon, carbon 14, and secular magnetic variations. Records for historic eruptions in compressional arcs (such as the Cascades) comprise 80 percent of the data for 530 volcanoes presently known to be active and more than 95 percent of the over 5000 eruptions tabulated. These dominantly andesitic volcanoes tend to erupt explosively, and their common location in belts along the borders of the Pacific Ocean put them near many population centers, making them a special danger to people. The average time interval between eruptions for those volcanoes that have had multiple eruptions in historic times is about 5 yr. The most catastrophic eruptions generally occur after the longest quiescent intervals. Data also show that there is no established pattern of timing of the paroxysmal phase of an explosive eruption. The paroxysmal phase can occur at the beginning of an eruptive cycle or at any time up to several years after the beginning—posing a serious problem in forecasting dangerous eruptions.

Some volcanologists have speculated that the rise of magma from subduction zones beneath arc volcanoes takes place in diapirs. The explosive nature of these andesitic volcanoes, in combination with the characteristic, relatively short eruptive period, suggests the presence of magma chambers at intermediate depth. The form of such chambers is not known; their detection by geophysical techniques has to date been equivocal.

The rate of eruption at volcanic loci in extensional tectonic regimes may be influenced by stresses that can provide avenues for magma rise and by convective movements in the mantle that supply both the heat and at least part of the melt. Cooling with partial crystallization and buildup of volatile pressure also must be important, especially for silicic volcanic systems with near-surface magma chambers.

ERUPTIVE MECHANICS

The mechanisms of volcanic eruptions can become better understood through theoretical and experimental studies as well as by more sophisticated field observations. New insights can be gained by consideration of data for volcanic eruptions on other planets and satellites, particularly Jupiter's moon, Io. Problems that are important from the viewpoints of both scientific progress and hazard evaluation include the interrelationships of physical factors that determine the explosive character of an eruption and the identification of sources of volatiles involved in explosive eruptions. The variety of sizes and shapes of volcanic jets associated with violent eruptions reflects differences in the geometric shapes of near-surface parts of the volcanic systems and, secondarily, the differences in thermodynamic properties and rheology of the erupting fluids.

The most substantial accelerations in a volcanic eruption occur when the volatiles, either juvenile or meteoric, have become a vapor or vapor-particulate mixture. A thermodynamic equation of state of the erupting fluid can be modeled by a pseudo-gas equation in which the vapor-particulate mixture is approximated by a gas of heavy molecular weight and low isentropic exponent (see Chapter 11). In volcanic systems the erupting fluid will accelerate to sonic conditions in narrow parts of a conduit, exiting with pressures that are high enough to erode surface craters. Pressure-balanced jets result if the pressure approximately equals ambient when the jet emerges at the surface. Overpressured or underexpanded jets occur if the jet pressure greatly exceeds atmospheric. Plinian eruption columns are typical terrestrial balanced jets because they commonly emerge through craters; on Io, jet 3 (Prometheus) may be an equivalent. The lateral blast of May 18, 1980, at Mount St. Helens, however, was an overpressured jet because it emerged directly at high pressure from the face of the mountain; on Io, jet 2 (Loki), which apparently emerges through a fissure, may be an equivalent. The accurate prediction of hazards from volcanic jets will require a detailed understanding of the pressures within jets, in terms of volcanic system geometry and fluid thermodynamic properties.

In some eruptions the volatiles are juvenile and exsolve as pressure on the magma is reduced. In terrestrial volcanic systems, however, meteoric water can be combined with lava in eruptive events either below the surface or subsequent to eruption. Experiments to model subsurface interactions of meteoric water and magma have been made with thermite and water encased in steel vessels (see Chapter 12). The thermite forms a basalt-like melt and its interaction with water can be controlled by varying the contact geometry of the container. Eruption modes ranging from strombolian and surtseyan to passive chilling with pillow formation have been produced. The explosivity has been found to be a function of both the water-to-melt mass ratio and the confining pressure.

The May 18, 1980, eruption of Mount St. Helens was recorded by infrared sensors on military satellites. Study of these records has shown that the eruption started with two major explosions about 2 min apart. The first explosion, near the summit, occurred when a giant avalanche relieved confining pressure on the intruding magma. The second explosion occurred 8 km to the north near the Toutle River and may have been caused by interaction of hot dacite with surface water in the river valley and possibly Spirit Lake (see Chapter 10).

Mud flows and floods are a special hazard for areas surrounding major volcanoes, where unstable terranes are produced by a combination of steep slopes, immense quantities of water (including snow and ice), and large masses of unconsolidated tephra and hydrothermally altered rock. These hazards persist long after the eruption. Substantial success in evaluating flood hazards for the environs of Mount St. Helens has been achieved by computer modeling of two danger zones (Jennings *et al.*, 1981). The interface between volcanology and hydrology offers important research opportunities, and vigorous pursuit of these opportunities will help alleviate risks to life and property.

EMERGENCY PLANNING

Most of the potentially dangerous volcanic loci in the western United States have been identified and can be monitored. Some volcanic systems other than Mount St. Helens, possibly including loci outside the Cascades, may become active in the next few decades. The youngest volcano in the conterminous United States outside of the Cascade Mountains is an alkali basalt cinder cone and associated flow that erupted only 390 yr ago in the eastern Mohave Desert, 180 km from Los Angeles (see Chapter 1). Eruptions from such vents are more likely to be moderately explosive or nonexplosive than catastrophic, but there is still a significant risk that a major tuff-flow eruption, like the Bishop tuff, might occur or that an ash fall might blanket crops and affect population centers in a number of states.

Clearly we should plan for such emergencies. Recent experiences with volcanic crises in the Caribbean (see Chapter 13) and at Mount St. Helens (Miller *et al.*, 1981) have provided insights into problems that can arise.

A common problem in all such crises is the effective flow of information from informed scientists to the public. This difficulty is compounded by a number of factors, including the following:

• Conflicting information about risks or conflicting predictions about future events undermines confidence in crisis leadership and promotes either panic or apathy. These conflicts can arise because individuals or groups of scientists disagree or appear to disagree about the probable future course of events. Differences of opinion are inevitable. However, in an emergency interested scientists who lack up-to-date information should refrain from making comments to the press.

• Confusion results from multiple official voices. Problems have developed during volcanic crises through a variety of governmental organizations issuing press releases. Even if all their releases are virtually the same, differences in emphasis and wording can lead to confusion and distrust.

• Conflicting reports in crises arise because of distortions or misinterpretations by the press. These also must be regarded as inevitable, but their damage can be minimized if the public and the press develop confidence in a scientific team that has the official responsibility for monitoring the eruption.

RECOMMENDATIONS

1. Physical and chemical processes interact in the generation and eruption of magma. The diversity and extent of these interactions, however, is poorly understood. In western North America there is a broad correlation of magma composition with tectonism in that andesites are primarily erupted in zones of tectonic compression, whereas eruption of basalt, sometimes accompanied by rhyolite, is characteristic of broad areas under tectonic extension. Why these correlations exist, however, is a matter of speculation.

Basalt is now generally believed to be the thermal parent of major extrusions of rhyolite. In what degree basalt is also the chemical parent of these rhyolites is a matter that has been debated for decades and remains an unsolved problem. The mechanics of formation of large silicic magma chambers with basaltic roots extending into the mantle are not known. Partial fusion and assimilation of crustal rocks are widely believed to be involved in the generation of these silicic lavas, but in such cases what is the physical nature of the interface between the basaltic magma and the crustal rocks? How is the geometric form of a major magmatic system influenced by tectonic forces? Partial fusion of wall rocks would be unlikely in the absence of volatiles, but what is the source of the volatiles and how do they migrate during formation of large magma chambers in the crust? These are important petrologic problems, all of which have physical as well as chemical aspects.

Research on petrologic processes, especially those that involve reactions between magma and crustal or mantle rocks, should be undertaken on the physical as well as the chemical aspects of the processes. We suggest that progress in understanding such processes may be enhanced if geochemists and geophysicists collaborate in a consideration of the physical as well as the chemical nature of magmatic evolution.

2. Precursory measurements of both a geophysical and a geologic nature are essential for reliable assessment of volcanic hazards. An inadequate base of such measurements for the Soufrière Volcano on Guadeloupe in 1976 resulted in great difficulty both for scientists who were called on to forecast the course of that eruption and for a government concerned with the continuing evacuation of 70,000 people (see Chapter 13). The catastrophic eruption of Mount St. Helens on May 18, 1980, was preceded by a 2-month period during which seismicity was high and minor explosions were intermittent. This period provided an opportunity for installation of geophysical instruments and development of a rudimentary measurement base. Of equal importance in assessing the Mount St. Helens hazard was the detailed field study of the nature and sequence of prehistoric eruptions (Crandall and Mullineaux, 1978). These authors forecasted eruptive events at Mount St. Helens with remarkable accuracy.

A sustained effort should be made to obtain a precursory geophysical and geologic data base for potentially dangerous volcanic loci. Without such a data base trustworthy forecasts are not possible.

Such data should include identification of the nature and extent of prehistoric flows and tephra deposits, construction of hazard maps, hydrologic characterization, and geodetic monitoring (trilateration as well as leveling observations). Passive seismic monitoring should be maintained in volcanic areas where the presence of magma is suspected. The data base at Kilauea Volcano has allowed fairly accurate short-term (days) predictions to be made; a similar capability for forecasting events at Mount St. Helens is now being obtained.

Active seismic studies, such as those carried out in Yellowstone (see Chapter 7), are important for theoretical as well as hazard assessment. Such studies may be of great value in helping us understand the physical form of magma chambers. Active seismic studies together with electrical and other geophysical measurements (e.g., heat flow, magnetics, groundwater behavior) may provide information on subsurface magma accumulation and should be actively pursued.

3. Remote monitoring data obtained by branches of the U.S. Department of Defense (DOD) have been made available in a few instances to volcanologists for application to problems of general geophysical interest. An exciting example of such collaboration was the use of infrared data of satellite origin to determine the initial sequence of events during the cataclysmic eruptions of Mount St. Helens on May 18, 1980 (see Chapter 10). Another example was the use of underwater sound data to detect and locate a number of active submarine volcanoes in the Pacific Ocean (Norris and Johnson, 1969).

A related circumstance in which use of remote monitoring might potentially be useful is in the occasional occurrence of pumice rafts in the Pacific Ocean. These are produced in submarine eruptions, but in some cases the locations of the sources are difficult to identify because the pumice rafts drift intact for vast distances. Triangulation with data from hydrophone arrays might, however, result in identification of these vents.

Another kind of data that would be useful to volcanologists is high-resolution aerial photography. Volcanoes exist for which no published photographs are available because of their remote location and frequent cloud cover; Reventador Volcano in Ecuador is an example.

Data obtained in the course of routine remote monitoring for the DOD may be retained for only short periods of time. Moreover, reduction of such data for volcanological purposes can be a time-consuming process for those few persons who have access to the raw data. For these reasons applications to theoretical problems will be most easily brought about through collaboration of research workers with interested scientists engaged in DOD operations.

Volcanologists in research positions should make every practical effort to build intellectual bridges with colleagues whose related work is within the DOD establishment. We urge the Department of Defense to develop policies wherein maximum use, including volcanological use, is made of data acquired during routine geophysical monitoring.

4. The USGS is the only organization in the United States with the available manpower, support, and expertise to assume the task of monitoring dangerous volcanic eruptions, of warning the public, and of providing analyses of events. A commendable job was done by the USGS during the Mount St. Helens crisis. Operational decisions to block access to an area or to evacuate people, however, must be the responsibility of other organizations. These will normally be local, either state or municipal, bodies or local representatives of federal agencies, including the Forest Service and the National Park Service. The Federal Emergency Management Agency (FEMA) is responsible for contingency planning and coordinating responses of all federal agencies to disasters; in the past FEMA has cooperated effectively with the USGS.

It would be very helpful during a volcanic crisis if the USGS team responsible for monitoring the event and providing information was able to establish communications with local officials and the press *in advance* of the crisis. The nature of volcanic eruptions is such that there will normally be ample time to lay a communications groundwork. Efforts to provide such a groundwork are under way at Mammoth Lakes.

The USGS should continue and extend these communication efforts with local, state, and federal agencies wherever signs develop of potential volcanic activity that could be a danger to the public.

REFERENCES

Crandall, D. R., and D. R. Mullineaux (1978). Potential hazards from future eruptions of Mount St. Helens, Washington, *U.S. Geol. Surv. Bull. 1383-C*, 26 pp.

Crosson, R. S. (1983). Review of seismicity in the Puget Sound region from 1970 through 1978, in *Proceeding of Workshop XIV: Earthquake Hazards of the Puget Sound Region, Washington*, J. C. Yount and R. S. Crosson, eds., U.S. Geol. Surv. Open-File Rep. 83-19, pp. 6-18.

Jennings, M. E., V. R. Schneider, and P. E. Smith (1981). Computer assessments of potential flood hazards from breaching of two debris dams, Toutle River and Cowliz River systems, in *The 1980 Eruptions of Mount St. Helens, Washington*, P. W. Lipman and D. R. Mullineaux, eds., U.S. Geol. Surv. Prof. Paper 1250, U.S. Government Printing Office, Washington, D.C., pp. 829-836.

Miller, C. D., D. R. Mullineaux, and D. R. Crandall (1981). Hazard assessments at Mount St. Helens, in *The 1980 Eruptions of Mount St. Helens, Washington*, P. W. Lipman and D. R. Mullineaux, eds., U.S. Geol. Surv. Prof. Paper 1250, U.S. Government Printing Office, Washington, D.C., pp. 789-802.

Norris, R. A., and R. N. Johnson (1969). Submarine volcanic eruptions recently located in the Pacific by sofar hydrophones, *J. Geophys. Res. 74*, 650-664.

U.S. Geological Survey (1982). Notice of potential volcanic hazard issued for eastern California, press release, May 25.

BACKGROUND

The Source Regions of Alkaline Volcanoes

1

ARTHUR L. BOETTCHER
University of California, Los Angeles

ABSTRACT

The spatial and temporal distributions of kimberlites and alkali basalts provide information on the mechanisms by which these important rocks originate. Kimberlites erupted throughout most or all of the Proterozoic and Phanerozoic eons, but they are restricted to Precambrian shield areas and surrounding fold belts. Abundant chemical and isotopic data support the concept that kimberlites originate from a source region in the mantle that was enriched by metasomatic fluids, the ingress of which appears to have been triggered by epeirogenic crustal and mantle doming. The source regions appear to have drifted coherently with the continents.

In contrast, alkali basalts that contain mantle-derived xenoliths appear to be restricted to the Phanerozoic eon, with the vast majority of known occurrences being less than 5 m.y. old. Most of these basalts were emplaced in active orogenic environments associated with subduction. It appears that metasomatism of the source regions of these rocks was caused by warping of the crust and underlying mantle as the result of subduction of topographic features such as ridges and transform faults.

The concept of a metasomatic fluid enriching the source regions of these alkaline rocks is consonant with that of a heterogeneous mantle proposed from petrographic, chemical, and isotopic studies of xenoliths and basalts. It seems clear that H_2O, Fe, Ti, and LIL elements such as K and LREE are among the most mobile components. The time interval between metasomatism and generation of magma might be short, as witnessed by the inception of alkali basalt magmatism all along the Pacific Coast of North America less than 10 m.y. ago.

INTRODUCTION

The distribution of extrusive and intrusive igneous rocks, if sufficiently constrained in space and time, provides information on the mechanisms by which such rocks originate. For example, the emplacement of the enormous volumes of anorthosite massifs, perhaps with attendant rhyolitic lavas and granites, during late Proterozoic time in two large arcuate belts (Herz, 1969) provides clues to the cooling and evolution of the crust-mantle system and perhaps to the inception of the process of modern-type plate tectonics (Goodwin, 1981). Along similar lines, the restriction of most peridotitic komatiites to Archean terranes has been interpreted as evidence for much higher crustal and mantle temperatures than have prevailed in post-Archean time (e.g., Green, 1975; Sleep, 1979).

If such temporal and spatial constraints can be established for explosive alkaline volcanoes, such as kimberlites and alkali basalts, this will place restrictions on the origin of alkaline magmas and shed light on the mantle source regions.

TEMPORAL AND SPATIAL DISTRIBUTIONS OF KIMBERLITES AND ALKALI BASALTS

Kimberlites

Kimberlite is difficult to define, and a liberal interpretation is adopted here. If the stringent definition of Skinner and Clement (1979) is used, the pipes on the Colorado Plateau and southeastern Australia would not qualify.

13

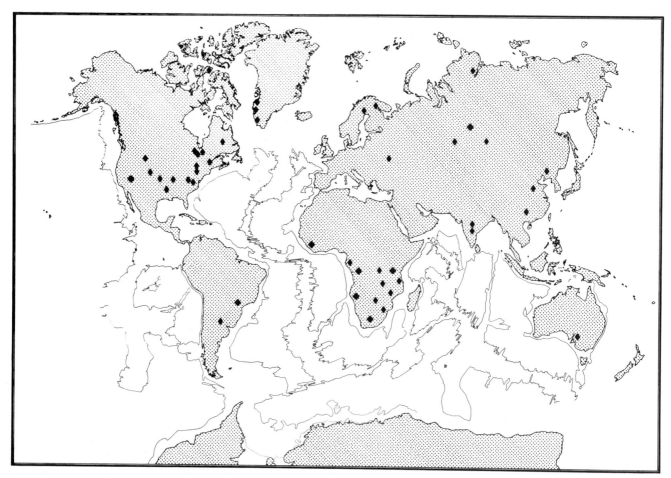

FIGURE 1.1 Distribution of kimberlites (modified from Dawson, 1980). The larger diamonds are kimberlite provinces; the smaller diamonds represent one or several kimberlites.

Kimberlites are concentrated in stable cratonic regions of the continental crust (see Figure 1.1). Some kimberlites also occur in the folded belts surrounding the cratons, such as those in Virginia, Tennessee, and Arkansas and around the Kaapvaal craton in southern Africa, but there is no evidence that these are other than postorogenic (Dawson, 1980). In fact, I am unaware of any kimberlites that show the effects of regional, dynamic metamorphism; this strengthens the argument that they develop in or near stable cratons, free from subsequent diastrophism. However, Haggerty et al. (1981) proposed that Precambrian dikes of graphitic schists in western Africa are metamorphosed kimberlite. Similarly, Bardet (1973) reported Precambrian dikes of "metakimberlite" in Ivory Coast and Gabon, Africa, that ostensibly contain diamonds. However, they have low K_2O (~ 0.5 wt.%) and Al_2O_3 (~ 1.0 wt. %) and high SiO_2 (~ 57 wt. %). Bailey (1964) observed that these regions in which kimberlites occur are also those of epeirogenic crustal doming. Kimberlites are not the only products of igneous activity associated with epeirogeny, but in many cratonic settings they are the only intracratonic igneous rocks indubitably associated with this process.

In addition, kimberlites are commonly spatially associated with fundamental fracture systems, such as in the Lesotho and

Kimberley areas of South Africa. In Greenland, kimberlite dikes were emplaced following regional doming and during the initial stages of rifting between Greenland and Canada during the Mesozoic (Andrews and Emeleus, 1975). In southeastern Australia, in a region of epeirogenic uplift beginning in the Cenozoic, Stracke et al. (1979) found a strong spatial correlation between the distribution of Permian to Quaternary kimberlites and the proposed landward extensions of neighboring oceanic transform faults. Following a concept developed earlier by Wilson (1965), they suggested that the zones of weakness resulting from predrift rifting determined the position and configuration of the subsequent transform faults. Thus, kimberlites appear to require deep-seated fractures as plumbing systems; however, there are other requirements, as documented by the conspicuous scarcity or absence of kimberlites in the Tertiary-Quaternary rift valleys of East Africa (e.g., Dawson, 1980).

Knowledge of the timing of the intrusion of kimberlites is an important parameter in understanding their origin. Many kimberlites are thought to be Cretaceous in age (e.g., Davis, 1977), but closer examination reveals a much more complex picture. For example, Paleozoic kimberlites are common in Siberia and North America. Precambrian ($\sim$950-1450 m.y. old), whole-rock, K-Ar ages have recently been determined for kimberlites in the Indian shield (Paul *et al.*, 1975). A particularly perplexing chronology occurs for the African kimberlites. Most of the kimberlites in southern Africa are younger than the Karroo sediments and lavas and are, therefore, post-Jurassic (Dawson, 1980). Radiogenic age determinations compiled by Dawson reveal nearly continuous emplacement of kimberlites throughout the Cretaceous. In addition, Hawthorne *et al.* (1979) and Dawson (1980) tabulated radiogenic ages for 6 South African kimberlites ranging from 150 to 730 million years (m.y.) old, in addition to the 1200-m.y.-old National and Premier pipes. Bardet and Vachette (1966) reported ages of about 90, 700, 1150, and 2200 m.y. for 4 kimberlites in the west African craton.

Dating kimberlites can be difficult; Zartman *et al.* (1967) reported that micas from such rocks commonly contain excess radiogenic Ar, and it is difficult, if not impossible in some instances, to distinguish primary from secondary phlogopite (e.g., Carswell, 1975). In addition, some minerals used for dating kimberlites may be xenocrysts derived from older crust; the 1750-m.y. age determined by Allsopp *et al.* (1967) on galena in the Premier kimberlite must be interpreted with caution. Kramers (1979) demonstrated that dating inclusions in diamonds also can yield widely unreliable ages, because at least some diamonds appear to be xenocrysts. For example, the Cretaceous Kimberley kimberlite contains diamonds with sulfide inclusions that yield ages in excess of 2 billion years (b.y.). Nevertheless, the old ages of some African kimberlites appear to be established, and Dawson (1980) suggested that Archean kimberlites may occur in several regions in Africa, as evidenced by the occurrence of diamond-bearing alluvial sedimentary rocks derived from Archean terranes.

Why has southern Africa been the site of emplacement of kimberlite for over a billion years? This is particularly puzzling considering the great depths (up to 250 km), indicated by geobarometric studies of xenoliths, from which kimberlites originate, together with the knowledge that the continent has drifted with respect to the surrounding lithosphere. One explanation might be that ancient continental areas extend to greater depths than ordinarily accepted. Alexander and Sammis (1975) showed that these ancient land masses are underlain by mantle whose geophysical properties extend to depths of about 400 km. Jordan (1978) developed the concept of tectosphere—thick root zones beneath shields that translate coherently with the continent during plate motion.

Further evidence for thick subcontinental mantle emerges from a study of the age of continental flood basalts. The Deccan basalts of India (Ghose, 1976) are a vast outpouring of tholeiitic lavas that erupted about 65 to 50 m.y. ago (Wellman and McElhinny, 1970). However, during the period of eruption of the Deccan basalts, India was migrating from the site of Gondwanaland toward Asia, and it is reasonable to assume that the source region migrated with it.

Alkali Basalts

When discussing the origin of alkali basalt, it is essential to define what criteria are used to categorize such rocks, as this term has been applied in various ways. The system developed by Macdonald and Katsura (1964) to distinguish tholeiites from

alkali basalts is based only on Hawaiian lavas, although it has been used extensively elsewhere. However, such a classification is too encompassing for present purposes. I will restrict this discussion to the flows and tephra with normative nepheline that commonly contain mantle xenoliths.

As shown in Figure 1.2, xenolith-bearing alkali basalts occur in a variety of structural settings and appear to have compositions that are independent of the environment of eruption, be it continental, oceanic, or island arc (e.g., Schwarzer and Rogers, 1974). However, systematic studies of Sr and Nd isotopes may reveal differences resulting from various degrees and kinds of crustal contamination. If the occurrence of mantle xenoliths, high-pressure megacrysts, and primitive $Mg/(Mg + Fe)$ ratios (e.g., Irving and Green, 1976) is taken as evidence that alkali basalts are primary, then undifferentiated, uncontaminated magmas occur in such diverse crustal environments as southeastern Australia (Kesson, 1973; Irving and Green, 1976); Mauritius in the Indian Ocean (Baxter, 1976); the eastern Mojave Desert (Wise, 1969; Wilshire *et al.*, 1971; Katz and Boettcher, 1980); the Colorado Plateau (Best and Brimhall, 1974); Anjouan, Comores Archipelago on the transition from oceanic to continental crust between Africa and Madagascar (Flower, 1973); Oahu, Hawaii (Jackson and Wright, 1970); Heimay, Iceland (Jakobsson, 1972); on the Pacific coast of Baja, California (Basu, 1977); and in Owens Valley, California, between the Sierra and Inyo mountains (Boettcher, unpublished data).

One requirement is a plumbing system that permits rapid ascent of these magmas from the mantle, and it is well established that some, not all, of the occurrences are in regions of demonstrable faulting. Forbes and Kuno (1967) stated that there is a high probability of finding mantle xenoliths in Cenozoic alkali basalts with $TiO_2 > 1.75$ wt % and with normative nepheline or leucite if they occur in uplifted and faulted (high-angle) terranes. However, the extensive alkali basalts of Ethiopia (Mohr, 1972), for example, meet these criteria but are apparently barren and are low in total alkalis. Alkali basalts commonly occur along offsets or transform faults; this contrasts with the preponderance of tholeiites along major rifts. Examples of this are the nepheline-normative lavas of the Cerros del Rio and El Alto fields along offsets of the Rio Grande rift, compared with the tholeiitic Servilleta and Albuquerque rocks on the main segment of the rift (Baldridge, 1978). Similar relationships occur in Iceland (Sigurdsson, 1970), Hawaii (Jackson and Wright, 1970), the Northern Pacific Ocean (Windom *et al.*, 1981), and the rift valley in Kenya (Wright, 1963).

Because of the enrichment of light rare-earth elements (LREE) in alkali basalts, it is commonly proposed that the source region is garnet lherzolite (e.g., Frey and Prinz, 1978). However, the depth of origin of most alkali basalts based on mineralogical geobarometers is less than the ~75-km depth required to stabilize garnet lherzolite (e.g., Green and Ringwood, 1967); in addition, garnet-bearing xenoliths are scarce in alkali basalts. I believe that we have much to learn about the effects of composition on the distribution coefficients of REE in garnet-bearing assemblages (see Apted and Boettcher, 1981).

Xenolith-bearing alkali basalts and kimberlites generally do not occur in the same areas. Possible exceptions are the diamond-bearing kimberlites and the xenolith-bearing alkali ba-

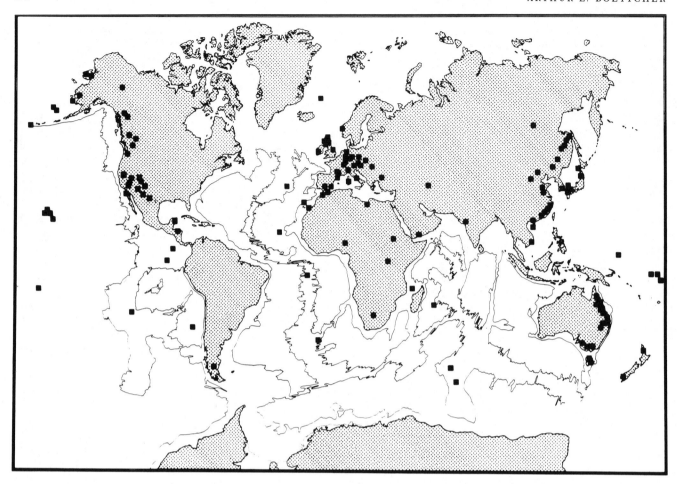

FIGURE 1.2   Distribution of mantle-xenolith-bearing alkali basalts (modified from Forbes and Kuno, 1967). Some squares represent more than one volcano. Antarctic locations are off map.

salts of eastern China, such as in the Shangdong and Jilin provinces (R. Cao, Institute of Geochemistry, People's Republic of China, personal communication, 1981), the Navajo-Hopi area of the southwestern United States where xenolith-bearing minettes and kimberlite-like diatremes occur in the same areas (Ehrenberg, 1979), and southeastern Australia (Ferguson and Sheraton, 1979; Ferguson et al., 1979; A. J. Irving, University of Washington, personal communication, 1981).

Another important aspect of the origin of alkali basalts is the timing of the volcanism. Explosive alkali basalts are generally assumed to be geologically young, and I am unaware of any Precambrian examples. Poldervaart (1962) could find no occurrences of any alkali basalts older than ~1 b.y. The oldest known mantle-xenolith-bearing alkali basalts include those of Devonian age in Seiland, northern Norway (Robbins, 1975); the late Paleozoic basanites of the Scottish Midland Valley (Chapman, 1976); the Delegate pipe of Jurassic age; and Kelly's Point of Permian age in New South Wales, Australia (see Wass and Irving, 1976). Engel et al. (1965) estimated the volume of tholeiitic and alkali basalts extruded during the last 3.5 b.y. (the abscissa of their Figure 4 should be in volume/unit time,

not volume). They show no alkali basalts older than ~250 m.y., and their production rate of tholeiites has progressively declined by about an order of magnitude since 3.5 b.y. before present (b.p.). The reader will certainly question the uncertainties that exist because of the incomplete and complex Precambrian record and because of the unknown volume of basalt consumed at plate boundaries, but the trends are probably reliable. The decrease in the production of tholeiites can be ascribed to a declining heat production in the mantle as K, U, and Th decay. For example, present radiogenic heat production is ~7 picowatts/kg, which is approximately one third the value that existed 3.5 b.y. ago. The extensive occurrences of Miocene and younger alkali basalts in Queensland, Australia (e.g., Wass and Irving, 1976), and the western United States are unknown in older terranes. The youngest volcano in the continental United States outside of the Cascade Mountains is an alkali basalt cone and associated flow (~390 yr old) in the eastern Mojave Desert, California (Katz and Boettcher, 1980). Numerous other examples of very young alkali basalts are known.

How then can we explain the abundance of young alkali basalts?

MANTLE METASOMATISM

Because of the nearly simultaneous inception of alkali basalt magmatism in western North America from Baja California, Mexico, to British Columbia, Canada, less than 10 m.y. ago, some catalytic mechanism must be invoked. Yoder (1976, p. 99) considered the heat budget in zones of magma genesis and calculated the time required to partially melt a garnet peridotite. He determined that 5 percent partial melting would require about 80 m.y., assuming that all radiogenic heat remains in the source area and that the rock was ready at the temperature of the beginning of melting. Yoder used a value for the enthalpy of melting of 686 J/g (estimated from Bradley, 1962) that incorporated the value of 357 J/g for diopside (from Robie and Waldbaum, 1968) and that is about one half the value suggested by Ferrier (1968), Weill *et al.* (1980), and Boettcher *et al.* (1982). Yoder also used a value of radiogenic heat production of 10^{-14} cal/cm^3/sec, which Schubert *et al.* (1980) suggested may be *high* by a factor of 2. For these reasons I believe that much longer times, perhaps 200 m.y., would be necessary to produce 5 percent liquid. Thus, for alkalic volcanism to begin nearly simultaneously in crustal environments as diverse as those on the Colorado Plateau, the Mojave Desert, Owens Valley, and coastal Baja California requires physicochemical conditions that must be considered extraordinary. I propose that the major governing factor was the introduction of volatile components into hot, relatively dry mantle peridotite.

Bailey (1964, 1974) proposed that epeirogenic crustal warping and concomitant pressure release have provided the focus for ingress of volatile components into the source regions of alkaline magmas. This proposal has received strong support from many geochemists and petrologists, and it is particularly appropriate to examine the role of such a mechanism in the origin of kimberlites and alkali basalts.

Structural Controls on Metasomatism

It is instructive to compare the worldwide distributions of kimberlites (Figure 1.1) and alkali basalts (Figure 1.2). The former are clearly associated with anorogenic or regional epeirogenic uplift; the latter are concentrated on oceanic islands and circumoceanic belts. However, as pointed out by Forbes and Kuno (1967), this circumoceanic belt does not coincide with the orogenic belt characterized by tholeiitic basalts and explosive andesitic volcanoes. Rather, the alkali basalt zones lie toward the interiors of continents and toward back-arc basins in the circum-Pacific arcs. Glazner (1981a, 1981b) proposed that late Tertiary faulting and volcanism in the southwestern United States resulted from warping of the crust as the North American plate passed over the Mendocino fault zone. This zone was a major discontinuity in the subducted Farallon plate, because it was the contact between the young lithosphere on the south and the thicker, older lithosphere on the north. The timing of this event (10 to 15 m.y. ago) agrees with the age of the oldest alkali basalt volcanoes in the eastern Mojave block (~ 10 m.y. old). This mechanism also may be applied to southeastern Australia (see Stracke *et al.*, 1979), to the Aleutian Islands, the

circum-Mediterranean zone, Antarctica, and elsewhere. It also may explain the absence of explosive alkali basalts in South America north of 33°S, which is the site of active subduction but which is removed from major ridges and transform faults. In contrast, at Palei-Aike at the southern tip of South America, alkali basalts that contain garnet lherzolites and other mantle-derived xenoliths occur along the subducted projection of the Chile Ridge, near the triple junction of the Nazca, Antarctic, and South American plates (Skewes and Stern, 1979).

Metasomatism of Magma Source Regions

Kimberlites and alkali basalts are relatively rich in K, Na, Ti, volatile components, and large-ion-lithophile (LIL) elements. Gast (1968) suggested that the alkali basalt clan originated by small degrees (2 percent or less; Kay and Gast, 1973) of melting of mantle peridotite. This certainly remains as a possible explanation of the "enriched" chemistry of these rocks, but a precursory metasomatic event conceivably could supply the requisite elements and eliminate the necessity of separation of such small percentages of anatectic liquid, which may be physically difficult because of a decreasing density contrast between liquid and crystals at depth (Stolper *et al.*, 1981). The large K/Na ratios of kimberlites relative to those in alkali basalts probably reflect the fact that the K is mostly ensconced in phlogopitic micas in the source regions of kimberlites, whereas the Na-rich amphiboles common in alkali basalts are stable only at shallower depths.

At the First International Kimberlite Conference in 1973, Lloyd and Bailey (1975) presented geochemical data on mantle-derived xenoliths in alkaline volcanic rocks from Uganda and West Germany. This clearly demonstrated that Ti, alkalis, volatile components, and LIL elements were introduced into the mantle, converting the peridotite into phlogopite-, amphibole-, and clinopyroxene-bearing assemblages, which were subsequently partially melted and also incorporated as xenoliths. At the same conference, Harte *et al.* (1975) described similarly metasomatized xenoliths from the Lesotho kimberlite. Boyd (1973) earlier reported that solutions rich in TiO_2, K_2O, and H_2O had altered the mantle above depths of 170 km beneath South Africa. Green (1971) concluded that magma or other fluid had similarly enriched the mantle low-velocity zone.

The consequences of chemically altered xenoliths are profound. Xenoliths provide the only samples of subcontinental mantle, and they have been used extensively as indicators of the chemistry and mineralogy of the mantle. In addition, if it can be demonstrated that mobile fluids have percolated through the mantle, much-needed information would be provided about the chemical and physical parameters in the interior of the Earth. (The term *fluid* is used here to indicate either silica-rich liquids (magmas) or volatile-rich hydrothermal solutions.) Toward these goals, Varne and Graham (1971), Wilshire and Trask (1971), Best (1974), Frey and Green (1974), and Francis (1976) investigated xenoliths in alkali basalts from throughout the world and concluded that metasomatism of the mantle sampled by these volcanic rocks resulted from hydrous anatectic liquids or deep-seated hydrothermal solutions. Erlank and Shimizu (1977) determined that $^{87}Sr/^{86}Sr$ ratios of xenoliths were significantly higher than those of the host kimberlites, suggesting that metasomatism was not the result of interaction with the kimberlite but rather occurred earlier. Particularly strong evidence for metasomatism is provided by the xenoliths from alkali basalts such as those at Victoria, Australia (Frey and Green, 1974), and San Carlos, New Mexico (Frey and Prinz, 1978), as well as the xenoliths from kimberlites examined by Shimizu (1975) that are highly enriched in incompatible (LIL) elements and "depleted" (Mg-rich) in terms of major elements.

Is there any direct observational evidence for the ingress of metasomatizing solutions? Wilshire and Trask (1971) first directed our attention to dikes and veins (commonly rich in amphibole and/or mica) in xenoliths in alkali basalts. Subsequently, Irving (1977), Stewart and Boettcher (1977), Frey and Prinz (1978), Boettcher et al. (1979), and Windom and Boettcher (1980) recognized that these features result from mobile liquids or fluids in the mantle. Wilshire and Pike (1975) reported analogous features in some alpine peridotites. Complexities in the chemistries of some of the large crystals (megacrysts) of phlogopitic mica in kimberlites have been shown by Farmer and Boettcher (1981) to originate from mobile solutions prior to incorporation into the host kimberlitic fluid.

Amphiboles and other phases also are common megacrysts in alkali basalts. Some researchers have proposed that these megacrysts are phenocrysts that crystallized from the alkali basalt at depth. However, there is considerable chemical and isotopic evidence (Boettcher and O'Neil, 1980) to support the suggestion of Wilshire and Trask (1971) that the megacrysts are fragments of disaggregated veins in mantle peridotite. This complements the experimental results of Stewart et al. (1979) that such amphiboles are not stable near the liquidus of alkali basalts and must result from subsolidus processes rather than crystallization from basaltic magma.

There are several interpretations of the significance of these veins. Wilshire et al. (1980) believed them to result from filter pressing of silicate magma, resulting in only local metasomatism. Conversely, Boettcher and O'Neil (1980) and Menzies and Murthy (1980) concluded that the metasomatism was pervasive and precursory to the alkali basalt magmatism.

To determine the provenance of the fluids, Boettcher and O'Neil (1980) performed chemical, isotopic, and petrographic studies of micas and amphiboles from kimberlites, alkali basalts, and their included xenoliths and megacrysts. In Figure 1.3 the micas from kimberlites plot in a rectangle with narrow limits of -60 to -80 δD and 5 to 6 $\delta^{18}O$, which is defined as a possible "field of primary values," because kimberlites originate at great depths (~ 150 km; Boyd, 1973). Conversely, the micas and amphiboles in alkali basalts have a much wider spread of δD and $\delta^{18}O$ values. The narrow range of the "primary values" from kimberlite may reflect the uniform isotopic composition of deep-seated fluids. However, this range of δD of -60 to -80 is the same as that for oceanic sediments (Savin and Epstein, 1970), and we cannot discount that the provenance of this mantle H_2O may be related to subducted materials.

Boettcher and O'Neil (1980) also examined the content of H_2O (as OH^-) in these hydrous phases. The phlogopites from the kimberlites all had normal H_2O contents of 3 to 4 wt. %,

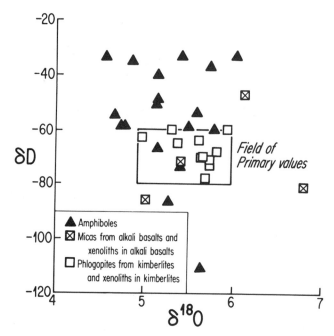

FIGURE 1.3 Value of δD and $\delta^{18}O$ for mantle-derived amphiboles and micas (from Boettcher and O'Neil, 1980, with permission of the American Geophysical Union). The rectangle is the range of values for micas in kimberlites and xenoliths in kimberlites and is the range proposed for primordial H_2O.

with 10 to 20 percent of the hydroxyl positions occupied by F and Cl. Surprisingly, the amphiboles and micas in the alkali basalts had extremely low H_2O contents; for example, some amphiboles had values as low as 0.06 wt. % H_2O (averaging ~ 0.6 wt. %) compared with >1.5 wt. % for most igneous amphiboles. Most samples contained only ~ 0.1 wt. % (F + Cl), with the remaining positions apparently occupied by oxygen. The wide range of δD, $\delta^{18}O$, and H_2O values for the hydrous minerals in alkali basalts may reflect complex interaction with aqueous fluids diluted with O, F, Cl, and components containing C and N in the source regions ($<\sim 70$ km). In contrast, the narrow range of values for kimberlites may result from the uniform composition of deep-seated primordial fluids in their source regions (100 to 250 km).

It is unknown whether the temperature-composition conditions of the source regions permit the existence of CO_2-H_2O fluids. However, it is now well established that fluids rich in H_2O will result in melting (Mysen and Boettcher, 1975) with disolution of the H_2O into the liquid. In addition, CO_2-rich fluids are unstable at depths below ~ 100 km, depending on the temperature and composition of the fluid; at greater depths, CO_2 reacts with olivine and pyroxenes to produce Mg-rich carbonates (e.g., Eggler, 1978; Wyllie, 1979). However, with the exception of several reports of Mg-rich carbonate inclusions in crystals from kimberlites, such phases have not been recovered from deep-seated rocks, even from diamond-bearing kimberlites. This strongly suggests that the stable forms of carbon in the source regions of kimberlites are chemically reduced species, such as diamond, graphite, carbides (Perchuk, 1976), or as carbon in solution in silicate minerals (Freund, 1981). This is supported by the widespread occurrence in kimberlites of such reduced phases as native metals and carbides (see Dawson, 1980; Pasteris, 1981, for a summary). The presence of ferric iron (Fe^{3+}) in mantle minerals such as amphiboles and micas should not be interpreted as evidence of high oxygen fugacities (fO_2), particularly in alkalic rocks where Na^+ can provide local charge balance when Fe^{3+} is in tetrahedrally coordinated sites in silicate structures (see Mysen *et al.*, 1980). Pressure and bulk composition will both have significant effects on the chemical potential of Fe^{3+}, largely independent of fO_2.

Additional evidence for a chemically reduced mantle has been provided by Arculus and Delano (1981), who measured the "intrinsic" oxygen fugacity of spinels in megacrysts and xenoliths from West Germany, Australia, and the United States. All of their samples were within $+0.25$ to -1.0 log fO_2 units of iron-wüstite values.

The question remains of whether chemically reduced fluids (exclusive of silicate-rich magmas) can exist in the source regions of alkali basalts and kimberlites. Recently, such species as H_2, N_2, and CH_4 have been detected as inclusions in diamonds (e.g., Melton and Giardini, 1981), but more work is required to verify this. In addition, the solubilities and transport capabilities of these species and their potentials as mantle fluids capable of pervasive metasomatism remain to be determined. To date, the only studies at temperatures and pressures of the subcontinental mantle are determinations of the solubility of potassium in H_2O coexisting with phlogopite (Ryabchikov and Boettcher, 1980), the partitioning of rare-earth elements between garnet peridotite minerals and H_2O (Mysen, 1979), the solubility of diopside in H_2O and in CO_2 (Eggler and Rosenhauer, 1978), and the composition of H_2O vapor in equilibrium with coexisting forsterite and enstatite (Nakamura and Kushiro, 1974). Equally necessary are additional studies such as those of crack propagation (Anderson and Grew, 1977) and the fluid dynamics of mantle metasomatism (Spera, 1981).

CONCLUSIONS

There are abundant geochemical and petrological data indicating that the mantle is chemically heterogeneous and that the source regions of many magmas, particularly alkaline ones, are enriched in volatiles, LIL elements, and other components such as Ti. In contrast to the generally accepted model that the upper mantle is peridotite, Anderson (1981) proposed a model of the evolution of the Earth in which a primitive magma ocean, rich in LIL elements, crystallizes to a mantle with a ~ 200-km-thick peridotite layer overlying a denser eclogite cumulate. Late-stage fluids from the eclogite layer rise and metasomatize the periodotite. This enriched peridotite layer is the source of continental basalts and alkaline ocean-island basalts. The depleted eclogite layer is the postulated source of mid-ocean ridge basalts.

The validity of the details of this model can be argued, but it is consonant with the suggestion of DePaolo and Wasserburg (1979) and Carlson *et al.* (1981) that ocean-island basalts and continental flood basalts have source regions with similar Sr and Nd isotopic compositions. Conversely, mid-ocean ridge basalts have lower $^{87}Sr/^{86}Sr$ and higher $^{143}Nd/^{144}Nd$ ratios, which suggests derivation from a depleted mantle region. This is in accord with the proposals of Gast (1968) and Tatsumoto *et al.* (1965) that abyssal basalts are depleted in LIL elements that may reflect a previous depletion event, such as partial melting, in the source region. The major difference between these models is that Anderson's (1981) source region for continental and ocean-island basalts requires an enriched peridotite, whereas the source in the recent Carlson *et al.* model is peridotite that need not have been metasomatized. This difference is an important one in terms of the concept of metasomatism in the source regions of alkaline volcanoes.

The Nd and Sr isotopic data by Menzies and Murthy (1980), O'Nions *et al.* (1979), and Basu and Tatsumoto (1980) revealed that significant isotopic heterogeneity occurs in the source regions of kimberlites and alkali basalts. The $^{143}Nd/^{144}Nd$ values of xenoliths from both rock types range from strongly depleted to highly enriched, suggesting that the originally depleted source regions have been enriched to various degrees by one or more metasomatic events. Menzies and Murthy (1980) summarized isotopic data indicating that such events have operated in the source regions of kimberlites and alkali basalts in Alaska, California, and South Africa during the last ~ 250 m.y.

Metasomatism provides several ingredients necessary for partial fusion:

1. Fe, Ti, and other elements that lower solidus temperatures of mantle assemblages;
2. Volatiles such as H_2O that lower melting temperatures and increase the mobility of other components; and
3. Heat-producing elements, primarily K, U, and Th.

If the host mantle is near the solidus temperature prior to metasomatism, the time interval between introduction of these elements and the generation of magma may be small. If not, a longer time will be required to accumulate radiogenic heat. For example, the former situation probably prevailed along the west coast of North America during the Neogene and Quaternary periods.

Based on evidence such as the eruption of kimberlites in South Africa for at least the past 1.5 b.y., the chemical and isotopic differences between subcontinental and suboceanic mantle, and the eruption of the Deccan flood basalts while India was rapidly drifting relative to surrounding mantle, it is tempting to suggest that old continents have keels at least 250 km thick that drift with the continents and are the source of much of the magmatism.

Because of metasomatism, xenoliths probably are not reliable samples of mantle. Rather than being depleted (by partial melting), as is commonly accepted, they probably are richer in volatiles and LIL elements than the bulk composition of the upper mantle. We must also remember that alkali basalts and kimberlites sample only a restricted range of mantle environments (Figures 1.1 and 1.2), and extrapolation to the entire mantle is unwarranted.

ACKNOWLEDGMENTS

I appreciate the invitation from Dr. F. R. Boyd to participate on the Panel on Explosive Volcanism. Conversations wth Dr. A. J. Irving on the volcanoes in New South Wales were valuable, as were those with Ronglong Cao for the occurrences in the People's Republic of China, with Drs. Steve Haggerty and Henry Meyer for western Africa, with Dr. Chris Scarfe for western Canada, and with Dr. Alexander McBirney for Central America. I greatly appreciated the in-depth reviews by Drs. F. R. Boyd, David Eggler, and Asish Basu. This work was supported by NSF Grant EAR78-16413 and is Institute of Geophysics and Planetary Physics Contribution No. 2271.

REFERENCES

Alexander, S. S., and C. G. Sammis (1975). New geophysical evidence on the driving mechanisms for continental drift, *Earth Miner. Sci.* 4, 25.

Allsopp, H. L., A. J. Burger, and C. Van Zyl (1967). A minimum age for the Premier kimberlite pipe yielded by biotite Rb-Sr measurements, with related galena isotopic data, *Earth Planet. Sci. Lett. 3*, 161-166.

Anderson, D. L. (1981). Hotspots, basalts, and the evolution of the mantle, *Science 213*, 82-89.

Anderson, O. L., and P. C. Grew (1977). Stress corrosion theory of crack propagation with applications to geophysics, *Rev. Geophys. Space Phys. 15*, 77-104.

Andrews, J. R., and C. H. Emeleus (1975). Structural aspects of kimberlite dyke and sheet intrusion in south-west Greenland, in *Physics and Chemistry of the Earth, Vol. 9*, L. H. Ahrens, J. B. Dawson, A. R. Duncan, and A. J. Erlank, eds., Pergamon, Oxford, pp. 43-50.

Apted, M. J., and A. L. Boettcher (1981). Partitioning of rare earth elements between garnet and andesite melt: An autoradiographic study of P-T-X effects, *Geochim. Cosmochim. Acta 45*, 827-837.

Arculus, R. J., and J. W. Delano (1981). Intrinsic oxygen fugacity measurements: Techniques and results for spinels from upper mantle peridotites and megacryst assemblages, *Geochim. Cosmochim. Acta 45*, 899-913.

Bailey, D. K. (1964). Crustal warping—a possible tectonic control of alkaline magmatism, *J. Geophys. Res. 69*, 1103-1111.

Bailey, D. K. (1974). Continental rifting and alkaline magmatism, in *The Alkaline Rocks*, H. Sorensen, ed., John Wiley & Sons, New York, pp. 148-159.

Baldridge, W. S. (1978). Petrology and petrogenesis of basaltic lavas of the central Rio Grande Rift, in *1978 International Symposium on the Rio Grande Rift*, K. H. Olsen and C. E. Chapin, eds., Los Alamos Scientific Laboratory Report LA-7487-C, pp. 16-17.

Bardet, M. G. (1973). *Géologie du Diamant*, Mem. Bur. Rech. Geol. Minire 83, 235 pp.

Bardet, M. G., and M. Vachette (1966). *Déterminations d'Ages de Kimberlites de l'Ouest Africain et Essai d'Interprétation des Datations des Diverses Venues Diamantifères dans le Monde*, Bur. Rech. Geol. Minières Rep. DS 66, 59 pp.

Basu, A. R. (1977). Olivine-spinel equilibria in lherzolite xenoliths from San Quintin, Baja California, *Earth Planet. Sci. Lett. 33*, 443-450.

Basu, A. R., and M. Tatsumoto (1980). Nd-isotopes in selected mantle-derived rocks and minerals and their implications for mantle evolution, *Contrib. Mineral. Petrol. 75*, 43-54.

Baxter, A. N. (1976). Geochemistry and petrogenesis of primitive alkali basalt from Mauritius, Indian Ocean, *Geol. Soc. Am. Bull. 87*, 1028-1034.

Best, M. G. (1974). Mantle-derived amphiboles within inclusions in alkalic-basaltic lavas, *J. Geophys. Res. 79*, 2107-2113.

Best, M. G., and W. H. Brimhall (1974). Late Cenozoic alkalic basaltic magmas in the western Colorado Plateau and the Basin and Range transition zone, U.S.A., and their bearing on mantle dynamics, *Geol. Soc. Am. Bull. 85*, 1677-1690.

Boettcher, A. L., and J. R. O'Neil (1980). Stable isotope, chemical and petrographic studies of high-pressure amphiboles and micas: Evidence for metasomatism in the mantle source regions of alkali basalts and kimberlites, *Am. J. Sci. 280-A*, 594-621.

Boettcher, A. L., J. R. O'Neil, K. E. Windom, D. C. Stewart, and H. G. Wilshire (1979). Metasomatism of the mantle and the genesis of kimberlites and alkali basalts, in *The Mantle Samples: Inclusions in Kimberlites and Other Volcanics*, F. R. Boyd and H. O. A. Meyer, eds., Am. Geophys. Union, Washington, D.C., pp. 178-182.

Boettcher, A. L., C. W. Burnham, K. E. Windom, and S. R. Bohlen (1982). Liquids, glasses and the melting of silicates to high pressures, *J. Geol. 90*, 127-138.

Boyd, F. R. (1973). A pyroxene geotherm, *Geochim. Cosmochim. Acta 37*, 2533-2546.

Bradley, R. S. (1962). Thermodynamic calculations on phase equilibria involving fused salts, Part II. Solid solutions and applications to the olivines, *Am. J. Sci. 260*, 550-554.

Carlson, R. W., G. W. Lugmair, and J. D. MacDougall (1981). Columbia River volcanism: The question of mantle heterogeneity or crustal contamination, *Geochim. Cosmochim. Acta 45*, 2483-2499.

Carswell, D. A. (1975). Primary and secondary phlogopites and clinopyroxenes in garnet lherzolite xenoliths, in *Physics and Chemistry of the Earth, Vol. 9*, L. H. Ahrens, J. B. Dawson, A. R. Duncan, and A. J. Erlank, eds., Pergamon, Oxford, pp. 417-429.

Chapman, N. A. (1976). Inclusions and megacrysts from undersaturated tuffs and basanites, East Fife, Scotland, *J. Petrol. 17*, 472-498.

Davis, G. L. (1977). The ages and uranium contents of zircons from kimberlites and associated rocks, in *International Kimberlite Conference 2, Santa Fe, New Mexico, Extended Abstracts*.

Dawson, J. B. (1980). *Kimberlites and Their Xenoliths*, Springer-Verlag, New York, 252 pp.

DePaolo, D. J., and G. J. Wasserburg (1979). Neodymium isotopes in flood basalts from the Siberian Platform and inferences about their mantle sources, *Proc. Natl. Acad. Sci. USA 76*, 3056-3060.

Eggler, D. H. (1978). Stability of dolomite in a hydrous mantle, with implications for the mantle solidus, *Geology 6*, 397-400.

Eggler, D. H., and M. Rosenhauer (1978). Carbon dioxide in silicate melts: II. Solubilities of CO_2 and H_2O in $CaMgSi_2O_6$ (diopside) liquids and vapors at pressures to 40 kb, *Am. J. Sci. 278*, 64-94.

Ehrenberg, S. N. (1979). Garnetiferous ultramafic inclusions in minette from the Navajo volcanic field, in *The Mantle Sample: Inclusions in Kimberlites and Other Volcanics*, F. R. Boyd, and H. O. A. Meyer, eds., Am. Geophys. Union, Washington, D.C., pp. 330-344.

Engel, A. E. J., C. G. Engel, and R. G. Havens (1965). Chemical characteristics of oceanic basalts and the upper mantle, *Geol. Soc. Am. Bull. 76*, 719-734.

Erlank, A. J., and N. Shimizu (1977). Strontium and stronium isotope distributions in some kimberlite nodule and minerals, in *International Kimberlite Conference 2, Santa Fe, New Mexico, Extended Abstracts*.

Farmer, G. L., and A. L. Boettcher (1981). Petrologic and crystal-chemical significance of some deep-seated phlogopites, *Am. Mineral. 66*, 1154-1163.

Ferguson, J., and J. W. Sheraton (1979). Petrogenesis of kimberlitic rocks and associated xenoliths of southeastern Australia, in *Kimber-*

lites, Diatremes, and Diamonds: Their Geology, Petrology, and Geo-chemistry, F. R. Boyd and H. O. A. Meyer, ed., Am. Geophys. Union, Washington, D.C., pp. 140-160.

Ferguson, J., R. J. Arculus, and J. Joyce (1979). Kimberlite and kim-berlitic intrusives of southeastern Australia: A review, *J. Austral. Geol. Geophys. 4*, 227-241.

Ferrier, M. A. (1968). Mesure de l'enthalpie du diopside synthétique entre 298° et 1885° K, *C. R. Acad. Sci. Paris, t. 267*, 101-103.

Flower, M. F. J. (1973). Evolution of basaltic and differentiated lavas from Anjouran, Comores Archipelago, *Contrib. Mineral. Petrol. 38*, 237-260.

Forbes, R. B., and H. Kuno (1967). II. Peridotite inclusions and basaltic host rocks, in *Ultramafic and Related Rocks*, P. J. Wyllie, ed., John Wiley & Sons, New York, pp. 328-337.

Francis, D. M. (1976). Amphibole pyroxenite xenoliths: Cumulate or replacement phenomena from the upper mantle, Nunivak Island, Alaska, *Contrib. Mineral. Petrol. 58*, 51-61.

Freund, F. (1981). Mechanism of the water and carbon dioxide solu-bility in oxides and silicates and the role of O^-, *Contrib. Mineral. Petrol. 76*, 474-482.

Frey, F. A., and D. H. Green (1974). The mineralogy, geochemistry and origin of lherzolite inclusions in Victorian basanites, *Geochim. Cosmochim. Acta 38*, 1023-1059.

Frey, F. A., and M. Prinz (1978). Ultramafic inclusions from San Carlos, Arizona: Petrologic and geochemical data bearing on their petrogenesis, *Earth Planet. Sci. Lett. 38*, 129-176.

Gast, P. W. (1968). Trace element fractionation and the origin of thol-eiitic and alkaline magma types, *Geochim. Cosmochim. Acta 32*, 1057-1086.

Ghose, N. C. (1976). Composition and origin of Deccan basalts, *Lithos 9*, 65-73.

Glazner, A. F. (1981a). Cenozoic Evolution of the Mojave Block and Adjacent Areas, Ph.D. thesis, Univ. of California, Los Angeles, 175 pp.

Glazner, A. F. (1981b). Detachment faulting, volcanism, and the pas-sage of the Mendocino Fracture Zone under the southwestern United States, *Geol. Soc. Am. Abstr. Progr. 13*, 460.

Goodwin, A. M. (1981). Precambrian perspectives, *Science 313*, 55-61.

Green, D. H. (1971). Composition of basaltic magmas as indicators of conditions of origin: Application to oceanic volcanism, *Phil. Trans. R. Soc. Lond. A268*, 707-725.

Green, D. H. (1975). Genesis of Archean peridotitic magmas and con-straints on Archean geothermal gradients and tectonics, *Geology 3*, 15-18.

Green, D. H., and A. E. Ringwood (1967). The stability field of alu-minous pyroxene peridotite and garnet peridotite and their relevance in upper mantle structure, *Earth Planet. Sci. Lett. 3*, 151-160.

Haggerty, S. E., P. B. Taft, and L. A. Tompkins (1981). Diamonds in graphite schists, *EOS 62*, 416.

Harte, B., K. G. Cox, and J. J. Gurney (1975). Petrology and geological history of upper mantle xenoliths from the Matsoku kimberlite pipe, in *Physics and Chemistry of the Earth, Vol. 9*, L. H. Ahrens, J. B. Dawson, A. R. Duncan, and A. J. Erlank, eds., Pergamon, Oxford, pp. 477-506.

Hawthorne, J. B., A. J. Carrington, C. R. Clement, and E. M. W. Skinner (1979). Kimberlites: Field relations, in *Kimberlites, Dia-tremes, and Diamonds: Their Geology, Petrology, and Geochemis-try*, F. R. Boyd and H. O. A. Meyer, eds., Am. Geophys. Union, Washington, D.C., pp. 59-70.

Herz, N. (1969). Anorthosite belts, continental drift, and the anor-thosite event, *Science 164*, 944-947.

Irving, A. J. (1977). Mantle pyroxenitic "liquids" and "cumulates": Geochemistry of complex xenoliths from San Carlos, Kilbourne's

Hole and Eastern Australia, in *International Kimberlite Conference 2, Santa Fe, New Mexico, Extended Abstracts*.

Irving, A. J., and D. H. Green (1976). Geochemistry and petrogenesis of the Newer basalts of Victoria and South Australia, *Geol. Soc. Austral. J. 23*, 45-66.

Jackson, E. D., and T. L. Wright (1970). Xenoliths in the Honolulu volcanic series, Hawaii, *J. Petrol. 11*, 405-430.

Jakobsson, S. P. (1972). Chemistry and distribution of Recent basaltic rocks in Iceland, *Lithos 5*, 365-386.

Jordon, T. H. (1978). Composition and development of the continental tectosphere, *Nature 274*, 544-548.

Katz, J., and A. L. Boettcher (1980). The Cima volcanic field, in *Tec-tonic Framework of the Mojave and Sonoran Deserts, California and Arizona, South Coast*, Geol. Soc. Am., Boulder, Colo., pp. 236-241.

Kay, R. W., and P. W. Gast (1973). The rare earth content and origin of alkali-rich basalts, *J. Geol. 81*, 653-682.

Kesson, S. E. (1973). The primary geochemistry of the Monaro alkaline volcanics, southeastern Australia—evidence for upper mantle het-erogeneity, *Contr. Mineral. Petrol. 42*, 93-108.

Kramers, J. D. (1979). Lead, uranium, strontium, potassium and ru-bidium in inclusion-bearing diamonds and mantle-derived xenoliths from southern Africa, *Earth Planet. Sci. Lett. 42*, 58-70.

Lloyd, F. E., and D. K. Bailey (1975). Light element metasomatism of the continental mantle: The evidence and the consequences, in *Physics and Chemistry of the Earth, Vol. 9*, L. H. Ahrens, J. B. Dawson, A. R. Duncan, and A. J. Erlank, eds., Pergamon, Oxford, pp. 389-416.

Macdonald, G. A., and T. Katsura (1964). Chemical composition of Hawaiian lavas, *J. Petrol. 5*, 82-137.

Melton, D. E., and A. A. Giardini (1981). The nature and significance of occluded fluids in three Indian diamonds, *Am. Mineral. 66*, 746-750.

Menzies, M., and V. R. Murthy (1980). Enriched mantle: Nd and Sr isotopes in diopsides from kimberlite nodules, *Nature 283*, 634-636.

Mohr, P. A. (1972). Regional significance of volcanic geochemistry in the Afar Triple Junction, Ethiopia, *Geol. Soc. Am. Bull. 83*, 213-222.

Mysen, B. O. (1979). Trace-element partitioning between garnet per-idotite minerals and water-rich vapor: Experimental data from 5 to 30 kbar, *Am. Mineral. 64*, 274-287.

Mysen, B. O., and A. L. Boettcher (1975). Melting of a hydrous mantle: I. Phase relations of natural peridotite at high pressures and tem-peratures with controlled activities of water, carbon dioxide, and hydrogen, *J. Petrol. 16*, 520-548.

Mysen, B. O., F. Seifert, and D. Virgo (1980). Structure and redox equilibria of iron-bearing silicate melts, *Am. Mineral. 65*, 867-884.

Nakamura, Y., and I. Kushiro (1974). Composition of the gas phase in Mg_2SiO_4-H_2O at 15 kbar, *Carnegie Inst. Wash. Yearbook 73*, 255-258.

O'Nions, R. K., S. R. Carter, N. M. Evensen, and P. J. Hamilton (1979). Geochemical and cosmochemical applications of Nd isotope analysis, *Ann. Rev. Earth Planet. Sci. 7*, 11-38.

Pasteris, J. D. (1981). Occurrence of graphite in serpentinized olivines in kimberlite, *Geology 9*, 356-359.

Paul, D. K., D. C. Rex, and P. G. Harris (1975). Chemical charac-teristics and K-Ar ages of Indian kimberlite, *Geol. Soc. Am. Bull. 86*, 364-366.

Perchuk, L. L. (1976). Gas-mineral equilibria and a possible geochem-ical model of the Earth's interior, *Phys. Earth Planet. Inter. 13*, 232-239.

Poldervaart, A. (1962). Aspects of basalt petrology, *J. Geol. Soc. India 3*, 1-14.

Robie, R. A., and D. R. Waldbaum (1968). *Thermodynamic Properties of Minerals and Related Substances at 298.15°K (25.0°C) and One Atmosphere (1.013 Bars) Pressure at Higher Temperatures*, U.S. Geol. Surv. Bull. 1259, 256 pp.

Robbins, B. (1975). Ultramafic nodules from Seiland, northern Norway, *Lithos 8*, 15-27.

Ryabchikov, I. D., and A. L. Boettcher (1980). Experimental evidence at high pressure for potassic metasomatism in the mantle of the Earth, *Am. Mineral. 65*, 915-919.

Savin, S. M., and S. Epstein (1970). The oxygen and hydrogen isotope geochemistry of ocean sediments and shales, *Geochim. Cosmochim. Acta 34*, 43-64.

Schubert, G., D. Stevenson, and P. Cassen (1980). Whole planet cooling and the radiogenic heat source contents of the Earth and Moon, *J. Geophys. Res. 85*, 2531-2538.

Schwarzer, R. R., and J. J. Rogers (1974). A worldwide comparison of alkali olivine basalts and their differentiation trends, *Earth Planet. Sci. Lett. 23*, 286-296.

Shimizu, N. (1975). Rare earth elements in garnets and clinopyroxenes from garnet lherzolite nodules in kimberlites, *Earth Planet. Sci. Lett. 25*, 26-32.

Sigurdsson, H. (1970). Structural origin and plate tectonics of the Snaefellsnes Volcanic Zone, western Iceland, *Earth Planet. Sci. Lett. 10*, 129-135.

Skewes, M. A., and C. R. Stern (1979). Petrology and geochemistry of alkali basalts and ultramafic inclusions from the Palei-Aike volcanic field in southern Chile and the origin of the Patagonian Plateau lavas, *J. Volcanol. Geotherm. Res. 6*, 3-25.

Skinner, E. M. W., and C. R. Clement (1979). Mineralogical classification of southern African kimberlites, in *The Mantle Sample: Inclusions in Kimberlites and Other Volcanics*, F. R. Boyd and H. O. A. Meyer, eds., Am. Geophys. Union, Washington, D.C., pp. 129-139.

Sleep, N. H. (1979). Thermal history and degassing of the Earth: Some simple calculations, *J. Geol. 87*, 671-686.

Spera, F. J. (1981). Carbon dioxide in igneous petrogenesis: II. Fluid dynamics of mantle metasomatism, *Contrib. Mineral. Petrol. 77*, 56-65.

Stewart, D. C., and A. L. Boettcher (1977). Chemical gradients in mantle xenoliths, *Geol. Soc. Am. Abstr. Progr. 9*, 1191-1192.

Stewart, D. C., A. L. Boettcher, and D. H. Eggler (1979). Phase relationships of kaersutite at upper mantle conditions: Implications for its subsolidus origin in ultramafic nodules, *EOS 60*, 418.

Stolper, E., D. Walker, B. H. Hager, and J. F. Hays (1981). Melt segregation from partially molten source regions: The importance of melt density and source region size, *J. Geophys. Res. 86*, 6261-6271.

Stracke, K. J., J. Ferguson, and L. P. Black (1979). Structural setting of kimberlites in south-eastern Australia, in *Kimberlites, Diatremes, and Diamonds: Their Geology, Petrology, and Geochemistry*, F. R. Boyd and H. O. A. Meyer, eds., Am. Geophys. Union, Washington, D.C., pp. 71-91.

Tatsumoto, M., C. E. Hedge, and A. E. J. Engel (1965). Potassium, rubidium, strontium, thorium, uranium, and the ratio of strontium-87 to strontium-86 in oceanic tholeiitic basalt, *Science 150*, 886-888.

Varne, R., and A. L. Graham (1971). Rare earth abundances in hornblende and clinopyroxene of a hornblende lherzolite xenolith: Implications for upper mantle fractionation process, *Earth Planet. Sci. Lett. 13*, 11-18.

Wass, S. Y., and A. J. Irving (1976). *XENMEG: A Catalogue of Occurrences of Xenoliths and Megacrysts in Basic Volcanic Rocks of Eastern Australia*, The Australian Museum, Sydney, Australia, 455 pp.

Weill, D. F., R. Hon, and A. Navrotsky (1980). The igneous system $CaMgSi_2O_6$-$CaAlSi_2O_8$: Variations on a classic theme by Bowen, in *Physics of Magmatic Processes*, R. B. Hargraves, ed., Princeton U. Press, Princeton, N.J., pp. 49-92.

Wellman, P., and M. W. McElhinny (1970). K-Ar age of Deccan Traps, India, *Nature 227*, 959-965.

Wilshire, H. G., and J. E. N. Pike (1975). Upper-mantle diapirism: Evidence from analogous features in alpine peridotite and ultramafic inclusions in basalt, *Geology 3*, 467-470.

Wilshire, H. G., and N. J. Trask (1971). Structural and textural relationships of amphibole and phlogopite in peridotite inclusions, Dish Hill, California, *Am. Mineral. 56*, 240-255.

Wilshire, H. G., L. C. Calk, and E. C. Schwarzman (1971). Kaersutite—a product of reaction between pargasite and basanite at Dish Hill, California, *Earth Planet. Sci. Lett. 10*, 281-284.

Wilshire, H. G., J. E. N. Pike, C. E. Meyer, and E. C. Schwarzman (1980). Amphibole-rich veins in lherzolite xenoliths, Dish Hill and Deadman Lake, California, *Am. J. Sci. 280-A*, 576-593.

Wilson, J. T. (1965). A new class of faults and their bearing on continental drift, *Nature 207*, 343-347.

Windom, K. E., and A. L. Boettcher (1980). Mantle metasomatism and the kimberlite-lamprophyre association: Evidence from an eclogite nodule from Roberts Victor Mine, South Africa, *J. Geol. 88*, 705-712.

Windom, K. E., K. E. Seifert, and T. L. Vallier (1981). Igneous evolution of Hess Rise: Petrologic evidence from DSDP Leg 62, *J. Geophys. Res. 86*, 6311-6322.

Wise, W. S. (1969). Origin of basaltic magmas in the Nojave Desert area, California, *Contrib. Mineral. Petrol. 23*, 53-64.

Wright, J. B. (1963). A note on possible differentiation trends in Tertiary to Recent lavas of Kenya, *Geol. Mag. 100*, 164-180.

Wyllie, P. J. (1979). Magmas and volatile components, *Am. Mineral. 64*, 469-500.

Yoder, H. S., Jr. (1976). *Generation of Basaltic Magma*, National Academy of Sciences, Washington, D.C., 265 pp.

Zartman, R. E., M. R. Brock, A. V. Heyl, and H. H. Thomas (1967). K-Ar and Rb-Sr ages of some alkalic intrusive rocks from central and eastern United States, *Am. J. Sci. 265*, 848-870.

Tectonic Influence on Magma Composition of Cenozoic Basalts from the Columbia Plateau and Northwestern Great Basin, U.S.A.

2

RICHARD W. CARLSON
Carnegie Institution of Washington
Department of Terrestrial Magnetism

INTRODUCTION

An important factor controlling the character of a volcanic eruption is the composition of the lava being extruded. This lava composition represents the end result of chemical differentiation processes acting on a magma formed by partial melting of a source material deep below the volcanic vent. Results from the field of experimental petrology (e.g., Yoder, 1979) have greatly expanded our understanding of how the major-element composition of a melt can be related to source characteristics and the degree and pressure-temperature conditions of the melting event. In addition, the application of trace-element and isotope geochemistry to the study of magma genesis allows determination of the relative importance of fractionation processes (e.g., partial melting, fractional crystallization) versus chemically heterogeneous sources in producing magmas of a certain composition.

This chapter discusses the chemical and isotopic characteristics of two distinct groups of Cenozoic basaltic volcanics from the northwestern United States: the Columbia Plateau basalts and the low-K, high-Al basalts of the Oregon and Modoc plateaus. Cenozoic volcanism in the northwestern United States shows a wide variety of eruptive styles, from the relatively quiescent fissure eruptions of the Columbia Plateau basalts to the highly explosive calc-alkaline Cascade volcanics and rhyolitic ash flows of the Oregon Plateau. The link between magma composition, eruptive style, and tectonic character of the eruption area of these lavas is believed to provide important information regarding the process of magma generation and eruption.

TECTONIC SETTING

The geographic extent of major Cenozoic volcanic provinces in the northwestern United States is outlined in Figure 2.1. Extensive Cenozoic volcanism in the western United States is

FIGURE 2.1   Geographic extent of major Cenozoic volcanic provinces of the northwestern United States.

believed to result directly from interaction between the colliding Pacific and North American plates (e.g., Christiansen and Lipman, 1972). Details of this interaction have been presented by a number of researchers, and the reader is referred to the discussion by Christiansen and McKee (1978) for a more thorough treatment than that presented below.

Beginning in the late Oligocene the nature of convergence between the Pacific and American plates changed markedly as a result of the intersection of the American plate with the East Pacific Rise. Accompanying this change in tectonic style of the continental border, the Basin and Range region was subjected to tensions resulting in the oblique extension now characterizing the area. In particular, the area of the Great Basin may have undergone up to 100 to 300 km of east-west extension beginning in the early Miocene (Hamilton and Myers, 1966). In contrast to this high degree of extension, the Columbia Plateau, which marks the northern margin of oblique extension, experienced perhaps less than a few tens of kilometers east-west extension (Taubeneck, 1970; Christiansen and McKee, 1978). Presumably as a result of the changing tectonic setting of this area, volcanism in the Basin and Range province shifted in character from predominantly calc-alkaline prior to the Miocene to bimodal basalt and rhyolite in the Neogene (Christiansen and Lipman, 1972).

ISOTOPIC CHARACTERISTICS

Sr and Nd isotopic data for a number of basalt types from the Columbia Plateau and the Oregon and Modoc plateaus are summarized in Table 2.1. The Sr isotopic data for these samples are compared with similar data for other volcanic provinces of the northwestern United States in Figure 2.2. As can be seen in Figure 2.2, these volcanic rocks exhibit a wide range in their Sr isotopic compositions. An important feature of these data is that, with some notable exceptions, the isotopic composition of a given sample depends more on its geographic locality than either the rock type or its age. For example, both andesites and high-Al tholeiites from the Cascade Range have $^{87}Sr/^{86}Sr$ values within the range of 0.7027 to 0.7048 (Church and Tilton, 1973); farther to the east, Snake River Plain basalts have much higher $^{87}Sr/^{86}Sr$ (>0.7060) (Leeman and Manton, 1971) similar to that measured for the initial $^{87}Sr/^{86}Sr$ of nearby Mesozoic batholiths (Armstrong et al., 1977). This correlation between isotopic composition of igneous rocks in the western United States and their emplacement locality has been noted by a number of workers (e.g., Doe, 1967; Noble et al., 1973; Zartman, 1974; Armstrong et al., 1977). The area of maximum isotopic variability seems to roughly follow the known exposures of pre-Cenozoic basement. That is, in general, rocks

TABLE 2.1 Trace Element (ppm) and Isotopic Composition of Selected Basalts from the Northwestern United States

	Wanapum			Saddle Mt.			High-Al Basalt	Mid-Ocean Ridge Basalt
	Picture Gorge	Grande Ronde	Robinette Mt.	Frenchman Springs	Ice Harbor	Umatilla		
Sc	40	35	38	36	40	25	39	61
Cr	113	33	148	56	198	8.8	226	297
Ni	77	20	—	75	90	60	165	97
Rb	10	37	—	34	13	46	2.6	1
Sr	246	300	—	318	251	293	241	130
Ba	320	593	160	560	500	3550	161	14
La	8.7	23	7.2	24	35	47	3.3	2.7
Sm	4.1	6.3	3.0	7.3	10	11	2.2	3.1
Yb	3.2	3.06	1.8	3.9	4.9	4.4	2.3	2.9
(La/Sm)[a]	1.36	2.35	1.54	2.11	2.25	2.74	0.94	0.56
(La/Yb)[a]	1.82	5.03	2.68	4.12	4.78	7.15	0.98	0.62
K/Rb	550	330	—	345	513	461	750	1300
Sr/Nd	17	15	—	10.3	5.2	5.6	37	14
$^{87}Sr/^{86}Sr$	0.7036	0.7042 0.7059	0.7042	0.7051	0.7077	0.7089	0.7032 0.7074	0.7028
$^{\varepsilon}Nd$	7.5	+4.0 +0.3	+3.1	+1.2	−4.6	−7.1	+7.4 −3.9	+10

[a] Chondrite normalized.

SOURCE: Elemental data for Picture Gorge, average of PG-1 and PG-14 from Carlson et al. (1981); Grande Ronde and Robinette Mt., average from Wright et al. (1978); Frenchman Springs, Ice Harbor, Umatilla, Carlson et al. (1981); High-Al Basalt, Hart (1981b); Mid-Ocean Ridge Basalt, Engel et al. (1965), Schilling (1972): Isotopic data for Columbia basalts from McDougall (1976) and Carlson et al. (1981); for High-Al Basalts from Hart (1981b).

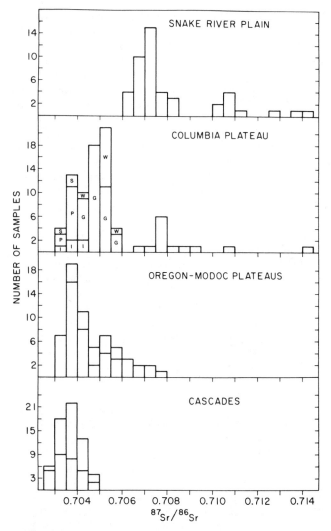

FIGURE 2.2 Histogram of $^{87}Sr/^{86}Sr$ measured for Cenozoic volcanics from the northwestern United States. Where divided columns are shown, lower division represents data for basaltic compositions, upper division for intermediate to silicic magmas. Units in the Columbia Plateau group are: I-Imnaha, P-Picture Gorge, G-Grande Ronde, W-Wanapum, S-Saddle Mt. All Columbia Plateau basalts with $^{87}Sr/^{86}Sr$ belong to the Saddle Mt. unit. Data for the Cascades from Peterman *et al.* (1970), Church and Tilton (1973), and Mertzman (1979); for the Oregon-Modoc Plateau from Noble *et al.* (1973), Mark *et al.* (1975), Hart (1981b), McKee *et al.* (1983), and Carlson (unpublished); for the Columbia Plateau from McDougall (1976), Carlson *et al.* (1981), and Carlson (unpublished); and for the Snake River from Leeman and Manton (1971).

erupted in the western portions of the region have rather low $^{87}Sr/^{86}Sr$, generally below 0.704, whereas those samples emplaced farther to the east have much higher $^{87}Sr/^{86}Sr$. Similar information is contained in the Pb (Doe, 1967; Zartman, 1974) and Nd (Carlson *et al.*, 1981; Hart, 1981a) isotopic composition of igneous rocks from the northwestern United States.

The question remains, however, as to whether the isotopic variability in these rocks is best explained by mixing between primary melts and materials of the continental crust or by systematically variable isotopic composition in the mantle source of these lavas. This distinction contains a number of important implications regarding the physical and chemical characteristics of the mantle underlying this region. For instance, in one extreme case where all isotopic variability is assigned to mixing between a primary, isotopically homogeneous magma and materials derived from the continental crust, the conclusion could be drawn that the mantle underlying this region has $^{87}Sr/^{86}Sr$ less than or equal to the lowest measured values for volcanics in the area (0.7027; Church and Tilton, 1973) and $^{143}Nd/^{144}Nd$ greater than or equal to the highest measured for these same lavas (0.51304; Carlson *et al.*, 1981). This conclusion arises because the continental crustal component would be expected, in general, to have higher $^{87}Sr/^{86}Sr$ and lower $^{143}Nd/^{144}Nd$ than that of a mantle-derived melt. These values of Sr and Nd isotopic composition for the mantle in this region are similar to those observed for basalts erupted in oceanic areas and imply that this mantle has had, on average, Rb/Sr and Sm/Nd, respectively, less than and greater than expected for undifferentiated bulk earth materials (bulk earth $^{87}Sr/^{86}Sr$ = 0.7045 to 0.7050; DePaolo and Wasserburg, 1976; O'Nions *et al.*, 1977; and $^{143}Nd/^{144}Nd$ = 0.51264; Lugmair *et al.*, 1975).

The extreme alternative to this crustal contamination scenario is that the isotopic composition measured for each basaltic rock in the area has a corresponding region within the mantle with similar isotopic composition. In this case the observed correlation between isotopic composition and geographic position of these eruptives implies that the mantle underlying this area becomes progressively more isotopically distinct from oceanic-type "depleted" mantle proceeding in the direction of the continental craton. One explanation for the presence of chemically and isotopically distinct mantle underlying a continental platform is the existence of a lithospheric "keel" beneath stable continental crust (Leeman, 1975). This "keel" would then translate with the overlying crust during plate movement. As discussed by Brooks *et al.* (1976) and Jordan (1978), such a lithospheric keel will provide a ready trap for incompatible-element-rich fluids rising from deeper in the mantle. In such a model, as a crustal section becomes older, its lithospheric keel will increase in size and become enriched in incompatible elements (e.g., Rb, Nd). Consequently, it will evolve, through time, a distinct isotopic composition from that found in the surrounding mantle. The important distinction between this model of an "enriched" mantle source region and crustal contamination of a lava is twofold. First, an incompatible-element-enriched, ultramafic mantle material would be capable of producing melts with high Mg/Fe but, at the same time, enriched in lithophilic elements. In contrast, any interaction between a primary magma and sialic crustal material would be expected to rapidly lower the Mg/Fe of the melt. Second, if a thick (100 to 300 km) lithospheric keel exists beneath continents and it is significantly enriched in incompatible elements over primordial or bulk earth mantle values, the volume of mantle from which incompatible elements have been extracted over Earth history to produce the continental crust and its enriched mantle keel is considerably greater than the mantle that would have been depleted if the crust alone represents the only "enriched" reservoir on the Earth.

To evaluate the two possibilities of crustal contamination or heterogeneous mantle sources to explain the isotopic variability of Cenozoic volcanics in the northwestern United States, consideration of the chemical composition of these lavas in league with the isotopic data is vital. The following discussion considers this information for the Columbia Plateau and Oregon-Modoc Plateau basalts.

COLUMBIA PLATEAU BASALTS

The Columbia Plateau province contains roughly 200,000 km^3 of basaltic rock erupted in the interval 17 to 6 million years (m.y.) ago with the main volume being extruded prior to 14 m.y. ago (Swanson et al., 1979). Detailed field work in the area (e.g., Swanson et al., 1979) has resulted in a breakdown of the Columbia Plateau basalts into five main stratigraphic units each of which contains a number of chemically distinct members. Of these, two of the older units, the Imnaha and Picture Gorge, which together constitute about 19 volume percent of the Columbia Plateau basalts, were erupted from vents confined to the area south of a major SW-NE trending structural feature—the Blue Mountains anticline (Fruchter and Baldwin, 1975; Camp and Hooper, 1981), which exposes rocks of Upper Triassic age (Armstrong et al., 1977). Basalts of the Imnaha and Picture Gorge sequences have lower ^{87}Sr/^{86}Sr and higher ^{143}Nd/^{144}Nd and tend to be somewhat less fractionated (i.e., higher Mg/Fe, CaO, and Al$_2$O$_3$) than the main volume of Columbia Plateau basalts belonging to the Grande Ronde unit (see Tables 2.1 and 2.2).

Basalts classified with the Grande Ronde unit account for approximately 75 volume percent of the Columbia Plateau lavas (Swanson and Wright, 1979). Feeder dikes with composition similar to Grande Ronde flows occur over a wide portion of the southeastern quarter of the plateau (Taubeneck, 1970). A large number of chemical analyses are available for Grande Ronde basalts (e.g., Wright et al., 1978; Swanson et al., 1979). Over 100 analyses of Grande Ronde lavas are averaged in Table 2.2. As can be seen, with respect to Picture Gorge basalts, Grande Ronde lavas are notably enriched in SiO$_2$, K$_2$O, and TiO$_2$ and the incompatible trace elements Rb, Ba, and La. This unit also exhibits a rather wide range in isotopic composition with ^{87}Sr/^{86}Sr from 0.7042 to 0.7059 and ^{143}Nd/^{144}Nd between 0.51265 and 0.51284 (McDougall, 1976; Carlson et al., 1981).

Following the cessation of extrusion of Grande Ronde-type basalts, lavas of the fourth stratigraphic unit, the Wanapum, were erupted. Volumetrically, Wanapum basalts constitute about 5 percent of the Columbia Plateau volcanic pile. Chemically, most units in the Wanapum group have slightly lower SiO$_2$ and Al$_2$O$_3$ contents and higher TiO$_2$ and P$_2$O$_5$ contents compared with the Grande Ronde unit (Table 2.2). For the most part, Wanapum basalts show a narrow range in isotopic composition (Table 2.1) with the exception of one chemical type—Robinette Mountain, which has a distinctive chemical composition compared with other Wanapum lavas (Swanson et al., 1979) and has much lower ^{87}Sr/^{86}Sr (0.70425) and higher ^{143}Nd/^{144}Nd (0.51279) than observed for other members of the Wanapum sequence (Carlson, unpublished data).

The final volcanic episode on the Columbia Plateau province was characterized by sporadic eruptions of rather small volumes until approximately 6 m.y. ago when volcanism ceased in this region (Swanson et al., 1979). These last eruptions constitute the Saddle Mountains sequence, which contains about 1 percent of the volume of Columbia Plateau lavas. These basalts form a chemically diverse group with different chemical types being erupted from widely spaced vent areas in the central and eastern extremes of the province (e.g., Swanson et al., 1979; Camp and Hooper, 1981). Two extremes in the compositional spectrum of Saddle Mountains basalts are given in Tables 2.1 and 2.2. Isotopically, the majority of Saddle Mountain basalts have evolved compositions with ^{87}Sr/^{86}Sr > 0.7075 and ^{143}Nd/^{144}Nd < 0.51245. In contrast, certain Saddle Mountain flows erupted in the southern extremes of the Plateau in northeastern Oregon (Taubeneck, 1979) have rather constant and relatively

TABLE 2.2 Major Element Composition (in weight %) of Selected Basalts from the Northwestern United States

	Wanapum			Saddle Mt.				
	Picture Gorge	Grande Ronde	Robinette Mt.	Frenchman Springs	Ice Harbor	Umatilla	High-Al Basalt	Oceanic Tholeiite
SiO^2	50.9	54.8	50.0	52.3	47.5	54.7	47.7	49.3
Al$_2$O$_3$	15.5	14.2	17.1	13.2	12.5	14.1	16.5	17.0
FeO	11.7	11.6	10.0	14.4	17.5	12.6	9.87	8.61
MgO	5.77	4.29	7.84	4.04	4.41	2.71	9.21	7.19
CaO	10.1	8.01	11.0	7.90	8.80	6.14	11.3	11.7
Na$_2$O	3.12	3.01	2.44	2.67	2.44	3.20	2.43	2.73
K$_2$O	0.66	1.47	0.27	1.41	1.23	3.68	0.23	0.16
TiO$_2$	1.68	2.00	1.00	3.17	3.79	2.80	1.04	1.49
P$_2$O$_5$	0.28	0.35	0.19	0.71	1.54	0.88	0.15	0.16

SOURCES: Picture Gorge, average of chemical types of 6 and 7 from Swanson et al. (1979); Grande Ronde, average from Wright et al. (1978); Robinette Mt., Frenchman Springs, Ice Harbor, Umatilla, Swanson et al. (1979); High-Al Basalt, Hart (1981b); Oceanic Tholeiite, Engel et al. (1965).

FIGURE 2.3 Nd versus Sr isotopic composition of Columbia Plateau basalts, Oregon-
Modoc Plateau basalts, and selected other
mantle-derived rocks and minerals. Symbols
represent ■, Imnaha; ▼, Picture Gorge; ●,
Grande Ronde; ○, Wanapum; □, Saddle Mt.;
and △, Oregon-Modoc low-K tholeiites. Also
shown are two mixing arrays that would result
from mixing between the two end-member
compositions given in the figure. Isotopic data
for the "rhyolitic" end-member were measured for a crustal xenolith in the Columbia
Plateau lavas by Carlson *et al.* (1981). Crosses
along the two curves show 10 percent increments in the addition of rhyolitic component.
Ratios of cumulate to assimilate mass (C:A) are
given at the ends of each curve and illustrate
the effect of combined crystal fractionation and
crustal assimilation on the isotopic evolution
paths. Fields for ocean-ridge (ORB) ocean-island (OIB) basalts and mantle xenolith min-

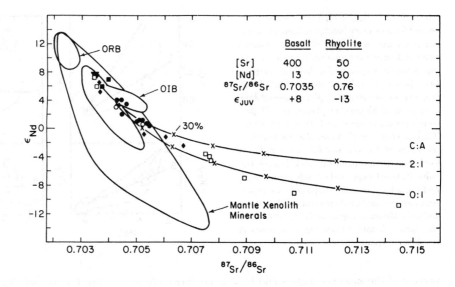

erals are from Carlson *et al.* (1981) as are data for the Columbia Plateau basalts. Data for Oregon-Modoc Plateau samples are from Hart (1981a)
and Carlson *et al.* (1981). ε_{Nd} is defined as $[(^{143}Nd/^{144}Nd)_{sample}/(^{143}Nd/^{144}Nd)_{chondrite} - 1] \times 10^4$. Chondritic $^{143}Nd/^{144}Nd = 0.51264$.

low $^{87}Sr/^{86}Sr$ (0.7035) and high $^{143}Nd/^{144}Nd$ (0.51293) (Carlson
et al., 1981) yet vary in major-element composition from oli-
vine-normative basalt to andesite and nepheline-normative basalt (Taubeneck, 1979). This difference in isotopic composition
between the northeast Oregon Saddle Mountain flows and the
majority of Saddle Mountain members erupted farther north
might be indicative of the general correlation between isotopic
composition and geographic position of igneous rocks in the
northwestern United States.

ORIGIN OF CHEMICAL CHARACTERISTICS

Qualitatively, the distinguishing chemical features of the Grande
Ronde unit (i.e., high SiO_2, K_2O, incompatible trace elements,
and $^{87}Sr/^{86}Sr$) are what would be expected to result from mixing
between old sialic crustal material and a primary magma similar
to the least fractionated Picture Gorge and Imnaha basalts.
However, some of the Grande Ronde chemical and isotopic
characteristics could also be explained by a mantle source that
has been metasomatically enriched in trace elements and that
contains a suitable amount of H_2O to promote the production
of SiO_2-rich melts (Mysen and Boettcher, 1975). To properly
evaluate the possibility of crustal assimilation as a method of
producing the Grande Ronde basalts, the effect of such mixing
on the elemental and isotopic composition of a primitive basalt
composition needs to be quantitatively examined.

Figure 2.3 plots the Nd and Sr isotopic compositions of Columbia Plateau basalts and shows the relation of these data to
those obtained from a variety of oceanic basalts and minerals
extracted from mantle xenoliths in alkalic and kimberlitic volcanics. The fields for oceanic basalts and xenoliths are included
to illustrate the range in isotopic composition observed in unquestionably mantle-derived rocks and minerals. Also shown
in Figure 2.3 are curves that would be expected to result from
mixing between primitive Imnaha magma and a crustal com-

ponent, which in this case is represented by the composition
shown in the figure. The effect of variations in the assumed
composition of the crustal component is discussed by Carlson
et al. (1981). For this example the isotopic composition of the
crustal component is that measured for a quartz-rich, presumably metasedimentary, crustal xenolith in the Columbia Plateau
lavas. Both Sm-Nd and Rb-Sr model ages for this xenolith
suggest that it was derived from material of early Proterozoic
age existing beneath the Columbia Plateau. The two curves in
Figure 2.3 show two cases of mixing—one simple binary mixing, the other includes the effects expected when the magma
is also precipitating crystals that are then removed from the
magma. This latter case offers the magma the opportunity to
regain some of the heat it loses in melting its wallrock by gaining
the heat of crystallization of the phases forming out of the
magma (Bowen, 1928). The differences between the binary
mixing and combined mixing-crystal fractionation curves result,
in this case, from the rapid increase in Nd concentration in the
magma caused by the removal of a crystallizing phase assemblage that strongly rejects Nd into the melt. Thus, as the Nd
content of the magma increases, a given amount of crustal
material will have less of an effect on the Nd isotopic composition of the melt, whereas its effect on the Sr isotopic composition is even slightly enhanced in this model because of
decreasing Sr content in the magma.

Both of these mixing curves satisfy the isotopic data of the
Picture Gorge, Imnaha, Grande Ronde, and Wanapum lavas,
which, it should again be stressed, constitute about 99 percent
of the Columbia Plateau basalts. On the other hand, only the
binary mixing model follows the data for the volumetrically
minor, high $^{87}Sr/^{86}Sr$, Saddle Mountains basalts. Another important feature of this mixing model is that to approach the
highest $^{87}Sr/^{86}Sr$ values seen for Grande Ronde lavas less than
about 20 percent of the magma's mass is derived from the
crustal component. On the other hand, to reach the high $^{87}Sr/$
^{86}Sr seen for some Saddle Mountain basalts, greater than 60

FIGURE 2.4 Major-element variation expected for a primary basaltic magma undergoing mass exchange with the crust. Curves
marked A show evolution paths in binary mixing between magma and a rhyolitic composition; those marked B include the effect of frac-
tionally crystallizing a phase assemblage
composed of 55 percent plagioclase, 35 percent low-Ca pyroxene, and 10 percent olivine
(Carlson *et al.*, 1981). A cumulate to assimilate
mass ratio of 2 is assumed for curves B. Open
fields encompass major-element data for Imnaha, Picture Gorge, and Grande Ronde lavas,
and cross-hatched and stippled fields represent the compositional spectrum of Wanapum
and high-Al basalts, respectively. The dots along
the curves denote 10 percent increments of
rhyolitic component.

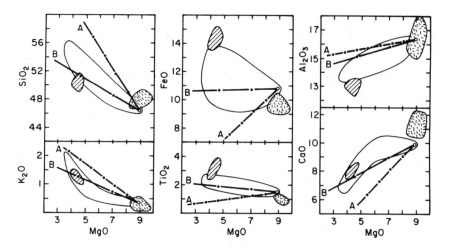

percent of the magma's mass would have to be derived from
the crust.

Extending the results of this crustal interaction model to the
major-element composition of the Columbia Plateau lavas, Figure 2.4 compares the compositional trends expected for mixing
between a primitive Imnaha basalt and a granitic material.
Chosen in this example to represent this granitic component
is a material similar in elemental, but not isotopic, composition
to rhyolites from the Newberry volcano of central Oregon (Higgins, 1973). Simple binary mixing between this rhyolite and
the Imnaha composition fails to produce a mixing array that
encompasses the Columbia Plateau compositions. Binary mixing of up to 60 percent crustal component is required to produce the lower MgO concentration seen for these basalts. In
contrast, if the magma is allowed to crystallize while it is
undergoing crustal assimilation and the crystallizing assemblage is removed from the magma, a much better fit to the
compositional data can be achieved given certain assumptions
regarding the composition of the phase assemblage being removed.

In Figure 2.4 the set of curves marked B represent the
compositional evolution of the primitive Imnaha magma when
it undergoes assimilation of the Newberry rhyolite coupled with
the removal of a crystallizing phase assemblage composed of
55 percent plagioclase, 35 percent low-Ca pyroxene, and 10
percent olivine. This phase assemblage is similar to that proposed by Cox (1980) to explain the compositional trends found
in continental flood basalts in general. The assumed ratio of
mass of removed crystals to assimilated crust is 2:1 in Figure
2.4. This model provides a reasonable match to the major-
element trends exhibited by Imnaha, Picture Gorge, Grande
Ronde, and Wanapum basalts and requires only 20 percent (by
mass) assimilation of the crustal component to produce the
compositions with lowest MgO, in accord with the relative
contribution of crustal material indicated by the isotopic data.
The sensitivity of this model to changes in the composition of
both the crystallizing phase assemblage and the ratio of cumulate to assimilate mass is discussed in greater detail by Carlson *et al.* (1981).

This model of crystallization-assimilation also tends to match

the Rb, Sr, and Nd contents of these lavas. Again, about 20
percent of a crustal component seems to be required to explain
the more evolved trace-element compositions. In this model
the low ratio of crystallized mass to melted crustal mass (2:1),
which provides the best solution to the isotopic and compositional data, indicates that the magma was not being rapidly
cooled by ingesting its wallrock. This suggests that the magma-
wallrock interaction, in this case, occurred at great depth, presumably in the lower crust where the temperature of the wallrock was near its melting temperature even before the injection
of the mafic lavas.

While this combined crystallization-assimilation model ap-
proximates the chemical trends seen in the majority of the
Columbia Plateau basalts, the composition of the Saddle Mountains basalts, which have high $^{87}Sr/^{86}Sr$, is not satisfactorily
explained. To reach an $^{87}Sr/^{86}Sr > 0.707$ as observed for these
samples, the model discussed above would predict that the
magma would approach an andesitic major-element composition but would have a high FeO (> 10 wt.%) and Rb (70 ppm)
and low MgO (1 wt.%) content. In contrast, the Pomona chemical unit of the Saddle Mountains basalts has an $^{87}Sr/^{86}Sr =
0.7075$ yet also has compositions of MgO = 7.1 wt.%, CaO =
10.6 wt.%, and Rb = 16 ppm (Carlson *et al.*, 1981). This
disparity between predicted and observed composition is difficult to overcome by minor variations in the model and may,
in the end, mean that these isotopically distinct Saddle Mountain basalts do not owe their characteristics to the influence of
sialic continental crust.

On the other hand, if instead of using the chemical and
isotopic compositions of a primitive Imnaha or Picture Gorge
basalt as the primary magma for all the Columbia Plateau lavas,
the high $^{87}Sr/^{86}Sr$ Saddle Mountain basalts are assumed to be
derived from a parental magma similar in composition to the
Pomona chemical unit, then the composition of other Saddle
Mountain basalts can be explained by using a crystallization-
crustal assimilation model similar to that described earlier.
Such a model of mixing between sialic crustal materials and a
primary magma with $^{87}Sr/^{86}Sr = 0.7075$ and $^{143}Nd/^{144}Nd =
0.51245$ can in fact account for much of the major- and trace-
element and isotopic variations seen for the remainder of the

Saddle Mountain basalts. The idea of an isotopically evolved parent magma for this series may also be supported by the fact that the majority of primitive Snake River Plain basalts have $^{87}Sr/^{86}Sr$ values clustering around 0.7075 (Leeman and Manton, 1971).

Another important factor is the lack of variation in isotopic composition for members of the Picture Gorge sequence and especially for the chemically diverse samples of Saddle Mountain volcanics from northeastern Oregon. Although these Saddle Mountain eruptives vary in composition from olivine normative basalt to andesite, their $^{87}Sr/^{86}Sr$ of 0.7035 is constant and similar to values measured for both Picture Gorge and Imnaha lavas. The lack of isotopic variability of the Picture Gorge, Imnaha, and these Saddle Mountains flows may imply that (1) these magmas did not undergo mass exchange with crustal materials, (2) no old and isotopically evolved crustal materials were present on the extreme southern edge of the province where these flows were erupted, or (3) the mantle underlying this area is distinctly less isotopically evolved than that to the north.

Combined crystallization and crustal assimilation can thus account for a number of the chemical and isotopic features of the main volume of the Columbia Plateau flood basalts. The predicted chemical trends from this model are somewhat sensitive to the composition of assimilated material and quite sensitive to the assumed crystallizing phase assemblage and to the ratio of cumulate to assimilate mass (Carlson *et al.*, 1981). However, the fact that the crystallization-assimilation model seems to satisfy both the chemical and isotopic trends observed in the major volume of the Columbia Plateau lavas strongly suggests that these basalts did indeed melt and mix with some of the crustal materials they came in contact with. This might not be unexpected because the manner in which the Columbia Plateau lavas were erupted, that being in large volumes continually erupted through the same limited geographic area over a prolonged time period, maximizes the probability that these magmas heated the surrounding crust beyond its melting point (Marsh, Chapter 5, this volume). Nevertheless, the Columbia Plateau basalts lack the physical evidence for extensive mixing with the crust; they are notably devoid of both phenocrysts and xenoliths. This might be explained by assuming that the parental lavas rose to the surface very rapidly from reservoirs deep in the crust. The rapid ascent would promote superheating of the magmas and possibly also the resorption of whatever phenocrysts and xenoliths were entrained with the rising magma. Although the lack of physical evidence of crustal assimilation need not necessarily contradict the geochemical evidence for such a process, it does make verification of the proposed fractionation scenario more difficult.

OREGON-MODOC PLATEAU

Post-Eocene volcanism south of the Columbia Plateau is composed predominantly of bimodal rhyolite and high-Al basalt. This section addresses the compositional characteristics of the low-K, high-Al basaltic lavas in this area. These basalts display a rather limited range in major- and trace-element composition despite their eruption over a wide geographic area and a long time span (0 to 11 m.y.; Hart, 1981a). They have a distinct composition, marked in particular by high Al_2O_3 content and low TiO_2 and alkali contents (see Tables 2.1 and 2.2 and Figure 2.4). Chemically, these basalts share many similarities with mid-ocean ridge basalts (MORB); however, slight, but important, differences exist. Most important are their higher and more variable $^{87}Sr/^{86}Sr$ and lower $^{143}Nd/^{144}Nd$ than MORB (Table 2.1). The lowest $^{87}Sr/^{86}Sr$ measured for the high-Al tholeiites on the Oregon-Modoc Plateau is about 0.7030 for those samples occurring in northern California and southern Oregon just east of the Cascade chain (Mertzman, 1979; Carlson *et al.*, 1981, Hart, 1981b; McKee *et al.*, 1983). Corresponding values of $^{143}Nd/^{144}Nd$ for these same samples are on the order of 0.51297 (Carlson *et al.*, 1981, and unpublished data). Slightly lower values of $^{87}Sr/^{86}Sr$, as low as 0.7027 (Church and Tilton, 1973), have been measured for high-Al basalts in the Cascade chain. In contrast to these relatively low values of $^{87}Sr/^{86}Sr$, samples of chemically similar basalts from northern Nevada, southeastern Oregon, and southwestern Idaho have higher and more variable $^{87}Sr/^{86}Sr$ in the range of 0.705 to 0.7076 (Mark *et al.*, 1975; Hart, 1981b) and $^{143}Nd/^{144}Nd$ values as low as 0.51237 (Hart, 1981a).

Again, as in the case of the Columbia Plateau basalts, these higher $^{87}Sr/^{86}Sr$ and lower $^{143}Nd/^{144}Nd$ values are what might be expected for a magma containing some component derived from old continental crust. However, unlike the case for Columbia Plateau basalts, the chemical composition of the high-Al basalts shows only a poor correlation with isotopic composition. In fact, the low alkali content of all the high-Al basalts in this area presents a strong argument against the presence of any sialic crustal component in these lavas. For example, to reach $^{87}Sr/^{86}Sr$ values of 0.707, about 35 percent of a component similar to that used for crustal contaminant for the Columbia Plateau lavas is required to be present in the high-Al basalt. Such a mixture should also then have SiO_2, K_2O, and Rb contents of 56 wt.%, 1.3 wt.%, and 57 ppm, respectively. Those high-Al basalts that actually have $^{87}Sr/^{86}Sr = 0.707$ have SiO_2, K_2O, and Rb concentrations of 48 wt.%, 0.2 wt.%, and 3 to 5 ppm, respectively (Mark *et al.*, 1975; Hart, 1981b). Thus, the chemical composition of these basalts does not allow any significant component of sialic, isotopically evolved crustal material. Rather their composition suggests that they are relatively unmodified partial melts of ultramafic mantle materials. Furthermore, judging from the high Mg/Fe of these basalts, they did not experience significant amounts of fractional crystallization and hence are likely to have been transported quickly from their mantle source to their eruption on the surface. Texturally, the majority of these low-K, high-Al olivine tholeiites are diktytaxitic equigranular and lack abundant phenocrysts—in support of their chemically unfractionated nature.

To explain the variable isotopic composition of the high-Al basalts it is necessary to consider the possibility of long-standing chemical heterogeneities in their mantle sources or perhaps mixing between a "depleted" or low $^{87}Sr/^{86}Sr$ mantle region and the isotopically evolved mantle source suggested for Snake River Plain basalts (Leeman and Manton, 1971) and for the Saddle Mountain basalts. Judging from the relatively minor variation in minor-element composition of these basalts, the heterogeneity in their sources seems to have been primarily restricted to incompatible trace-element abundances. It must be pointed out, however, that there is no well-defined correlation between isotopic composition and incompatible trace-element abundance ratios such as Rb/Sr or Sm/Nd (Mark et al., 1977; Hart, 1981b). Hence these data do not define meaningful "pseudoisochrons." Furthermore, because most of the high-Al basalts have Rb/Sr values that are too low to have evolved to even their lowest $^{87}Sr/^{86}Sr$ values over the age of the Earth, it must be concluded that the low-Rb contents of these lavas reflect a previous differentiation event that extracted alkalies and other incompatible elements from the materials that would later melt to produce the high-Al basalts (e.g., McKee et al., 1983). In such a scenario the isotopic heterogeneity in the sources of these basalts must have existed before the depletion event, because the Sr isotopic evolution of these materials essentially stopped at the time of the depletion because of the low Rb/Sr ratios left after this event. Thus, the history of the source materials for the high-Al basalts must be marked by perhaps several events of both enrichment and depletion.

HETEROGENEOUS MANTLE

One method of imposing heterogeneity on an originally homogeneous mantle is that of extracting partial melt. The residue left behind in such an event will be highly depleted in its most incompatible elements and hence have lower Rb/Sr and higher Sm/Nd than it had before the melt extraction. Models of melt extraction, leaving behind depleted mantle reservoirs, have long been proposed (Tatsumoto et al., 1965) to explain the low $^{87}Sr/^{86}Sr$ measured in MORB. Variations in the degree of melt extraction also have been proposed as one explanation for the observed negative correlation between the Sr and Nd isotopic compositions of oceanic basalts (e.g., DePaolo, 1979).

Previous melt extraction offers a satisfactory explanation for lower $^{87}Sr/^{86}Sr$ and higher $^{143}Nd/^{144}Nd$ than expected for the bulk mantle; however, such extraction does not explain the respectively higher and lower $^{87}Sr/^{86}Sr$ and $^{143}Nd/^{144}Nd$ relative to bulk earth values of some high-Al basalts and the high $^{87}Sr/^{86}Sr$ Saddle Mountain basalts. If chemical heterogeneities in the mantle result from the migration of partial melts, examination of the possibility of "enriching" an area in the mantle by adding to it melt derived from some other region in the mantle is necessary. In such a model a fluid originating by the partial melting of a deeper portion of the mantle ascends and either solidifies within or equilibrates with overlying mantle (e.g., Lloyd and Bailey, 1975). Evidence for such metasomatic processes in the mantle can be found in the relatively high abundance of ultramafic xenoliths found to contain hydrous phases. Figure 2.5 shows the effect of the addition of various types of melt on the isotopic evolution of a mantle region. In this model 5 parts melt are combined with 95 parts of a primitive mantle composition (Jagoutz et al., 1979), and the mixture is allowed to evolve its isotopic composition with its new Rb/Sr and Sm/Nd.

To produce in this manner a mantle region that will evolve the high $^{87}Sr/^{86}Sr$ for a given $^{143}Nd/^{144}Nd$ (as observed in the high $^{87}Sr/^{86}Sr$ Saddle Mountain basalts), the addition of a component to the mantle that has a high Rb/Sr for a given Sm/Nd is necessary. One possibility for this high Rb/Sr component, and one that is commonly observed in mantle xenoliths, is phlogopitic mica. As shown in Figure 2.5, a 1 percent addition of phlogopite to the mantle will lead to nearly horizontal isotopic evolution on a Nd versus Sr isotopic plot because of the high Rb/Sr of the phlogopite and the fact that this mineral has very low abundances of the rare-earth elements (REE). Because of the rapid rise of $^{87}Sr/^{86}Sr$ in the phlogopite-bearing mantle, the addition of a phase enriched in the light REE for this mantle region would also be necessary to approach the Nd and Sr isotopic evolution trend observed for the Saddle Mountain basalts. One possible way to do this is to have the invading fluid be similar in composition to K- and light-REE-rich kimberlite magmas. An important constraint on models of mantle metasomatism as an explanation of the isotopic variability of northwest U.S. basalts is that in order to approach the high $^{87}Sr/^{86}Sr$ for a given $^{143}Nd/^{144}Nd$ observed for these lavas the involvement of phases in their source region that strongly enrich Rb over Sr seems necessary. Because likely phases with high Rb/Sr, such as micas, are stable only above depths of ∼ 150 km, the conclusion can be drawn that only those regions

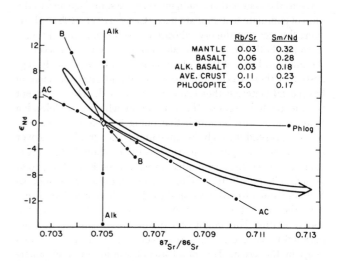

FIGURE 2.5 Present-day isotopic composition expected for a mantle region formed at various times in the past by the addition or subtraction of the listed melt compositions from a primitive mantle reservoir (B = basalt; Alk = alkali basalt; AC = average crust). Dots along each curve represent 10^9-yr intervals since the enrichment ($\varepsilon_{Nd} > 0$) or depletion ($\varepsilon_{Nd} < 0$) event that altered the Rb/Sr and Sm/Nd of the mantle reservoir. The melt addition curves assume that 5 percent melt was added to 95 percent primitive mantle to produce the "enriched" reservoirs, with the exception of the phlogopite composition where only 1 percent phlogopite was added to 99 percent primitive mantle. Isotopic evolution of "depleted" reservoirs assumes 1 percent melt (or 5 percent for the basaltic composition melt) was extracted from a volume of primitive mantle to leave behind the "depleted" material. The open, concave-upward field shows the range of data for northwest U.S. volcanics. Present-day isotopic composition of the model primitive mantle is shown by the diamond.

of the mantle above $\sim$ 150 km, when metasomatized with a fluid of appropriate composition, can evolve to the high $^{87}Sr/^{86}Sr$ observed for some basalts from the northwest United States.

As Figure 2.5 shows, the introduction of a melt of average crustal composition to a region of the mantle will eventually produce isotopic values within the mantle that approximate those observed for northwest U.S. basaltic lavas. However, such "enriched" regions of the mantle, whether enriched by a kimberlitic or a sialic composition liquid, would be expected to produce melts with considerably different trace-element characteristics than those observed in the high-Al basalts. In particular, these melts would be expected to have high Rb/Sr and light-REE enrichments (chondrite-normalized) in contrast to the low Rb/Sr and flat or slightly light-REE-depleted REE abundances of the high-Al basalts. For models of an enriched mantle source to explain the low incompatible trace-element contents but evolved isotopic compositions of the high-Al basalts, a multistage history of enrichment followed by depletion must be assumed for the source regions of these basalts. This last depletion event must have occurred close enough to the present day (i.e., less than a few hundred million years ago) that the isotopic composition of the area did not have time to change significantly. A mechanism capable of depleting this enriched mantle source involves the extraction of a small amount of partial melt. One possibility for the recent depletion of the high-Al basalt source regions is that they participated in the production of the numerous Mesozoic batholiths in the area as implied by the similarity of initial $^{87}Sr/^{86}Sr$ of these batholiths and nearby Cenozoic volcanics (Armstrong *et al.*, 1977). In such a model the mantle sources of the high-Al basalts would essentially represent the residues left behind when the melts, which now make up the Mesozoic batholiths, were extracted from this originally "enriched" subcontinental lithosphere.

SUMMARY AND CONCLUSIONS

To produce a model that explains the characteristics of volcanism in the northwest, consideration of both the chemical data for these rocks and the tectonic conditions of the region in which they were erupted is important. Observations in this regard are that the majority of Columbia Plateau basalts show chemical and isotopic evidence for assimilation of a crustal component, the Columbia Plateau contains an insignificant amount of rhyolitic composition rocks, and the tectonic condition of the crust in the area where these basalts were erupted is one of limited extension. In contrast, basaltic volcanism in the Oregon-Modoc Plateau shows no chemical indication of a significant crustal component, rhyolitic volcanism in the area is common and of large volume, and the region has experienced severe extension throughout its post-Eocene history. Using these observations, the following model is presented as a possible scenario for the volcanic history of the northwest United States.

Resulting perhaps from both an attempt to create an ensialic marginal basin and from differential stress caused by the new character of plate interaction along the west coast of North America, the region now forming the Oregon-Modoc Plateau began to undergo extension. Diapiric upwelling of the mantle beneath this crustal section led to its partial melting. Whether such "passive" mantle upwelling also occurred beneath the Columbia Plateau is difficult to establish. The limited extension and the much larger volumes of basalt present in the Columbia Plateau area may suggest a "driven" system where the diapiric upwelling of a deeper mantle section causes uplift and rifting of the overlying crust. Presumably in both the Columbia Plateau and the Oregon-Modoc Plateau regions, the intrusion of primary basic magmas into the crust led to the production of rhyolitic melts through anatectic melting of the lower crust. After this point, however, the evolutionary paths followed by the Columbia Plateau and Oregon-Modoc Plateau lavas must have diverged. Large block faults caused by the severe extension of the northern Great Basin seemingly provided low impedance paths for the rhyolitic and basaltic melts to follow, allowing the eruption of these melts on the Oregon-Modoc Plateau soon after they were formed. The relatively small volumes of single basaltic flows in this area may also mean that a short interval between melt production and eruption did not allow the formation of large-volume magma chambers.

In contrast to the Oregon-Modoc lavas, primary magmas under the Columbia Plateau did not encounter a severely extended crust. Hence, these magmas may have ponded at the crust-mantle interface for a considerably longer time than the Oregon-Modoc lavas. A long residence time at the crust-mantle interface would be expected to increase the total amount of heat loss from the magma, causing it to begin to crystallize; would allow for a higher degree of mixing between the magma and the crust above it; and would allow the large accumulations of magma required to explain the large volumes of single flows in the Columbia Province (Shaw and Swanson, 1970). Eruption of individual flows was possibly triggered by lowering the density of the magma by both fractionation and mixing with the crust, thus allowing its ascent through the crust and/or the creation of magma flow channels by fracturing of the crust. The scarcity of rhyolitic eruptions in the Columbia Plateau region can be explained in this model by assuming that mixing between any rhyolitic magma produced and the intruding basalt occurred more rapidly than the rhyolite rose through the crust.

The model presented here is obviously not a unique solution to the nature of volcanism in the northwest United States. In particular, if an isotopically heterogeneous mantle can explain the isotopic variability of all the volcanics in this region, as may be indicated by the Oregon-Modoc Plateau basalts, then the conclusion that the Grande Ronde unit of the Columbia Plateau basalts has undergone mixing with crustal materials may no longer be necessary. However, the fact that much of the isotopic and chemical data for the Grande Ronde basalts seems to fit the assimilation-fractional crystallization model is a strong argument in its favor. Regardless, the apparent correlation between magma type, eruption style, and tectonic condition of the eruption area of lavas in the northwest United States illustrates the importance of the interaction between physical and chemical processes in controlling the production, transport, and eruption of magmas.

ACKNOWLEDGMENTS

The major part of the work on Columbia Plateau basalts was completed as part of my Ph.D. thesis at the University of California, San Diego. The advice and assistance given by G. Lugmair and J. D. Macdougall is sincerely appreciated. I thank R. Stern, W. H. Hart, T. Wright, and D. Swanson for numerous helpful discussions. Helpful reviews by F. R. Boyd, B. Doe, W. Leeman, and T. Wright are much appreciated.

REFERENCES

Armstrong, R. L., W. H. Taubeneck, and P. O. Hales (1977). Rb-Sr and K-Ar geochronometry of Mesozoic granitic rocks and their Sr isotopic composition, Oregon, Washington, and Idaho, *Geol. Soc. Am. Bull. 88*, 397-411.

Bowen, N. L. (1928). *The Evolution of the Igneous Rocks*, Princeton U. Press, Princeton, N.J., 334 pp.

Brooks, C., D. E. James, and S. R. Hart (1976). Ancient lithosphere: Its role in young continental volcanism, *Science 193*, 1086-1094.

Camp, V. E., and P. R. Hooper (1981). Geologic studies of the Columbia Plateau: Part l, Late Cenozoic evolution of the southeast part of the Columbia River basalt province, *Geol. Soc. Am. Bull. 92*, 659-668.

Carlson, R. W., G. W. Lugmair, and J. D. Macdougall (1981). Columbia River volcanism: The question of mantle heterogeneity or crustal contamination, *Geochim. Cosmochim. Acta 45*, 2483-2499.

Christiansen, R. L., and P. W. Lipman (1972). Cenozoic volcanism and plate-tectonic evolution of the western United States, II. Late Cenozoic, *Phil. Trans. R. Soc. London A271*, 249-284.

Christiansen, R. L., and E. H. McKee (1978). Late Cenozoic volcanic and tectonic evolution of the Great Basin and Columbia Intermontane regions, in *Cenozoic Tectonics and Regional Geophysics of the Western Cordillera*, R. B. Smith and G. P. Eaton, eds., Geol. Soc. Am. Mem. 152, pp. 283-311.

Church, S. E., and G. R. Tilton (1973). Lead and strontium isotope studies in the Cascade Mountains: Bearing on andesite genesis, *Geol. Soc. Am. Bull. 84*, 431-454.

Cox, K. G. (1980). A model for flood basalt volcanism, *J. Petrol. 21*, 629-650.

DePaolo, D. J. (1979). Implications of correlated Nd and Sr isotopic variations for the chemical evolution of the crust and mantle, *Earth Planet. Sci. Lett. 43*, 201-211.

DePaolo, D. J., and G. J. Wasserburg (1976). Inferences about magma sources and mantle structure from variations of $^{143}Nd/^{144}Nd$, *Geophys. Res. Lett. 3*, 743-746.

Doe, B. R. (1967). The bearing of lead isotopes on the source of granitic magma, *J. Petrol. 8*, 51-83.

Engel, A. E. J., C. G. Engel, and R. G. Havens (1965). Chemical characteristics of oceanic basalts and the upper mantle, *Geol. Soc. Am. Bull. 76*, 719-734.

Fruchter, J. S., and S. F. Baldwin (1975). Correlations between dikes of the Monument swarm, central Oregon, and Picture Gorge basalt flows, *Geol. Soc. Am. Bull. 86*, 514-516.

Hamilton, W., and W. B. Myers (1966). Cenozoic tectonics of the western United States, *Rev. Geophys. 4*, 509-550.

Hart, W. K. (1981a). Chemical, geochronologic and isotopic significance of low K, high-alumina olivine tholeiite in the northwestern Great Basin, U.S.A., Ph.D. thesis, Case Western Reserve U., Cleveland, Ohio.

Hart, W. K. (1981b). Spatial variation of $^{87}Sr/^{86}Sr$ ratios in low K olivine tholeitte, N.W. Great Basin, U.S.A., *Carnegie Inst. Washington Yearb. 80*, 475-479.

Higgins, M. W. (1973). Petrology of Newberry Volcano, central Oregon, *Geol. Soc. Am. Bull. 84*, 455-488.

Jagoutz, E., H. Palme, H. Baddenhausen, K. Blum, M. Cendale, G. Dreibus, B. Spettel, V. Lorenz, and H. Wanke (1979). The abundances of major, minor, and trace elements in the Earth's mantle as derived from primitive ultramafic nodules, in *Proceedings of the 10th Lunar and Planetary Science Conference*, Pergamon, New York, pp. 2031-2050.

Jordan, T. H. (1978). Composition and development of the continental tectosphere, *Nature 274*, 544-548.

Leeman, W. P. (1975). Radiogenic tracers applied to basalt genesis in the Snake River Plain-Yellowstone National Park region—Evidence for 2.7 b.y. old upper mantle keel, *Geol. Soc. Am. Abstr. Progr. 7*, 1165.

Leeman, W. P., and W. I. Manton (1971). Strontium isotopic composition of basaltic lavas from the Snake River Plain, southern Idaho, *Earth Planet. Sci. Lett. 11*, 420-434.

Lloyd, F. E., and D. K. Bailey (1975). Light element metasomatism of the continental mantle: The evidence and the consequences, in *Physics and Chemistry of the Earth, Vol. 9*, L. H. Ahrens, J. B. Dawson, A. R. Duncan, and A. J. Erlank, eds., Pergamon, Oxford, pp. 385-416.

Lugmair, G. W., N. B. Scheinin, and K. Marti (1975). Sm-Nd age and history of Apollo 17 basalt 75075: Evidence for early lunar differentiation of the lunar exterior, in *Proceedings of the 6th Lunar Science Conference*, Pergamon, New York, 1419-1429.

Mark, R. K., C. Lee Hu, H. R. Bowman, F. Asaro, E. H. McKee, and R. R. Coats (1975). A high $^{87}Sr/^{86}Sr$ mantle source for low alkali tholeiite, northern Great Basin, *Geochim. Cosmochim. Acta 39*, 1671-1678.

McDougall, I. (1976). Geochemistry and origin of basalt of the Columbia River group, Oregon and Washington, *Geol. Soc. Am. Bull. 87*, 777-792.

McKee, E. H., W. A. Duffield, and R. J. Stern (1983). Late Miocene and early Pliocene basaltic rocks and their implications for crustal structure, northeastern California and south-central Oregon, *Geol. Soc. Am. Bull. 93*, 292-304.

Mertzman, S. A., Jr. (1979). Strontium isotope geochemistry of a low potassium olivine tholeiite and two basalt-pyroxene andesite magma series from the Medicine Lake Highland, California, *Contrib. Mineral. Petrol. 70*, 81-88.

Mysen, B. O., and A. L. Boettcher (1975). Melting of a hydrous mantle: II. Geochemistry of crystals and liquids formed by anatexis of mantle peridotite at high pressures and temperatures as a function of controlled activities of water, hydrogen, and carbon dioxide, *J. Petrol. 16*, 549-593.

Noble, D. C., C. E. Hedge, E. H. McKee, and M. K. Korringa (1973). Reconnaissance study of the strontium isotopic composition of Cenozoic volcanic rocks in the northwestern Great Basin, *Geol. Soc. Am. Bull. 84*, 1393-1406.

O'Nions, R. K., P. J. Hamilton, and N. M. Evensen (1977). Variations in $^{143}Nd/^{144}Nd$ and $^{87}Sr/^{86}Sr$ ratios in oceanic basalts, *Earth Planet. Sci. Lett. 34*, 13-22.

Peterman, Z. E., I. S. E. Carmichael, and A. L. Smith (1970). $^{87}Sr/^{86}Sr$ ratios of Quaternary lavas of the Cascade Range, northern California, *Geol. Soc. Am. Bull. 81*, 311-318.

Schilling, J. G. (1972). Sea floor evolution: Rare-earth evidence, *Phil. Trans. Roy. Soc. London A268*, 663-706.

Shaw, H. R., and D. A. Swanson (1970). Eruption and flow rates of flood basalts, in *Proceedings, Second Columbia River Basalt Symposium*, E. H. Gilmour and D. Stradling, eds., Eastern Washington State Coll. Press, Cheney, Wash., pp. 271-299.

Swanson, D. A., and T. L. Wright (1979). The regional approach to studying the Columbia River basalt group, in *International Symposium on Deccan Volcanism and Related Basalt Provinces in Other Parts of the World*, pp. 13-16.

Swanson, D. A., T. L. Wright, P. R. Hooper, and R. D. Bentley (1979). Revisions in stratigraphic nomenclature of the Columbia River Basalt group, *U.S. Geol. Surv. Bull. 1457-G*, G1-G59.

Tatsumoto, M., C. E. Hedge, and A. E. J. Engel (1965). Potassium, rubidium, strontium, thorium, uranium, and the ratio of strontium-87 to strontium-86 in oceanic tholeiitic basalt, *Science 150*, 886-888.

Taubeneck, W. H. (1970). Dikes of Columbia River basalt in northeastern Oregon, western Idaho, and southeastern Washington, in *Proceedings, Second Columbia River Basalt Symposium*, E. H. Gilmour and D. Stradling, eds., Eastern Washington State Coll. Press, Cheney, Wash., pp. 73-96.

Taubeneck, W. H. (1979). Columbia River basalt between the Minam River Canyon and the Grande Ronde Graben, northeast Oregon (abstr.), *Pacific Division AAAS Annual Meeting*, Moscow, Idaho.

Wright, T. L., D. A. Swanson, R. T. Helz, and G. R. Byerly (1978). Major oxide, trace element, and glass chemistry of Columbia River basalt samples collected between 1971 and 1977, *U.S. Geol. Surv. Open-File Rep. 79-711*, 146 pp.

Yoder, H. S., Jr., ed. (1979). *The Evolution of the Igneous Rocks, Fiftieth Anniversary Perspectives*, Princeton U. Press, Princeton, N.J., 588 pp.

Zartman, R. E. (1974). Lead isotopic provinces in the Cordillera of the western United States and their geologic significance, *Econ. Geol. 69*, 792-805.

Subduction Geometry and Magma Genesis

3

I. SELWYN SACKS
Carnegie Institution of Washington
Department of Terrestrial Magnetism

ABSTRACT

In normal subduction zones the slab sinks into the mantle at some approximately constant dip angle. There is asthenosphere between this slab and the continental lithosphere above, and the line of volcanoes has a simple relationship to the subducted plate. Between the slab and the surface, magma paths must be fairly confined, since the bulk properties of the asthenosphere seem laterally constant. In particular, the anelasticity, a measure of partial melt, is rather uniform down to depths of approximately 250 km and is similar beneath Japan, northern Chile, and normal oceans.

However, beneath central Peru (and central Chile) the (young) Nasca plate subducts initially at a "normal" angle of 30° and then at a depth of about 100 km deforms and travels essentially horizontally for about 300 km before interacting with the thicker continental lithosphere of the Precambrian shield and resubducting. The seismicity tends to be sparse except at the initial subduction and resubduction areas; the continuity of the slab was determined from anelasticity considerations. There is no recent volcanism in these regions. There is no low Q zone (asthenosphere) between the subducted and continental lithosphere. Young oceanic lithosphere has lower density and may be buoyant to considerable depths; after being forced down by the initial interplate contact, it may contort and underlay the continental lithosphere.

The tectonics (seismicity and lack of volcanism) are similar in the anomalous South American regions and the eastern Philippine Sea subduction region. It appears that volcanoes occur only when there is (hot, low Q) asthenospheric material above the subducting slab.

INTRODUCTION

The oceanic lithosphere and crust, generated at mid-ocean ridges, moves laterally (at velocities of 1 to 15 cm/yr) for considerable distances. In many regions this oceanic plate then subducts, presumably because of its interactions with an adjacent plate. The path of this subducting plate into the upper mantle is usually delineated by a zone of earthquakes that occur in the upper part of the plate. The seismicity is not necessarily continuous—the activity varies widely in adjacent regions. There is a line of volcanoes, commonly 90 to 150 km (occasionally up to 300 km) above the seismic plane.

Numerous studies have shown that there is little absorption of higher-frequency seismic waves traveling in this subducted plate. Indeed this was the primary technique used to identify a subducted plate. Q_p values seem to be well in excess of 1000 in the subducting plate but only a few hundred in the asthenosphere. This is commonly ascribed to the presence of a small degree of partial melt, which is known to cause high absorption. The path of the subducted plate is generally simple. Below the initial bend near the surface to a dip of 25 to 45° the plate sinks at the same angle with only minor contortions. (In the Marianas region, however, the dip increases with depth.)

Subducted Plate—Overlying Asthenosphere Interaction

One major effect of subduction must be to cool the asthenosphere into which it subducts. The oceanic plate, having been at the Earth's surface, is necessarily colder than the asthenosphere and therefore more dense. This is probably why it keeps sinking once it has been subducted. Its exact path is affected by possible motion in the upper mantle and by trench migration. The motion of the subducted plate affects the asthenosphere in two ways: (1) the heating of the subducted plate is at the expense of cooling the adjacent asthenosphere, which becomes more dense and will tend to sink, and (2) the drag of the plate on the asthenosphere will tend to force it to rotate, the adjacent area moving down with the plate. The two effects are cumulative. Figure 3.1 shows the disturbance to the temperature field. If the asthenosphere system (above the subduction zone) were closed, i.e., if there was a cooling of the asthenosphere in the wedge between the subducting and continental lithospheres without any involvement of the adjacent upper mantle material, the asthenospheric wedge would eventually cool so much that magma genesis would be inhibited, the viscosity would increase, and the subduction process would be modified.

Indications That the Asthenosphere Wedge Is Open

The observation that subduction and magmatism can continue for hundreds of millions of years (e.g., Japan) indicates that freezing out of the asthenospheric wedge clearly does not happen.

To get a quantitative estimate of the change in asthenospheric properties in this region, I compare the anelasticity of the region above the slab with that seaward of the trench. If, as is customarily assumed, the anelastic absorption of seismic waves in the asthenosphere is due mainly to partial melt, this comparison should be a sensitive check for cooling effects. Figure

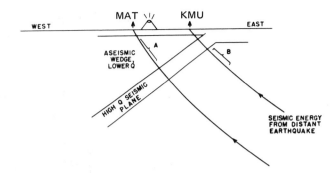

FIGURE 3.2 Sketch showing ray paths and seismic geometry beneath Japan. The two ray paths shown were found to have identical average Q, even though the path to MAT is much longer in the aseismic wedge than that to KMU. Therefore, the absorption in section A in the aseismic wedge must be exactly compensated by that beneath the seismic plane B, i.e., the average Q structure on either side of the high-Q seismic plane must be the same.

3.2 shows the seismic-ray paths studied, and Figure 3.3 compares, over a wide frequency range, the spectra of two paths from the same earthquake, which sample different regions of the asthenosphere. It is apparent that the attenuation of the two paths is essentially identical (over a frequency range of 4 to 0.03 Hz). The asthenosphere wedge (above the subducted plate) does not, therefore, appear to be cooling significantly. I have quantitative determinations of Q (a measure of the anelastic attenuation) for Japan and northern Chile. The ray paths studied are shown in Figure 3.4, and some results appear in Table 3.1 and Figure 3.5. The values determined for the asthenosphere in northern Chile are similar. The attenuation for compressional waves in the asthenosphere above the subducted plate ($Q_p = 200$) is equivalent to a value of 100 for shear waves. This is similar to values obtained by a number of workers from the attenuation of surface waves across oceanic paths and confirms the earlier observation (Figure 3.3) that the absorption was identical above the slab to that seaward of the trench. In this discussion the effects of shear heating and of emanations from the slab affecting the (asthenosphere) material above are ignored. Release of water from the slab has the effect of increasing the degree of partial melt—the opposite of the effect of the cooling. The shear heating effect is in the same direction. Either the system is in perfect balance, which is unlikely because the cooling effect seems very dominant, or the system is open. This means that fresh asthenosphere material is constantly being added to the system—the cooling and contamination effects then being small and noncumulative.

Subduction Zones and Volcanoes

Since many subduction zones are characterized by a volcanic front, and since this bears a fairly simple geometrical relation to the subducted plate, a direct causative relationship is suggested. What that may be is beyond the scope of this paper. However, observations in subduction zones that do not have present-day volcanism can indicate some parameters that ap-

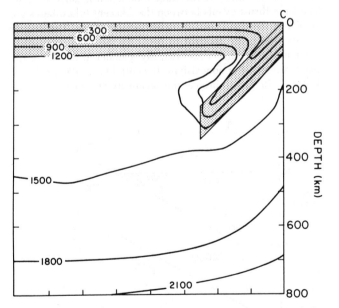

FIGURE 3.1 The temperature field of the asthenosphere and the subducting lithosphere 5 m.y. after the onset of subduction (after Hsui and Toksoz, 1979). The shading indicates the lithosphere. Note that the asthenosphere temperature above the slab is substantially reduced.

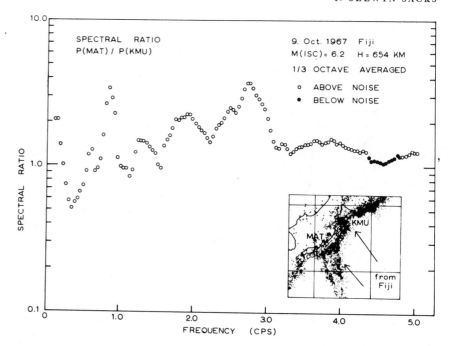

SPECTRAL RATIO
P(MAT) / P(KMU)

9. Oct. 1967 Fiji
M(ISC) = 6.2   H = 654 KM

1/3 OCTAVE AVERAGED

○ ABOVE NOISE
● BELOW NOISE

FIGURE 3.3   The spectral ratio as a function of frequency of the compressional wave (P) arrival at MAT to KMU from a deep earthquake in the Fiji region. The ray-path geometry is shown in Figure 3.2. A gross slope of the spectral ratio would indicate a difference in the Q values of the two paths compared. Undulations in the curve are caused by dissimilarities in the crustal structure of the two seismograph stations.

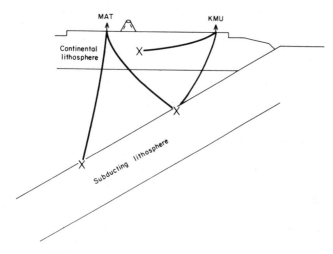

FIGURE 3.4   Ray paths studied to determine the Q structure of the asthenosphere. Crosses indicate earthquakes. The two paths from the shallower earthquake in the subducting plate gave higher Q to KMU than to MAT (after correcting for the effects of continental lithosphere). The deep event gave the same Q value to MAT as did the shallower event. See Table 3.1.

pear to be important. The geometry and seismicity of subduction zones in such anomalous regions are discussed below.

ANOMALOUS REGIONS

Central Peru and southwest Honshu, Japan, were studied to determine the geometry in anomalous subduction zones. Beneath central Peru, central Chile, and southwest Honshu the only clear seismicity delineating subduction is fairly shallow. The large thrust events between the adjacent lithospheres are similar to those in "normal" neighboring regions. The deeper events, which are thought to occur in the subducting plate itself and have tension axes (usually) parallel to the dip of the plate, are sparse or nonexistent. In central Peru, Barazangi and Isacks (1979) had to project the seismicity of a zone over 1000

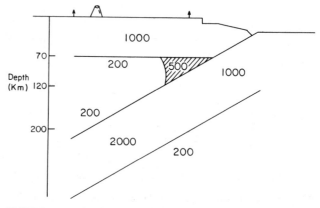

FIGURE 3.5 Q model for the Japan region. The hatched region has an anomalously high Q for the asthenosphere, $Q_p = 500$; normal oceanic asthenosphere has $Q_p = 200$.

TABLE 3.1   Asthenosphere Anelasticity Results for Paths Shown in Figure 3.4

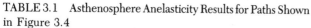

Station	Depth	Q_p Apparent	Q Asthenosphere
KMU	110	700	413
KMU	115	850	718
MAT	116	320	200
MAT	284	—	220

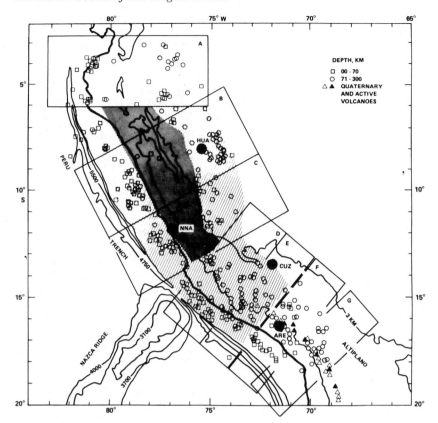

DEPTH, KM
□ 00 - 70
○ 71 - 300
△ ▲ QUATERNARY
AND ACTIVE
VOLCANOES

FIGURE 3.6 Seismicity of Peru, showing earthquakes with reliable locations selected by Barazangi and Isacks (1979). The cross-hatched area shows the large region with low seismicity. Intermediate-depth (circles) earthquakes in the hatched region have high-Q paths to the inland stations (solid dots)—HUA in the north and CUZ farther south. The anomalous zone is in sections B, C, D, and some of E.

km wide in order to have enough events to indicate a trend. Figure 3.6 shows this region of low activity. In southwest Honshu the seismic plane could be traced to a depth of 60 km or so, far less than the region to the south (Figure 3.10).

In southern Peru, northern Chile, and Kyushu (Japan) the volcanic front is well developed, but in the adjacent anomolous region there is no recent volcanic activity (Sykes, 1972; Isacks and Barazangi, 1977; Sacks, 1977; Barazangi and Isacks, 1979).

The absorption structure of the anomalous region is also different (Sacks and Okada, 1974; Chinn et al., 1980). Q_p values in the plate are well in excess of 1000 but are only about a few hundred in the asthenosphere. These low Q values are commonly ascribed to the presence of a small degree of partial melt. In comparing the Q structure of Japan and western South America, Sacks and Okada (1974) pointed out differences in paths from earthquakes in the slab at depths greater than 100 km to seismographs above them, i.e., far inland from the trench. In Japan these were low Q paths (as expected), but at stations CUZ and HUA (Figure 3.6) in central Peru these were high Q paths. Assuming that these earthquakes are in the subducted slab, the implication is that there is no low Q material (asthenosphere) between the subducted slab and the overlying continental lithosphere in central Peru.

SUBDUCTION GEOMETRY

Additional techniques allow delineation of the subducted plate in this region despite the inadequate control from traditional seismicity studies. The subduction geometry can be considered in three parts:

1. The initial subduction angle and depth can be determined under some conditions from observations of seismic wave energy converted at the upper surface of the subducted plate. If the dip angle is in the range of 20 to 35°, near-vertically traveling shear waves (reflected from the Earth's core) convert efficiently to compressional waves, which, being faster, arrive at the seismograph ahead of the shear wave (ScS) and are therefore observable. This phase (ScSp) has been studied both in Japan (Okada, 1977, Hasegawa et al., 1978) and in South America (Snoke et al., 1977, 1979). To observe the phase its amplitude must be larger than that of the direct S-coda. Recordings of small earthquakes by high-magnification seismographs are most useful. Such records were not available in the critical locations in South America, but sufficient clear arrivals (see Figure 3.7) were obtained on seismographs from the Worldwide Standardized Seismograph Network (WWSSN) in central Peru (NNA), southern Peru (ARE), and central Chile (PEL) and showed that the subduction had to have a dip of about 25 to 30° down to a depth of at least 100 km. This general trend is supported by the fault-plane orientation of earthquakes that occurred in this region. Stauder (1975) and Hasegawa and Sacks (1981) determined that the tension axes dipped at about 30°.

2. The horizontal region can be observed by projecting a very wide seismic zone (see Figure 3.6, zones B and C). However, this assumes that the zone is continuous (see the crosshatched region in Figure 3.6). Precision seismicity (see Figure

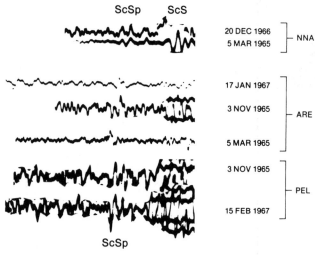

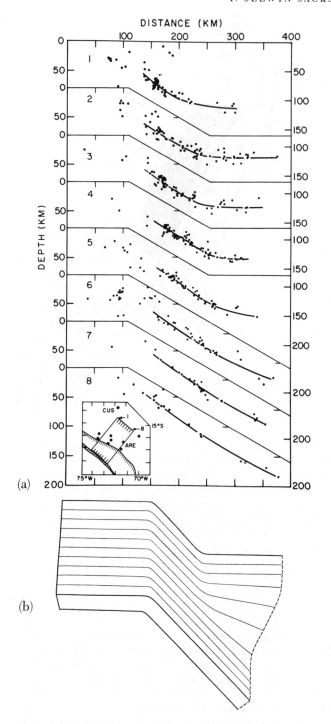

FIGURE 3.7   Photographs of vertical-component seismograms showing ScSp at each of three WWSSN South American stations: NNA
(central Peru), ARE (southern Peru), and PEL (central Chile). The
traces are aligned at the onset of ScS for each station—at NNA on the
vertical component with its clear onset and for the other stations (not
shown) on the horizontal components. Note the similarities of the
waveforms of the two NNA examples and the three ARE examples.
There is a time mark just before ScS on the NNA record of 1966
December 20 and one during ScSp on the ARE record of 1965 March
5. The ScSp phase is as clear in the anomalous regions—NNA and
PEL—as it is at ARE. The implication is that the slab must dip at 25
to 30° down to at least 100 km.

3.8) available near the southern boundary shows the development of a horizontal segment. Focal mechanisms of events
in this region (Hasegawa and Sacks, 1981) indicate essentially
horizontal focal mechanisms.

3. The resubduction zone is indicated by the easternmost
intermediate-depth earthquakes (150 km deep). These have
focal mechanisms with a 30° dip of the tension axis.

One interpretation of the subduction geometry is shown in
Figure 3.9, and the data and techniques are discussed more
fully in Hasegawa and Sacks (1981).

In the East Philippine Sea area the subduction in the volcano-free region (southwest Honshu) is delineated by earthquakes down only to about 60 km in depth (see Figure 3.10).
Nakanishi (1980) studied ScSp-converted phases in this region.
His assumed conversion interfaces are shown in Figure 3.11.
Hirahara (1981), using the method of Aki et al. (1977), studied
the velocity structure of this region. His results for the southwest Honshu region as well as the adjacent (normal) Kyushu
region are shown in Figure 3.12. Both Hirahara's and Nakanishi's results are consistent with a subducted plate traveling
approximately horizontally under the volcano-free region.

FIGURE 3.8   (a) Divided vertical cross sections of relocated earthquakes located inside the rectangles from regions 2 to 8 on the insert
map. Solid curves show the shape of the seismic zone (below 50-km
depth) inferred from the seismicity in each vertical section. (b) Schematic model of the descending Nazca plate near the boundary between
the nearly horizontal subduction zone beneath central Peru and the
more steeply dipping subduction zone beneath southern Peru. In both
regions the initial subduction angle is 30° and persists to at least 100
km.

SUMMARY AND IMPLICATIONS

Initial (shallow) subduction can have normal (~ 30°) dip but
has relatively sparse seismicity. At some depth, which I believe

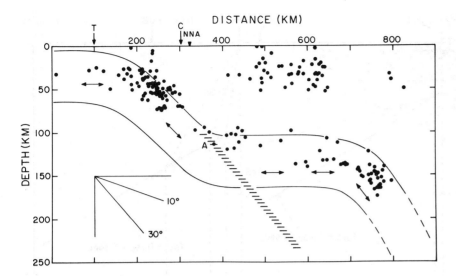

FIGURE 3.9   Cross section showing the inferred geometry of the descending Nazca plate beneath central Peru. Seismicity is from Barazangi and Isacks (1979). This is a 1000-km-wide section and therefore assumes the deformation in the slab constant over this width. The hatched zone denotes the possible conversion plane of the *ScSp* arrivals at station NNA (triangle) estimated by Snoke *et al.* (1977). Tension axes, shown by arrows, for focal mechanisms of intermediate-depth earthquakes are approximately parallel to the local dip of the descending Nazca plate.

is governed by the thickness of the adjacent continental lithosphere, the slab deforms and underlays the lithosphere. The seismic activity in the horizontal plate is very sparse and may even be nonexistent.

One indication of the forces involved comes from observations of the slab deformation between the normal and anomalous regions. A high-sensitivity local seismic network straddled this region in south-central Peru. The data indicate that the plate is continuous even though it deforms from a normally dipping plane to a horizontal one over a distance of only 80 km or so. Figure 3.8 shows the data and resultant model. This region is fully discussed by Hasegawa and Sacks (1981).

To sustain this deformation there must be an upward force of considerable magnitude on the subducted plate in the central Peru (anomalous) region. A likely possibility is that the central Peru portion is buoyant, at least at its western extremity and at a depth of 100 to 125 km. To understand the behavior of the subducted plate the density structure of oceanic plates before subduction needs to be considered.

Density of the Oceanic Plate

Being considerably colder, the subducting lithosphere is generally assumed to be more dense than the surrounding asthenosphere. The initiation of the subduction process is presumably due to the interaction of a moving oceanic plate with an adjacent (continental or oceanic) plate. The oceanic plate is forced to sink, at least to a depth equal to the thickness of the adjacent plate. Whether it continues to sink or it contorts and underlays the adjacent nonsubducting plate depends mainly on its density relative to that of the asthenosphere. The two possibilities are shown in Figure 3.13. The oceanic plate is essentially made up of a low-density basaltic crust (~ 2.9 g/cm^3) overlying a lithosphere with density similar to that of the asthenosphere from which it was derived (~ 3.4 g/cm^3). Because the subducting lithosphere is always colder than that of the asthenosphere surrounding it and is essentially of the same material, it will be denser. This is counteracted by the less dense crust. As oceanic crust has a remarkably constant thickness independent of age and region, and the lithosphere thickens proportionally to age, there will be some age (thickness) at which the subducting lithosphere is neutrally buoyant. Molnar and Atwater (1978), following calculations by McKenzie (1969), estimated that lithosphere less than 50 km thick will

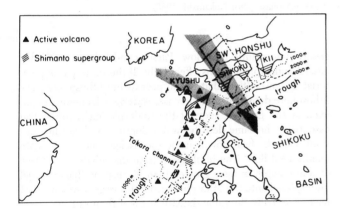

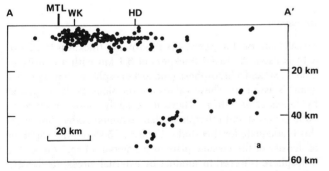

FIGURE 3.10   The top map shows the northwestern Philippine Sea plate. The western part, subducting under Kyushu and southward, is a normal subduction zone that includes volcanoes. The eastern part (Shikoku Basin) subducts under southwest Honshu and has no associated volcanoes. The seismicity of this section (bottom figure) shows normal dip but extends to less than 60-km depth.

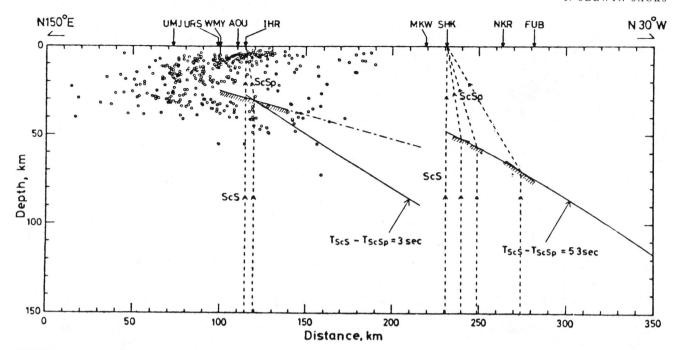

FIGURE 3.11 Vertical cross section of seismicity in the Shikoku, southwest Honshu region. Hypocenters indicated by open circles are for the period from January 1975 to December 1976 as determined by the Kochi Earthquake Observatory. The solid curves indicate the contours representing the theoretical differential travel times of 3 sec and 5.3 sec for the IHR and SHK stations, respectively. The solid line on the hatched portion represents the conversion interface corresponding to the vertical incidence of the ScS phase (after Nakanishi, 1980).

be buoyant. Pilger (1981) discussed the effects of subducting aseismic ridges.

In its path from the ridge to subduction the oceanic plate cools continuously through the seafloor. The density of all its constituents therefore increases with age. Any calculation of the relative density of the plate and the underlying asthenosphere needs to take this into account. Fortunately, it appears that the asthenosphere boundary layer "attached" to the moving lithosphere is considerable. To explain the bathymetry and heat flow as a function of age, Parsons and Sclater (1977) calculated the cooling of an equivalent "thermal plate" in which only vertical heat transport occurs. A thickness of about 130 km satisfies the data. This is, of course, considerably greater than the thickness of the lithosphere except for the very oldest seafloor. This implies that not only does the oceanic lithosphere cool and becomes denser but so does the asthenosphere below it. Since this is the asthenosphere into which the oceanic plate subducts, it is adequate in the buoyancy calculation to assume a constant density for both elements.

The thickening of the oceanic lithosphere with age has been determined by a number of workers using surface waves, as shown in Figure 3.14. Most are based on independent modeling of Love and Rayleigh wave data, although some fit both sets of data with the same model. There is general agreement that the lithosphere thickens rapidly in the first 30 million yr (m.y.) to about 50 km, then continues to thicken to about 100 km at 120 m.y. Note that the water depth does not increase following a square-root-of-age law beyond 80 m.y. or so because at greater ages the heat flow is essentially the background

steady-state mantle heat flow. Parsons and Sclater (1977) discussed and explained this effect fully. If the underplating material is denser (than that of the preexisting lithosphere), the lithosphere density would increase, causing a deepening ocean even at the greatest ages. If the underplating material had essentially the same density, water depth would remain constant once the heat flow became the background steady-state mantle heat flow. The implication from the relative constancy of water depth at ages greater than 80 m.y. is that material underplated onto the lithosphere from the asthenosphere must have essentially the same density as the asthenosphere.

Buoyancy

To calculate the buoyancy, the lithosphere thickness in Figure 3.14 is used. A crustal thickness of 6.1 km with a density of 2.83 g/cm^3 and a lithosphere (and asthenosphere) density of 3.4 g/cm^3 is assumed. These values are the same as those quoted by Leeds *et al.* (1974). There is certainly more structure in both the crust and lithosphere than assumed above; however, this is adequate for this study. Figure 3.15 shows the variation of density of the oceanic plate as a function of age. Once this lithosphere is forced to subduct by external forces, its density relative to that of the adjacent asthenosphere is controlled by the temperature difference. (This is a function of subduction velocity and plate age.) We can calculate an effective temperature difference that will make the subducted oceanic plate neutrally buoyant, as shown in Figure 3.16. For the youngest seafloors (less than 20 m.y. old) a temperature difference in

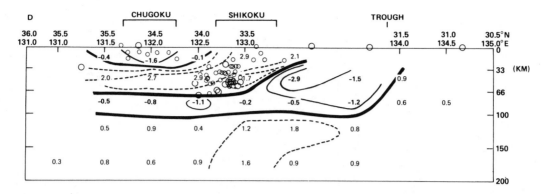

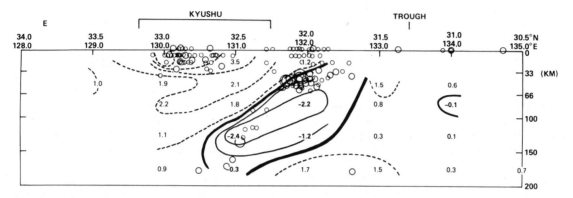

FIGURE 3.12　Three-dimensional seismic structure beneath southwest Honshu (upper) and Kyushu (lower). The slowness perturbations and hypocenters are projected onto vertical cross sections approximately normal to the Nankei trough. Thicker lines and negative signs indicate higher velocities, which are assumed to show the subducted plates. The younger East Philippine Sea plate dips down to about 60 km depth then flattens to horizontal (upper figure). The older, denser, West Philippine Sea plate has a dip that is maintained to at least 200 km depth (lower figure) (after Hirahara, 1981).

excess of 750°C is required for the plate to sink. For a 100-m.y.-old plate only a 350°C difference is required.

The calculation of the actual temperature structure is not straightforward since the plate as it is heated cools the asthenosphere, and the cooled asthenosphere, being denser, will tend to move down with the plate. Viscous drag also causes asthenosphere material to move with the plate. This has the effect of reducing the heating of the plate. Shear heating has the opposite effect. Hsui and Toksoz (1979) modeled young subduction. The temperature differences used here are based on their

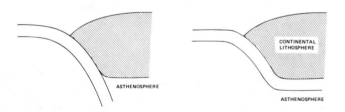

FIGURE 3.13　Subduction geometries that may result from (left) a sinking slab and (right) a buoyant slab. The initial subduction dip is controlled by the interplate foces and may be the same in both cases. If sufficiently buoyant, the slab may deform at depth and underlay the "continental" lithosphere.

model. Figure 3.1 shows the effective temperature difference for a subduction zone 5 m.y. after initiation. From Figure 3.16 we see that a seafloor 50 m.y. old is approximately neutrally buoyant. Whether it sinks or underlays the adjacent "continental" lithosphere (Figure 3.13) depends on whether the crust is anomalously thick or thin. Assuming isostatic equilibrium, a 0.25-km change in ocean depth requires a change in crustal thickness of about 1 km—hence the large effect shown in Figure 3.11. This result is consistent with observations. The "non-sinking" subduction under central Peru is of an ocean floor (including the Nazca ridge) that has a shallower depth (Pilger, 1981) than that for a "normal" ocean, i.e., as shown in Parson and Sclater's (1977) Figure 13. In the West Philippine Sea, a lithosphere of similar age subducts normally (i.e., sinks). The water depth here is 5.5 to 6 km (Murauchi *et al.*, 1968); "normal" is 5 to 5.5 km.

For an oceanic lithosphere 40 to 70 m.y. old the eventual subduction path seems to be strongly influenced by crustal thickness, which is reflected in the bathymetry. Beneath southern Peru the subduction is normal. To the north the thickened (lower-density) crust caused by the Nazca Ridge causes subduction to be of the buoyant variety. The West Phillipine Sea, of similar age, has a thinner than normal crust (the lithosphere is more dense) and it subducts normally.

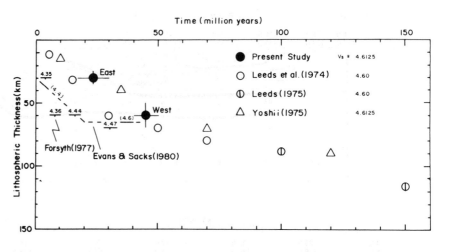

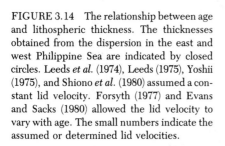

FIGURE 3.14 The relationship between age and lithospheric thickness. The thicknesses obtained from the dispersion in the east and west Philippine Sea are indicated by closed circles. Leeds *et al.* (1974), Leeds (1975), Yoshii (1975), and Shiono *et al.* (1980) assumed a constant lid velocity. Forsyth (1977) and Evans and Sacks (1980) allowed the lid velocity to vary with age. The small numbers indicate the assumed or determined lid velocities.

The Kinetics of the Basalt-Eclogite Phase Change

The previous discussion is based on a simple static model and is appropriate only for initial subduction. Once subduction has been operative for some time, the lower density (ρ = 2.9) basaltic crust may be transformed into eclogite, a very dense phase (ρ = 3.5). If this occurs, an ocean floor of any age will sink. The transformation to eclogite depends on the oceanic crust reaching sufficient temperature and pressure. Ito and Kennedy (1971) have made the most complete study of this transformation (see Figure 3.17). It is necessary, however, to consider the kinematics of the process in order to understand its behavior in different subduction zones.

For the transformation of basalt to eclogite to occur, there must be diffusive transport, both into and out of appropriate species to grow new minerals. At higher temperatures this process is fairly rapid. For 1200°C, Ito and Kennedy reported

that the process is complete in about 5 min. At 800°C, however, a week or more is required for the attainment of equilibrium. At even lower temperatures, very long times are required for the transformation, since the process is controlled by solid diffusion. Ito and Kennedy estimated that 10^6 to 10^7 yr are required at 400°C.

Ahrens and Schubert (1975) discussed the reaction rate of the gabbro-eclogite transformation in detail. While solid-state diffusion under completely dry conditions will not produce the transition in geologically meaningful times at temperatures less than about 600°C, these authors pointed out that interstitial water can accelerate the process greatly. Lambert and Wyllie (1968) and Essene *et al.* (1970) reported laboratory experiments that showed that eclogite will indeed form at moderate temperatures (<700°C) at sufficiently high (10 to 30 kbar) water pressure—equivalent to 5 to 10 percent (by weight) of included water. Measurements of water content in ocean-floor basalts, however, show far lower values. Engel *et al.* (1965) in their compilation of oceanic tholeiites found concentrations of 0.3 to 1.0 percent for Pacific Ocean samples. Dymond and Hogan (1973) and Craig and Lupton (1981) reported water contents in deep-sea basalts of about 0.3 percent, including one sample (DSDP-34 drill hole) that showed slight contamination with seawater from 30-m.y.-old crust. These observed low concentrations, relative to what is required to influence the basalt-eclogite transformation, suggest that this transformation in subduction zones is governed by solid diffusion rates and that temperature is therefore the important controlling parameter for the reaction rate. These temperatures are low in regions of anomalous subduction.

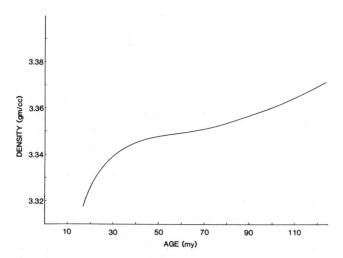

FIGURE 3.15 Density of the oceanic plate as a function of age. The structure, density data, and lithosphere thickness are the same as in Leeds *et al.* (1974). The specifics of the model are not important, only the general pattern of density increasing with age is. This trend is not sensitive to the details of the model, although the actual values are.

Heat Flow

My estimate of temperatures within the Earth is derived from observations of heat flow at the surface. Some results for western South America (Henry, 1981) are shown in Figure 3.18. These data cover both the southern Peru region, where subduction is "normal" (i.e., at 30° dip down to the deepest events), and the central Peru regions, where the subducted lithosphere appears to underlay the continental lithosphere without any intervening asthenosphere. At a similar position relative to the

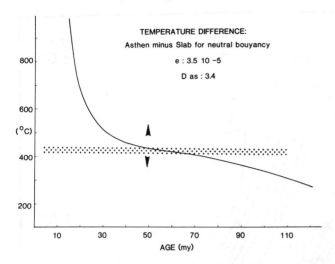

FIGURE 3.16　The excess temperature of the asthenosphere relative to the subducted plate in order that the plate be neutrally buoyant. If the temperatures are above the curve the plate will sink. Note that for ages less than 20 m.y., unreasonably high temperature differences would be required, and therefore the plate will probably be buoyant. The dotted zone is the effective mean temperature difference calculated from the model of Hsui and Toksoz (1979). The geometry and temperature field are shown in Figure 3.12. For the particular models, assumptions, and data used, the implication is that oceanic slabs less than about 40 m.y. old will be buoyant and slabs older than 80 m.y. will sink. However, the effect of anomalous crustal structure can be very great. The arrowheads indicate the shift of the stability curve for a crustal structure change causing a bathymetry anomaly of 0.15 km. In the case of central Peru (40 to 60 m.y. old) the seafloor is shallower (Pilger, 1981) and, assuming this is an indication of thicker crust, the slab is buoyant. In the case of the West Philippine Sea (Kyushu) the water depth is greater than expected for its age (40 to 50 m.y. old)—5.5 to 6 km in the Philippine basin rather than the expected 5 to 5.5 km. This implies a thinner (or lighter) crust and the slab should sink. In the case of the East Philippine Sea, its age (~20 m.y.) suggests that the slab must be buoyant.

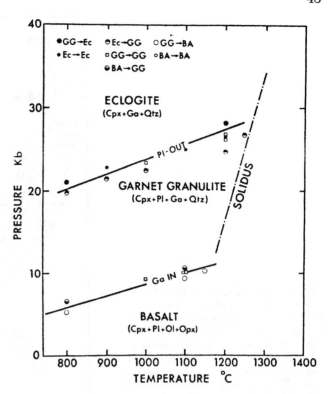

FIGURE 3.17　Pressure-temperature diagram of NM5. Ba is basalt, GG is garnet granulite, EE is eclogite, Ga is garnet, Cpx is clinopyroxene, Opx is orthopyroxene, Pl is plagioclase, Ol is olivine, and Qz is quartz, (after Ito and Kennedy, 1971, with permission of the American Geophysical Union).

subduction, above the 100-km contour, the mean heat flow in central Peru is ~30 mW/m², whereas in southern Peru the mean heat flow is about 60 mW/m².

　　To convert these surface values to temperature at depth, I follow the calculations (and assumptions) of Chapman and Pollack (1977). Their results are shown in Figure 3.19. For the heat flow observed in central Peru (~30 mW/m²), the expected temperature at a depth of 100 km is about 400 to 500°C. At this temperature the basalt-eclogite transformation will be retarded, possibly for 10^6 to 10^7 yr. At subduction rates in excess of 10 cm/yr (e.g., for the Nazca plate) the subducted plate may remain buoyant for a length of 100 to 1000 km, which is consistent with the geometry determined by Hasegawa and Sacks (1981).

　　Similar, though less extreme, conditions apply in the Philippine Sea subduction. The eastern and western parts of the Philippine Sea are of considerably different ages. The average age, determined from Deep Sea Drilling Project (DSDP) cores, of the Philippine basin just south of the Kyushu-Palau ridge is

about 50 m.y. The age of the Shikouku basin region just north of the ridge is about 20 m.y. Shiono *et al.* (1980), analyzing both Love and Rayleigh waves, determined lithosphere thicknesses of 60 and 30 km for the older and younger regions, respectively. These are in general agreement with thicknesses determined by other workers in different oceans of the same age (see Figure 3.14). The western part, subducting under Kyushu, is older (50 m.y.), thicker (60 km), and has a thinner than "normal" crust. It subducts normally. The eastern part, subducting under southwest Honshu, being younger (20 m.y.) and thinner (30 km), is buoyant and therefore underlays the adjacent continent, traveling essentially horizontally after it penetrates to the base of the continental lithosphere.

Normal Subduction

Most of the present-day subduction is normal, i.e., the subducted slab continues sinking as a more or less planar body below the initial near-surface deformation from a horizontal to an inclined slab. In the case of old lithosphere, its relatively high density (low temperature) ensures that it will sink (see Figure 3.16) irrespective of the transition, whereas for young lithosphere the buoyancy is controlled by the basalt-eclogite transformation. This in turn is controlled by the temperature. For young lithosphere any of three conditions ensure normal

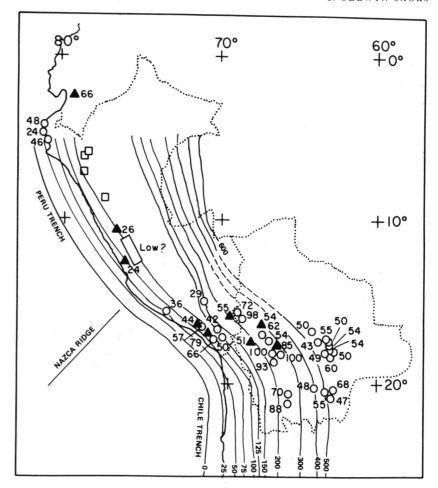

FIGURE 3.18 Location map of heat-flow sites in Peru and Bolivia with fully corrected heat-flow values shown in milliwatts per square meter. Open circles are sites established by the University of Michigan; open rectangle in central Peru represents five sites with questionable results but are believed to have low heat flow. Solid triangles are sites established by other investigators, which have been corrected for topography and uplift and erosional history (Henry, 1981). Contours are depths in kilometers to the top of the subducted Nazca plate (after Hasegawa and Sacks, 1981, with permission of the American Geophysical Union).

subduction: (1) contact of the subducted crust with the asthenosphere, (2) contact with a sufficiently hot (800°C) adjacent lithosphere, (3) part of a normal subduction region. Once normal subduction has been established, the negative buoyancy of the deeper part of the slab will ensure that the slab keeps sinking no matter what the buoyancy of the newly subducting material is. If subduction is interrupted (e.g., detached slab) it may restart either way. If the oceanic plate is less than about 70 m.y. old, bathymetry anomalies (e.g., aseismic ridges) may have a controlling role. This is presumably why the South American subduction is so segmented. In both central Peru and central Chile, buoyant subduction is flanked by normal subduction. The buoyant zones contain aseismic ridges.

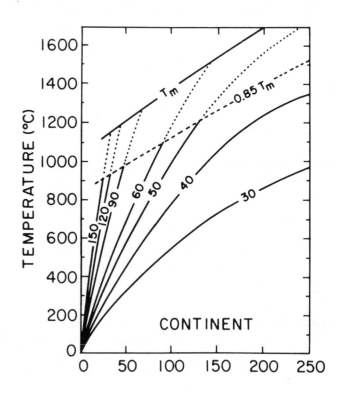

FIGURE 3.19 Geotherm families for oceanic and continental regions; family parameter is heat flow in mW/m². T_m is the mantle solidus; geotherms are dotted above 0.85 T_m to indicate provisionality (after Chapman and Pollack, 1977). For central Peru, with an observed heat flow of less than 30 above the 100-km contour (see Figure 3.18), the subducted crust temperature may be about 400 to 500°C. At this temperature the basalt-eclogite transition may take 1 to 10 m.y.

CONCLUSIONS

Observations have been described here that lead to models of subduction beneath western South America, Kyushu, and southwest Honshu. These regions are anomalous in that the subducted plate deforms at depth and has an essentially horizontal segment. The data show that initial subduction, being controlled by the interaction between the subducting and adjacent lithosphere, is at an angle of about 30°. Because it is buoyant the plate deforms and travels horizontally once it is below the continental lithosphere. For the buoyancy to be maintained, however, the transformation of the basalt in the crust to the denser eclogite must be retarded. This requires that the subducted plate be in contact with a sufficiently cold continental lithosphere.

The important parameters are as follows:

1. The age of the subducting ocean floor can determine the style of subduction. Oceanic plates younger than about 70 m.y. may subduct anomalously if the other conditions are met. Older oceanic plates are sufficiently dense that normal subduction is most likely.

2. The thickness of the oceanic crust can be the controlling factor for ocean floor ages that are 40 to 70 m.y. Aseismic ridges tend to cause buoyant subduction, whereas a thinner than normal crust (which will be associated with anomalously deep water) can ensure normal subduction of a young (~40 m.y.) plate.

3. Buoyant subduction can persist only if the subducting slab is in contact with a cool continental lithosphere. This is manifested by low heat flow above the region where the slab deforms to a near-horizontal aspect. Only in a cool environment will the basalt-eclogite transformation be retarded, so that the subducted plate can remain buoyant.

While there is a clear geometric relationship between the subducted plate and the volcanic front, there is no generally accepted model that explains the mechanism. Some insight is possible from the observations that there are no volcanoes in the anomalous regions. There the subducted lithosphere is in contact with the overlying continental lithosphere. The two consequences are that there is no hot mobile asthenosphere in contact with the subducted crust, and the temperature of the (subducted) crust is substantially lower. The relative importance of these two effects depends on whether volcanic magma is derived mainly from the asthenosphere or whether melting of the subducted crust is significant.

REFERENCES

Ahrens, T. J., and G. Schubert (1975). Gabbro-eclogite reaction rate and its geophysical significance, *Rev. Geophys. Space Phys. 13*, 383-400.

Aki, K., A. Christofferson, and E. S. Husebye (1977). Determination of the three-dimensional seismic structure of the lithosphere, *J. Geophys. Res. 82*, 277-296.

Barazangi, M., and B. Isacks (1979). Subduction of the Nazca plate beneath Peru: Evidence from spatial distribution of earthquakes, *Geophys. J. R. Astron. Soc. 57*, 537-555.

Chapman, D. S., and H. N. Pollack (1977). Regional geotherms and lithospheric thicknesses, *Geology 5*, 265-268.

Chinn, D. S., B. L. Isacks, and M. B. Barazangi (1980). High-frequency seismic wave propagation in western South America along the continental margin, in the Nazca plate and across the Altiplano, *Geophys. J. R. Astron. Soc. 60*, 209-244.

Craig, H., and J. E. Lupton (1981). Helium-3 and mantle volatiles in the ocean and the oceanic crust, in *The Sea, Vol. 7: The Oceanic Lithosphere*, C. Emiliani, ed., Wiley-Interscience, New York, pp. 391-428.

Dymond, J., and L. Hogan (1973). Noble gas abundance patterns in deep-sea basalts—primordial gases from the mantle, *Earth Planet. Sci. Lett. 20*, 131.

Engel, A. E. J., C. G. Engel, and R. G. Havens (1965). Chemical characteristics of oceanic basalts and the upper mantle, *Geol. Soc. Am. Bull. 76*, 719-734.

Essene, E. J., B. J. Hensen, and D. H. Green (1970). Experimental study of amphibolite and eclogite stability, *Phys. Earth Planet. Inter. 3*, 378-384.

Evans, J. R., and I. S. Sacks (1980). Lithospheric structure in the North Atlantic from observation of Love and Rayleigh waves, *J. Geophys. Res. 85, B12*, 7175-7182.

Forsyth, D. W. (1977). The evolution of the upper mantle beneath mid-ocean ridges, *Tectonophysics 38*, 89-118.

Hasegawa, A., and I. S. Sacks (1981). Subduction of the Nazca plate beneath Peru as determined from seismic observations, *J. Geophys. Res. 86*, 4971-4980.

Hasegawa, A., N. Umino, and A. Takagi (1978). Double-planed deep seismic zone and upper-mantle structure in the northeastern Japan Arc, *Geophys. J. R. Astron. Soc. 54*, 281-296.

Henry, S. G. (1981). Terrestrial heat flow overlying the Andean subduction zone, Ph.D. thesis, U. of Michigan, Ann Arbor.

Hirahara, K. (1981). Three-dimensional seismic structure beneath southwest Japan: The subducting Philippine Sea plate, *Tectonophysics 79*, 1-44.

Hsui, A. T., and M. N. Toksoz (1979). The evolution of thermal structures beneath a subduction zone, *Tectonophysics 60*, 43-60.

Isacks, B., and M. Barazangi (1977). Geometry of Benioff zones, in *Island Arcs, Deep Sea Trenches and Back-Arc Basins*, M. Talwani and W. C. Pitman, eds., Maurice Ewing Series I, Am. Geophys. Union, Washington, D.C.

Ito, K., and G. C. Kennedy (1971). An experimental study of the basalt-garnet granulite-eclogite transition, in *The Structure and Physical Properties of the Earth's Crust*, J. G. Heacock, ed., Geophys. Monogr. 14, Am. Geophys. Union, Washington, D.C., pp. 303-314.

Lambert, I. B., and P. J. Wyllie (1968). Stability of hornblende and a model for the low velocity zone, *Nature 219*, 1240-1241.

Leeds, A. R. (1975). Lithospheric thickness in the Western Pacific, *Phys. Earth Planet. Inter. 11*, 61-64.

Leeds, A. R., L. Knopoff, and E. G. Kausel (1974). Variations of upper mantle structure under the Pacific Ocean, *Science 186*, 141-143.

McKenzie, D. P. (1969). Speculations on the consequences and causes of plate motion, *Geophys. J. R. Astron. Soc 18*, 1-32.

Molnar, P., and T. Atwater (1978). Interarc spreading and Cordilleran tectonics as alternates related to the age of subducted oceanic lithosphere, *Earth Planet. Sci. Lett. 41*, 330-340.

Murauchi, S., N. Den, S. Asano, H. Hotta, T. Yoshi, T. Asanuma, K. Hagiwara, K. Ichikawa, T. Sato, W. J. Ludwig, J. I. Ewing, N. T. Edgar, and R. E. Houtz (1968). Crustal structure of the Philippine Sea, *J. Geophys. Res. 73*, 3143-3191.

Nakanishi, I. (1980). Precursors to ScS phases and dipping interface in the upper mantle beneath southwestern Japan, *Tectonophysics* 69, 1-35.

Okada, H. (1977). Fine structure of the upper mantle beneath Japanese Island arcs as revealed from body wave analyses, D.Sc. thesis, Hokkaido Univ., Sapporo, Japan, 129 pp.

Parsons, B., and J. G. Sclater (1977). An analysis of the variation of ocean floor bathymetry and heat flow with age, *J. Geophys. Res.* 82, 803-827.

Pilger, R. H., Jr. (1981). Plate reconstructions, aseismic ridges, and low-angle subduction beneath the Andes, *Geol. Soc. Am. Bull.* 91, 448-456.

Sacks, I. S. (1977). Interrelationships between volcanism, seismicity, and anelasticity in western South America, *Tectonophysics* 37, 131-139.

Sacks, I. S., and H. Okada (1974). A comparison of the anelasticity and structure beneath western South America and Japan, *Phys. Earth Planet. Inter.* 9, 211-219.

Shino K., I. S. Sacks, and A. T. Linde (1980). Preliminary velocity structure of Japanese Islands and Philippine Sea from surface wave dispersion, *Carnegie Inst. Washington Yearb.* 79, 498-505.

Snoke, J. A., I. S. Sacks, and H. Okada (1977). Determination of subducting lithosphere boundary by use of converted phases, *Bull. Seismol. Soc. Am.* 67, 1051-1060.

Snoke, J. A., I. S. Sacks, and D. James (1979). Subductions beneath western South America: Evidence from converted phases, *Geophys. J. R. Astron. Soc.* 59, 219-225.

Stauder, W. (1975). Subduction of the Nazca plate under Peru as evidenced by focal mechanisms and by seismicity, *J. Geophys. Res.* 80, 1053-1064.

Sykes, L. R. (1972). Seismicity as a guide to global tectonics and earthquake prediction, *Tectonophysics* 13, 393-414.

Yoshii, T. (1975). Regionality of group velocities of Rayleigh waves in the Pacific and thickening of the plate, *Earth Planet. Sci. Lett.* 25, 305-312.

Potentially Active Volcanic Lineaments and Loci in Western Conterminous United States

4

ROBERT L. SMITH *and* ROBERT G. LUEDKE
U.S. Geological Survey

ABSTRACT

Speculation about potentially active volcanoes in the conterminous United States has usually focused on the Cascade region. The reason for this focus is simply that Cascade volcanoes have erupted in historic time, are solfatarically active, and are morphologically similar to active cones around the world. The recent eruption of Mount St. Helens has confirmed this view.

Maps showing the distribution of Upper Cenozoic volcanoes and volcanic rocks in the western United States provide abundant evidence for the existence of many potentially active volcanic loci outside the Cascades region. Portrayed in time increments of 0 to 5, 5 to 10, and 10 to 16 million years (m.y.) the loci reveal both changing and static distribution patterns. For the last 5 m.y., most loci have been concentrated in linear zones. These zones record both waxing and waning eruptive frequencies within that time span.

Distribution patterns alone suggest that any volcanic locus that has appeared within the last 5 m.y. reflects a viable mantle source today and has potential for volcanic eruption. This observation is reinforced by a consideration of the life spans and periodicities of older volcanic loci, some of which span 10 m.y. or more. Almost the entire distribution pattern of volcanic loci 0 to 5 m.y. old is reflected by the distribution pattern of basaltic eruptions less than 10,000 yr old. We conclude that, in addition to some 15 volcanic loci in the Cascade province, more than 60 other identifiable loci in 11 western states have potential for future eruptions. Further, the general concentrations of loci suggest that new loci may form within a zone at any time.

INTRODUCTION

That many of the High Cascade volcanoes have potential for future eruptions has never been seriously questioned. They are geomorphically youthful, are easily equated with active volcanism of the circum-Pacific margin, and several have had historic eruptions. However, until the last few years, because of a flurry of interest in geothermal energy, toxic-waste disposal, and natural-hazards assessment programs, the potential for future eruptions from the numerous other sites of late Cenozoic volcanism in the western conterminous United States has seldom been considered. No comprehensive survey of these potential sites has been made. In this report we identify volcanic fields, exclusive of the Cascades, that have potential for future volcanic eruptions. We emphasize that this is a prelim-

inary study, and a great amount of work must be done before any meaningful level of confidence for long-term prediction will be attained.

This study is an outgrowth of the U.S. Geological Survey's (USGS) Geothermal Energy Research Program, and our analysis and conclusions are based on USGS Miscellaneous Investigations Series Maps I-1091-A through I-1091-E by Luedke and Smith (1978a, 1978b, 1981, 1982, 1983). This map series (scale of 1:1,000,000) shows the distribution, composition, and age of volcanic centers with vents dating from about 16 m.y. ago to the present time. The maps also illustrate five common rock compositions that occur within three different increments of time: 0 to 5, 5 to 10, and 10 to 16 m.y. ago. The maps represent the collective works of hundreds of individuals, and the reader is referred to this map series for references to specific source materials. Original sources rather than generalized map compilations were consulted wherever possible in the construction of these maps. However, our own interpretations were used concerning controversial problems and, where necessary, to maintain a standard treatment for meaningful comparisons of data.

With the completion of the 1:1,000,000 map series, a composite map of the western conterminous United States was compiled at 1:2,500,000 and separated into three time-increment maps, representing time spans of 0 to 5, 5 to 10, and 10 to 16 m.y. These maps, modified and reduced to a scale of about 1:3,750,000, are plates 1, 2, and 3 in the pocket in the back of this volume. Time-space-composition patterns as shown on these composite maps were then interpreted for their relevance to future volcanic activity.

From the map patterns displayed by our three arbitrarily chosen time increments, it was evident that in any discussion of future volcanism from existing volcanic fields the significant data base is primarily within the 0- to 5-m.y. time increment. The general pattern for this time increment was established about 10 m.y. ago and, except for minor subsequent modifications, has persisted to the present.

Knowledge of the past eruptive frequencies within these volcanic fields is essential for meaningful analysis of future activity. Because this kind of information on intraplate extensional-tectonic volcanic fields is virtually unknown in any detail, we had to rely on existing fragmental data from many fields, probable eruption frequencies relative to better-known types of volcanism, and deductive models. To give some order and perspective to this complex volcanic array, we introduce here (see Figure 4.1) a simplistic scale for periodicity and episodicity of volcanic eruptions from volcanic systems most likely to be found in the United States. This scale is essentially a simplified version from Smith (1979, Figure 12) modified to give emphasis to the basaltic component of extensional systems rather than to the derivative rhyolitic components. This scale suggests that future eruptions cannot arbitrarily be ruled out from any of the

less than 5-m.y.-old loci as long as the present distribution pattern of zones and loci prevails.

For interpretation of viability of the silicic components of the 0- to 5-m.y. loci, i.e., those that have greatest potential for future explosive activity, we have leaned heavily on the model that Smith and Shaw (1975, 1978) developed for estimating the igneous-related geothermal resource base. Their model provides a first approximation of the solidification state of a magmatic system based on estimates of magma-chamber volumes and age of the last eruption.

Migrating volcanic fields are of particular interest and give insight into plate and mantle motions. We have not attempted a sophisticated analysis of these migrations because of fragmentary data; although the subject is relevant, it is somewhat tangential to this paper. We feel compelled to offer some comments, however, based entirely on distribution patterns, longevity patterns, and motions as deduced from the map compilations. We have not made an extensive search of the plate-tectonic literature or the pertinent geophysical literature, but we think the omissions emphasize the power of the volcanic component of geodynamics to speak for itself.

TERMINOLOGY

The following terms are defined so as to clarify our usage of them here rather than as an attempt to apply rigorous definitions. Some of these terms as applied to volcanological concepts are relatively new; others are very old. Space does not permit an etymological review.

Volcano A recognizable source of primary volcanic materials or of secondary volcanic or nonvolcanic materials that may be inferred to have been distributed by volcanic explosion. Most volcanoes are constructional geomorphic features. A volcano may be short lived (<10 yr) or it may be long lived ($>10^6$ yr). Most long-lived volcanoes are better classified as *volcanic systems*.

FIGURE 4.1 Periodicity and episodicity of common volcanic systems. Periodicity is implied where not otherwise stated, although the frequency of caldera formation within some systems may be better termed *episodicity*. Given our present state of knowledge, many such designations are indeterminate.

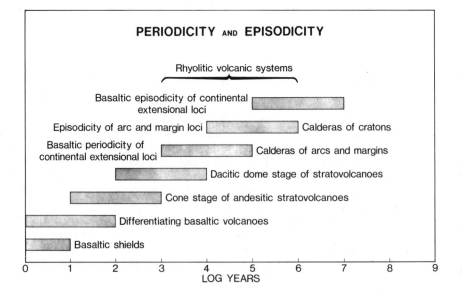

Volcanic system Long-lived differentiated volcanoes, volcanoes having complex plumbing systems, and volcano clusters clearly related in space and time are, because of their complexity, best thought of as volcanic systems. Most volcanoes, even simple forms, are parts of larger volcanic systems in a genetic sense or volcanic loci in a descriptive sense.

Volcanic locus A geographically isolated volcano or group of volcanoes identifiable as a spatial and/or temporal unit of volcanism. A locus may consist of 1 volcano or more than 500 volcanoes. A locus may be isolated from other loci but more commonly is part of a chain of loci termed a *volcanic zone*. Within a volcanic zone loci may coalesce to form poorly definable areas of time-space overlap. For some areas the terms *volcanic system* and *volcanic locus* are synonymous; however, for others, *loci*, as we use the term, contain multiple, perhaps independent, volcanic systems (e.g., Snake River Plain).

Volcanic zone Concentrations of volcanic loci are termed *volcanic zones*, which commonly are linear where well defined. Volcanic zones may often be more than 100 km wide, may reflect known tectonic elements of the crust in other examples, or may reveal unexplained tectonic elements in others. Some zones clearly reflect old, perhaps reactivated, tectonic boundaries, whereas some zones appear to reflect relatively new configurations of the crustal stress field.

Periodicity The eruptive frequency of a volcano or a volcanic system may be termed its *periodicity*. During the life span of a volcano or volcanic system, its periodicity may change as a function of several variables, such as volume of magma storage, changing magma composition, or rate of magma rise. These variables probably all depend on the rate and style of thermal input to the system, which in turn is probably a function of the overall activities and nature of the tectonic regime.

Episodicity Some of the volcanic systems or volcanic loci included in this paper appear to have experienced episodes (>1 m.y.) of "periodic" activity separated by long intervals (>1 m.y.) of inactivity or reduced activity. In some loci an episode may result in a compositional sequence that may be defined as a *cycle*. Silicic nuclei in these sequences are superimposed in some loci and transgressive in others. Linear transgressions may define a volcanic zone. Lateral transgressions mark migrations of entire zones.

Potentially active volcano (volcanic system) A volcano considered dormant by reason of its known or presumed eruptive periodicity is potentially active. In this paper we extend the term *potentially active* to include many volcanic systems not heretofore designated as potentially active or dormant. Neither the periodicity nor the episodicity of these systems is known, but from a consideration of map-distribution patterns that show change over a 15-m.y. period, future eruptions in many volcanic loci seem inevitable. Map patterns that show migrating volcanic loci and waxing and/or waning volcanic intensity may be powerful indicators of probable future sites of eruptions when viewed in the context of volcano-evolution models.

PERIODICITY AND EPISODICITY OF VOLCANOES

The periodicity of volcanoes is little known or understood. Few systematic studies have been made, and modern textbooks curiously avoid the subject. Any attempt to systematically analyze periodicity requires that the distinction between a volcano and a volcanic system be clearly recognized. For example, the Paricutin Volcano in Mexico was born and grew to a height of 360 m in 3 yr. Lava was discharged at a nearly constant rate for 9 yr, after which activity ceased (Williams and McBirney, 1979). In all probability, Paricutin Volcano is now extinct. However, Paricutin is only one of many similar cinder cones (monogenic volcanoes) that make up the volcanic field or locus. Thus, we would view the eruption of Paricutin as one "periodic" expression of the eruptive periodicity of the volcanic system that includes Paricutin and scores of other volcanoes.

Exclusive of the Cascades, most of the potentially active volcanic systems, which are the subject of this paper, are analogous to the Paricutin example; thus, in such fields we need to know the frequency of occurrence of individual cones and lava flows. Such information is not available, in any detail, for any volcanic locus known to us.

Volcanic loci of this type are found largely in areas of crustal extension and thus differ in their general behavior pattern and chemistry from volcanic systems directly related to convergent plate boundaries as found in island-arc and continental-margin environments (Cascade volcanoes). The differences in the behavior of volcanic systems from these two major environments will probably ultimately be found to be related to the major tectonic controls as they affect the rate and amount of melting, magma rise, and the amount of magma storage at both mantle and crustal levels.

Smith (1979) suggested that the eruption frequency of volcanic systems that produce large pyroclastic eruptions is a function of the volume of stored magma irrespective of tectonic setting. This is a simplistic view; nevertheless, there is an order-of-magnitude correlation between volumes of caldera-forming eruptions and the time between calderas formed in succession in the same volcanic system. The implication here is that it takes longer to generate the larger volumes of the differentiated magma that is usually involved in caldera-forming eruptions. The production rate of this magma—usually silicic in composition (rhyolitic to dacitic)—is approximately 10^{-3} km³/yr. Smith further suggested that the volume-time relationship also holds for smaller-volume systems that do not produce calderas.

If, as many workers now think, transfer of heat to melt crustal material is accomplished by injections of basalt into the crust, the longer-lived intraplate systems may have more time to generate larger volumes of differentiated magma by whatever processes such magmas are formed, if the injected basalt can be stored. Crustal extension may provide the storage space (Bacon, 1982). The efficiency of the extension mechanism for storage would be quite variable. If the system is too open, the basalt leaks out and little heat is stored (e.g., Columbia River lavas). If the system allows storage at depth but is not open in the upper crust, dominantly silicic magmas may be erupted

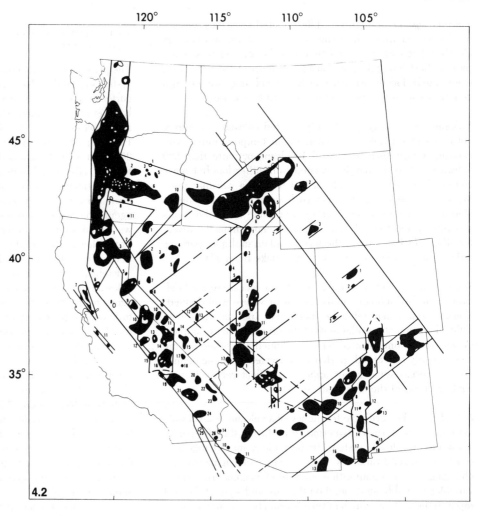

FIGURE 4.2   Volcanic loci 0 to 5 m.y. old. Black pattern indicates basaltic and andesitic volcanism; white spots within black are rhyolitic and dacitic volcanism. The grid pattern shows arbitrary boundary constraints on volcanic zones and loci. (Numbered loci are given in Table 4.1. See Figure 4.5 and text for details.)

FIGURE 4.3   Volcanic loci 10 to 16 m.y. old superimposed on a 0- to 5-m.y. grid pattern. Black pattern includes all volcanic rock compositions.

FIGURE 4.4   Volcanic loci 5 to 10 m.y. old superimposed on a 0- to 5-m.y. grid pattern. Black pattern includes all volcanic rock compositions.

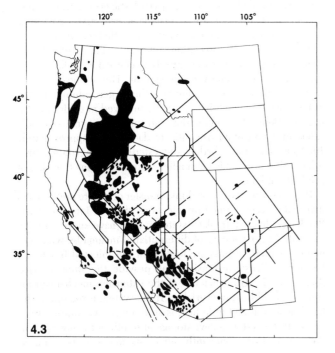

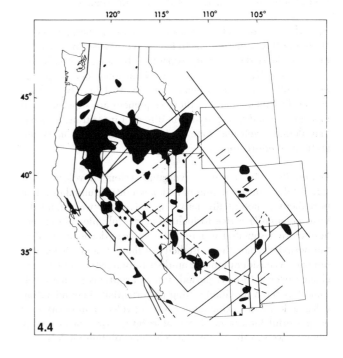

and in larger volumes; the basalt may never appear at the surface or may appear only in minor quantities (e.g., western United States in Oligocene-Miocene time).

Crustal thickness may help determine storage time and may influence ratios and volumes of mantle-derived/crustal-derived magmas. Evidence presented (see Chapter 2) for notable crustal contamination of the Columbia River lavas suggests to us that in thicker crust, such as that found in the Rocky Mountain provinces, the Columbia River magmatism might have been silicic rather than mafic. The implication here may be that in the incipient stages of crustal extension ponding of mantle-derived magma at the crust-mantle interface may be the dominant mode of thermal storage. In later stages of crustal extension the rise of mantle-derived melts is facilitated perhaps in part by earlier lower-crustal melting and thinning, and magma storage in higher-level fractures becomes the dominant mode of thermal storage and local melting.

In any case, volcanic systems related to arcs and margins erupt at higher frequencies and have shorter life spans than do intraplate volcanic systems. They also tend to be more focused, thereby giving rise to longer-lived "central vent" volcanoes. It seems probable that the phenomenon of focused thermal energy in these arc and margin systems leads to an earlier thermal storage and differentiation in the upper crust and hence to an earlier, albeit smaller-volume, central vent complex than is usually seen in intraplate loci. The intraplate loci (e.g., the Jemez Mountains locus), although longer lived, may require a special tectonic situation to focus thermal storage, even after several million years of crustal preheating in the locus. In the absence of focused thermal storage the intraplate volcanic system may never evolve beyond a field of basaltic cinder cones, including perhaps a rare, short-lived silicic nucleus. Depending on local crustal tectonics, all transitions between fields of mafic cinder cones and complex silicic volcanic systems can and do occur. Compare the 0- to 5-m.y.-old products of the Springerville, San Francisco Mountains, and the Jemez Mountains loci (see Figure 4.2 and Table 4.1).

Figure 4.1 shows in a general way the relative periodicities and episodicities of the more common kinds of volcanoes and volcanic systems in the United States. However, Figure 4.1 is a simplistic portrayal of extremely complex phenomena, and its usefulness for providing insight into any specific volcanic system should not be overinterpreted. The highest eruption frequencies are found in basaltic volcanoes, and it seems probable that large basaltic shields, such as the Hawaiian type, are built largely by eruption frequencies in the 1- to 10-yr range. As these shields grow larger and older, they either die because their heat supply has migrated to a new location or they change composition with longer repose periods between eruptions, until at some advanced state of differentiation their thermal supply finally shuts off. Little is known about this end-stage periodicity, but by the time trachytic composition is reached (rare in Hawaiian volcanoes), repose times may be in the 1000-yr range.

During their cone-building stages, andesitic stratovolcanoes provide the best base for the study of periodicity because they are numerous, erupt with frequencies that are a part of the historic record, and commonly evolve to dome and caldera

TABLE 4.1 Numbered Loci in Each State, as Shown in Figure 4.2

Arizona
1. Western Grand Canyon
2. San Francisco Mountains
3. Mormon Mountain
4. Hackberry Mountain
5. Spotted Mountain
6. Springerville
7. Sentinel
8. Florence
9. San Carlos
10. Raven Butte
11. Pinacate
12. Rucker School
13. Bernardino
14. Yuma Wash

California
1. Medicine Lake
2. Tuledad Canyon
3. Eagle Lake
4. Black Butte Lake
5. Chilcoot
6. Sutter Buttes
7. Clear Lake-Sonoma
8. Jackson Buttes
9. Ebbetts
10. Mono-Long Valley
11. Santa Clara (No. & So.)
12. Monarch Divide
13. Big Pine
14. Saline Range
15. Kern
16. Coso
17. Greenwater Range
18. Death Valley
19. Lava Mountains
20. Cima
21. Amboy-Pisgah
22. Goffs
23. Stepladder
24. Eagle Mountains
25. Salton Sea
26. Black Mountain

Colorado
1. Eagle
2. Woody Creek
3. Specie Mesa

Idaho
1. East Snake River Plain
2. Central Snake River Plain
3. West Snake River Plain
4. Pocatello
5. Blackfoot
6. Oneida

Montana
1. Sweetwater Basin
2. Soap Creek

Nevada
1. Sheldon-Antelope
2. Seven Troughs
3. West Humboldt
4. Winnemucca
5. Battle Mountain
6. Tahoe
7. Fallon
8. Schurz
9. Table Mountain
10. Candellaria
11. Silver Peak
12. North Pancake
13. South Pancake
14. Peak 5476
15. Timber Mountain
16. Halfpint
17. Gold Butte

New Mexico
1. Brazos
2. Taos Plateau
3. Raton
4. Mora
5. Jemez Mountains
6. Mount Taylor
7. Zuni-Bandera
8. Puertocito
9. Albuquerque
10. Quemado
11. Socorro
12. Mesa Redonda
13. Carrizozo
14. Jornada
15. Paxton Siding
16. Hermanas
17. Potrillo
18. Scott Peak

Oregon
1. Beech Creek
2. Crooked River
3. South John Day
4. Bear Valley
5. Logan Valley
6. Newberry-Malheur
7. Yamsay Mountain
8. Tenmile Butte
9. North Sheep Rock
10. Jordan Valley
11. Drake Peak

TABLE 4.1 Numbered Loci in Each State, as Shown in
Figure 4.2—Continued

Utah	9. Escalante
1. Wildcat Hills	10. Enterprise
2. Bear Lake	11. Kolob
3. Gray Back Mountain	12. Kanab
4. Gold Hill	
5. Honeycomb Hills	Wyoming
6. Fumarole Butte	1. Yellowstone
7. Mineral-Black Rock	2. Crescent
8. Loa	3. Leucite Hills

stages through a compositional heirarchy (i.e., basalt, andesite, dacite, rhyolite). From this record it is evident that most andesitic stratovolcanoes build their main cones by eruption frequencies of tens to hundreds of years (International Association of Volcanology, 1951-1975; Simkin *et al.*, 1981; see also Chapter 8 of this volume). The longer repose times probably mean that the compositions are changing toward those of the dacitic-dome and caldera stages, in which repose times in large-volume systems may be tens of thousands of years. For the intracratonic extensional volcanic systems little definitive data are available, but we may deduce that basaltic cinder-cone frequency within a locus is of the order of thousands to tens of thousands of years, whereas episodes of basaltic volcanism may occur with frequencies of hundreds of thousands to millions of years. For many loci the available data are insufficient to determine whether a true distinction exists between periodicity and episodicity or whether each basaltic eruption is simply a reflection of a random tectonic event. However, the distributional patterns of loci, packing density of volcanoes, and geologic groupings of cones and lavas all suggest at least a semisystematic control of events within a locus and within a volcanic zone. In other loci the episodic grouping of events is clear, allowing us to define petrogenetic cycles and providing insight into rates of formation of certain types of rocks and their host volcanoes.

Recent studies, such as Kennett *et al.* (1977), discuss episodes of high-intensity volcanism in a global tectonic context, with peak intensities of volcanism at intervals of as much as 5 m.y. Major pulses, referred to as Cascadian and Columbian, are approximately 15 m.y. apart. Such studies refer to regional or global episodicity and should not be confused with the episodicity within individual loci or volcanic systems that we refer to here.

DESCRIPTION OF SPECIFIC ZONES AND LOCI

Plates 1, 2, and 3 show the late Cenozoic volcanic fields in their entirety. Whereas this is appropriate for information concerning the distribution of volcanic rocks, it is less meaningful if we wish to focus on the distribution of the eruption sites (vents) of the volcanic rocks. Inspection of plates 1, 2, and 3, especially plate 3, shows many large and small areas of volcanic rocks. Some fields shown include many vents (see Luedke and Smith, 1978a, 1978b, 1981, 1982, 1983), whereas other fields

shown have few vents for comparable areas of volcanic rock, although the *areas* of vent distribution may be virtually the same. In other areas very little volcanic material may have been erupted at the surface, but the area of the vent distribution may be comparable to that of a larger volume field. Clearly, these different types of fields have different meanings, but for any meaningful time-space comparison among fields the area of vent distribution, rather than total outcrop area, is a better indicator of the area of the thermal anomaly at depth.

In Figures 4.2, 4.3, and 4.4 we have drawn boundaries for loci, based on the smallest area, to include vents judged to belong to the same thermal system. For many loci the cogenetic relationship of vents in a cluster is self-evident. For other loci the relationships are problematical and our judgments subjective, but we submit that this method of illustrating loci enhances the similarities among loci of different zones.

Inspection of plate 3 suggests that the volcanic fields shown are distributed in a rectilinear pattern, more obvious in some areas than in others. If it could be shown that there is a systematic distribution pattern, both genetic interpretation and hazard analysis would be much simplified. After delineating the loci as shown in Figure 4.2 we drew best-fit boundaries on the more obvious linear zones and found that all but 3 small loci out of the 112 shown can be accommodated in the rectilinear grid shown in Figure 4.2 and subsequent figures. Thus, regardless of whether the grid has significant tectonic integrity in its entirety, it serves as a basis for discussion. It also serves to outline the distribution of volcanism over the last 5 m.y. and, we conclude, serves to illustrate the most likely distribution of potentially active volcanism today. In Figures 4.3 and 4.4 the distributions of loci are shown in relation to the 0- to 5-m.y. grid to illustrate the changing distribution patterns; the loci are totally different for the 10- to 16-m.y. time increment (Figure 4.3) and are very similar for the 5- to 10-m.y. increment (Figure 4.4), but with large specific differences because of the 0- to 5-m.y. migration patterns.

In Figure 4.5 informal names have been introduced for the various linear zones as shown in Figure 4.2. These names aid in the cursory descriptions of the zones discussed in the following section. Two major types of zones are shown: (1) those that illustrate some control of loci distribution related to known rifting of the crust in an east-west extensional mode, as expressed by geomorphic evidence for faulting, and (2) those probably more profound zones that trend northeast and northwest, that appear to be related to a deeper tectonic control of melting in the mantle, and that are not necessarily influenced by the upper-crustal stress field.

Rocky Mountain Hinge Zone (RMH)

The Rocky Mountain hinge zone extends from central-western Montana across Wyoming and Colorado to northeastern New Mexico. This broad zone is the present approximate eastern limit of igneous activity in the western United States. This zone has occupied a position near its present one at least as far back as Paleozoic time but has fluctuated in exact location and has occasionally been breached or has shifted position as far to the northeast as the Black Hills, South Dakota, and formerly ex-

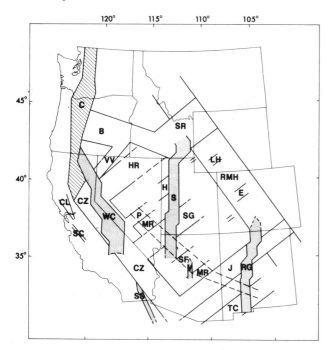

FIGURE 4.5 Grid constraining 0- to 5-m.y.-old volcanic loci. White, volcanic zones and loci exclusive of the Cascades; ruled pattern, Cascade volcanic zone; stippled patterns, volcano-tectonic rifts with active east-west extension.

C	Cascade zone	S	Sevier rift
CL	Clear Lake locus	SG	St. George zone
SC	Santa Clara locus	E	Eagle locus
SS	Salton Sea rift	RMH	Rocky Mountain hinge zone
CZ	California zone		
WC	Western Cordilleran rift	SF	San Francisco Mountains locus
B	Brothers zone		
VV	Virgin Valley zone	V	Verde rift
HR	Humboldt River zone	MR	Mogollon Rim lateral volcanic transgression
SR	Snake River zone		
P	Pancake zone	J	Jemez zone
H	Honeycomb Hills locus	RG	Rio Grande rift
LH	Leucite Hills locus	TC	Truth or Consequences zone

tended southward across west Texas into Mexico. We think the general position of this zone at the present time is as shown in Figure 4.5. Dominant vent alignments in the Yellowstone region (Wyoming), Leucite Hills loci (Wyoming), Eagle locus (Colorado), and the Raton and Mora loci (New Mexico) are parallel or subparallel to this zone. The youngest eruption on this zone is probably the 4000-yr-old basalt of the Eagle locus (Figures 4.5 and 4.8), and most workers think that the Yellowstone system contains magma at high crustal level at the present time (see Chapters 6 and 7 of this volume).

Rio Grande Rift (RG)

The Rio Grande rift is a major tectonic feature that had weak volcanic expression at least as far back as 15 m.y. ago. Much

has been written about the rift (Riecker, 1979) and its similarities to the other rift systems (Olsen and Chapin, 1978). Our analysis suggests that volcanism in the rift is, for the most part, more related to the volcanic zones that cross it than to the deep structure of the rift itself.

East-west extension across the rift is indicated by its graben-like geomorphic expression and north-south fractures that control volcanic vent alignments. Two young basaltic loci, the Brazos locus north of the Jemez zone and the Puertocito locus on the southern edge of the Jemez zone, lie to the west and outside of the present geomorphic expression of the Rio Grande graben. North-south vent alignment in these loci suggests that the western tectonic margin of the Rio Grande rift and its volcanic expression are moving westward toward the Colorado Plateau. Although there may have been mantle upwelling in the Rio Grande rift during its long history of development, it is doubtful that these upwellings represent a rise-and-spreading system in the sense of the African rifts. Perhaps the system will ultimately evolve to this stage, but for the present we think it best to regard the Rio Grande rift as a major extensional boundary in the upper crust between the more stable continental interior and the unstable western part of the continent. If the Colorado Plateau is ultimately consumed by Basin and Range-type faulting and volcanism, the Rio Grande rift will simply become the eastern margin of the Basin and Range province, a position it occupies today at least in its southern reaches. Conceivably the Sevier and Rio Grande rifts will ultimately merge.

Within the Rio Grande rift at least 4 loci have erupted within the last 1 m.y. (see Figure 4.6) and at least two loci have been

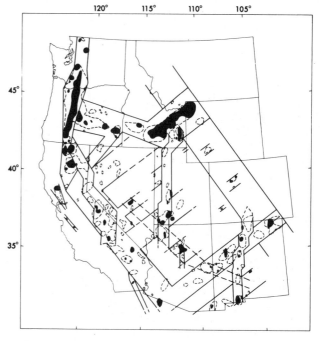

FIGURE 4.6 Volcanic loci active within the last 1 m.y. (black pattern), shown on the 0- to 5-m.y. grid pattern; 0- to 5-m.y. loci shown as dashed lines.

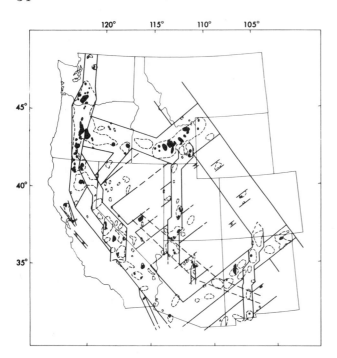

FIGURE 4.7 Volcanic loci active within the last 100,000 yr (black pattern) shown on the 0- to 5-m.y. grid pattern; 0- to 5-m.y. loci shown as dashed lines.

active within the last 100,000 yr (see Figure 4.7). Of particular interest is the geophysical anomaly in the vicinity of Socorro, New Mexico, which from its seismic activity, heat flow, and other characteristics is thought to be a magma body at a depth of about 6 km (Chapin *et al.*, 1980).

Jemez Zone (J)

The Jemez volcanic zone is the most spectacular phenomenon of its kind in the United States and is not easily understood. This zone lies in a region with virtually no seismicity and yet is clearly the zone of greatest volcanic potential in the southwest. Except for a lava flow dated at 11.8 m.y. in the Springerville locus, the earliest known indication of volcanism in the zone occurred about 10.5 m.y. ago at the intersection of the zone with the Rio Grande rift. Older basalts probably occur here but are not yet dated. The Jemez volcanic locus, which includes the Valles caldera, evolved at this intersection and is by far the most complex locus of either the Jemez zone or the Rio Grande rift and is, in fact, the most complex volcanic field in the entire southwestern United States within the last 15 m.y. During the early evolution of the Jemez field, there was a north-northeastward migration of the silicic foci comprising three volcanic cycles, but the vent distributions for nearly all Jemez precaldera vents were controlled by north-south fractures (Smith and Bailey, 1968; Smith *et al.*, 1970). Thus, in the Jemez Montains locus the combined tectonic controls of the Jemez zone and the Rio Grande rift are superposed. This superposition is also shown by the two major fault systems, the

Jemez and the Pajarito, both of which have been active throughout much of the life of the volcanic field. The Jemez Mountains locus undoubtedly owes its unique complexity, its long life (> 10 m.y.), and its caldera evolution to its location on this tectonic intersection. This is almost surely the result of greater magma (thermal) storage at high crustal level at the intersection than elsewhere within either the Jemez zone or the Rio Grande rift. This tectonic intersection has allowed a focusing of mantle-derived basalt and its storage, which in turn has increased the efficiency of crustal preheating and melting and has allowed a large volume of magma to accumulate and differentiate to give the diversity of volcanic rock compositions. Without such tectonic focusing of basalt into a restricted zone of the crust the Jemez Mountains locus would probably be just another field of cinder cones. We are seeing here an outstanding and demonstrable example of the effects of tectonic control on the diversity of igneous rocks.

The Jemez volcanic zone probably began to form before 10 m.y. ago (see plates 1 and 2), with volcanism in the Springerville locus (11.8 m.y. ago), in the Jemez Mountains locus (10.4 m.y. ago), and in the Raton locus (8.2 m.y. ago). These three loci probably continue to be active magmatic systems at depth, with the last dated eruptions at 800,000 yr ago, 130,000 yr ago, and <10,000 yr ago, respectively (see Figures 4.6 through 4.8). Most of the other loci in the Jemez zone came into existence with surface volcanism less than 4 m.y. ago. It would be presumptuous to assume that any of these loci are extinct. More detailed study is needed of all loci to attempt definition of the periodicities and episodicities of each. Within the Jemez vol-

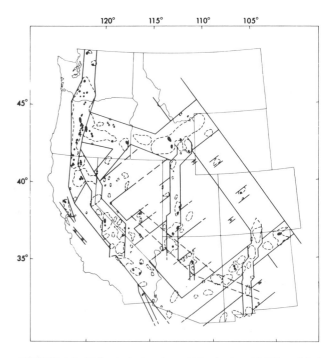

FIGURE 4.8 Volcanic loci active within the last 10,000 yr (black pattern) shown on the 0- to 5-m.y. grid pattern; 0- to 5-m.y. loci shown as dashed lines.

canic zone every locus is unique, ranging from (at present) small fields of few basaltic vents to large fields with hundreds of cinder cones and from fields of isochemical basalt to fields ranging in composition from basanites to high-silica rhyolites through a variety of intermediate rock compositions.

Whatever controls the position of the Jemez zone, whether it is an old Precambrian suture, a reflection of a more modern stress field, or both, we think that the zone is a more profound tectonic expression than is the Rio Grande rift, if intensity of volcanism is any measure. The superposition of volcanic and tectonic elements as seen in the Jemez Mountains locus and the general absence of geomorphic expression of tectonics in the Jemez zone suggest that such features as the Rio Grande rift reflect a dominantly upper-crustal stress field, that features like the Jemez zone reflect a lower-crustal or mantle stress field, and that these two stress fields are separated by a "décollement" zone(s) along which translation may or may not take place.

Truth or Consequences Zone (TC)

This tentative designation for a volcanic zone that appears to be subparallel to the Jemez zone includes several loci outside the Rio Grande rift, and their distribution seems best interpreted as an expression of a northeast-trending zone. A similar, not identical, conclusion was reached by Chapin *et al.* (1978). The major Potrillo locus common to both the Truth or Consequences zone and the Rio Grande rift contains some of the youngest, the largest volume of, and the most vents of basanite of any continental U.S. volcanic field yet mapped. North-south vent distribution in the Potrillo field suggests tectonic control by the east-west extension of the rift. However, within the Bernardino locus in southeastern Arizona the orientation of vents is northeast-southwest, more in accord with directions of the Jemez zone to the north.

The Truth or Consequences zone probably extends into Mexico, terminating either against the Gulf of California rift system or a zone of north-northwest-trending volcanic loci in Baja California. This Baja zone is either subparallel to the California zone or is an extension of it. The Truth or Consequences zone contains at least three loci active in the last million years (Figure 4.6).

San Francisco Mountains Locus (SF)

The San Francisco Mountains locus illustrates in a dramatic way the systematic migration of volcanism across geomorphic and presumed major tectonic boundaries. Sunset Crater erupted about 900 yr ago and is one of the youngest volcanic features of the United States exclusive of the Cascades. For more than 15 m.y. (see plates 1, 2, and 3 and Figures 4.2 through 4.4), basaltic volcanism has progressed northeastward to its present position within the locus. Within the last 5 m.y. this progression is recorded by migrating silicic foci that appear to show both a northeastward and eastward sense of relative motion (see Figure 4.9).

The major silicic foci of this transgression increases in area with time, suggesting an increase in the volume of silicic melt

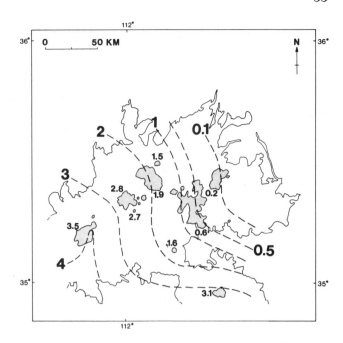

FIGURE 4.9   San Francisco Mountains locus, showing northeastward migration of silicic foci within the outlined volcanic field. Isochrons (dashed lines) drawn on silicic volcanic systems (stippled). Numbers represent millions of years. Data from McKee *et al.* (1974).

with time within the locus, as if the system were moving toward a higher-level larger-volume magma chamber that could ultimately reach a caldera state. A similar but earlier time-space-composition progression leading ultimately to caldera formation has been noted in the Jemez Mountains locus (Smith *et al.*, 1978).

Extending south from the San Francisco Mountains is an incipient rift zone (Figure 4.5), the Verde rift (V), that is probably active today and probably in some way connected with both the complexity of the present San Francisco field and lingering "remnants" of volcanism to the south in the area of 5- to 10-m.y. activity.

Sevier Rift (S)

We have provisionally termed the eastern margin of the Great Basin and the southwestern margin of the Colorado Plateau the Sevier rift. This complex rift system is terminated on the north by the Snake River Plain and on the south by the Mogollon Rim tectonic zone. In the southern part of the rift there is a distinct migration of volcanism (see Figure 4.10) northeastward and eastward onto the Colorado Plateau (Best and Brimhall, 1974). This volcanism appears spatially related to the zone of major faults in southwestern Utah and northwestern Arizona, such as Grand Wash, Hurricane, and Sevier. However, specific vent locations appear to be completely independent of the faults, suggesting a new stress regime.

In the western Grand Canyon region vent alignment is strongly north-south and, like the north-trending vents in the previously

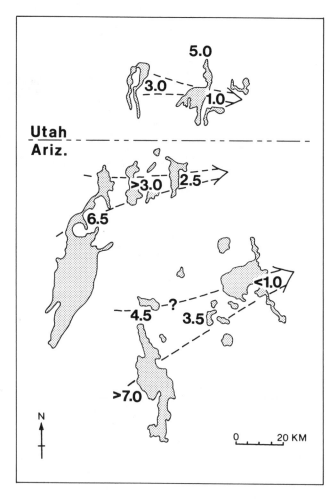

FIGURE 4.10 Western Grand Canyon locus showing eastward migration of hawaiite fields (patterned). Numbers represent millions of years. After Best and Brimhall (1974).

discussed Rio Grande rift, is probably indicative of east-west extension during formation of vents as old as 7.5 m.y. and as young as 10,000 yr (Luedke and Smith, 1978b).

In the central and northern parts of the Sevier rift in Utah the few 0- to 5-m.y.-old loci indicate that present activity is 75 to 100 km west of the Basin and Range-Colorado Plateau tectonic boundary. However, in Idaho, where the rift merges with the Snake River Plain zone, 0- to 5-m.y.-old activity coincides with the major tectonic boundary; here the rift also appears to merge with the Rocky Mountain hinge zone, as suggested by the distribution of vents in the Blackfoot locus. In the northern part of the Blackfoot locus the distribution of vents seems to be controlled by stress of the Rocky Mountain hinge, whereas the central part of the locus seems to be dominated by Snake River Plain stress (as suggested by the alignment of the Blackfoot domes), and the southern part of this locus shows the north-south vent alignment more typical of the Sevier rift. The position of the Blackfoot locus appears more related to the Snake River zone, which suggests that the zone may be migrating eastward as well as propagating northeastward.

Within the Sevier rift at least 5 loci have erupted in the last 1 m.y. (Figure 4.6) and at least 4 have erupted in the last 10,000 yr (Figure 4.8). Most of the activity is associated with the intersection of the rift and the St. George volcanic zone.

St. George Zone (SG)

In southwestern Utah there are several loci whose distribution patterns, considered together with the distribution of vents within each locus, suggest a northeast-trending volcanic zone, the St. George zone, subparallel to the Jemez zone to the south. Within this zone there has been an intensification of basalt eruption in the last 5 m.y., relative to the preceding 5 m.y., and an apparent eastward migration of volcanism.

To the northeast the St. George zone terminates in the Colorado Plateau and to the southwest probably joins the California zone but is obstructed in southern Nevada by a block of crust that has resisted volcanism at least since Mesozoic time. Leaks around this block more than 5 m.y. ago suggest that the St. George zone appeared at an earlier time.

Pancake Zone (P)

The Pancake zone is a designation for a hypothetical northeast-trending zone that includes the Pancake locus, the Honeycomb Hills, and Gold Hill loci in west-central Utah and perhaps the Leucite Hills locus in southwestern Wyoming; the position of these loci may all be rationalized in alternative ways. However, the Pancake locus, which includes the Lunar Crater volcanic field in central Nevada, shows a convincing pattern of migration eastward and northeastward from its beginning 5 to 10 m.y. ago. The 0- to 5-m.y.-old configuration suggests two semi-independent volcanic fields, where only one existed during the 5- to 10-m.y. interval. Geochronological and petrochemical data are needed to better understand this locus.

California Zone (CZ)

The California volcanic zone (Figure 4.2) contains between 20 and 25 loci, mostly in California and western Nevada. The zone terminates northwest against the Cascade zone and southeast probably in Mexico against the southwestern termination of the Truth or Consequences zone. The zone is thus about 1300 km long and more than 150 km wide over most of its length. The California zone, as defined, lies east of the active parts of the San Adreas fault system, except for the Garlock fault, which seemingly has no effect on the present distribution of loci other than perhaps a slight relative displacement of older loci on opposite sides of the fault zone.

Volcanism during the 5- to 10-m.y. interval was largely east of the present boundary of the California zone (as shown in Figures 4.2 and 4.5), although sporadic volcanism occurred within the zone and in southern California to the west of the zone. However, during the last 5 m.y. volcanism throughout the region has shifted westward and increased in intensity throughout the zone.

South of the Garlock fault, waxing volcanism in the Mojave Desert region includes a basanitic volcano that may represent the youngest (about 390 yr) eruption known in the conterminous United States outside the Cascade zone (see Chapter 1 of this volume).

Throughout at least the northern part of the California zone volcanoes of each locus are probably positioned by north-south rifting associated with an east-west extension of the upper crust (see the next section). The California zone contains silicic volcanic systems important to both geothermal interest and volcanic hazards. Except for Yellowstone, the Mono-Long Valley system may have the greatest potential for catastrophic eruption of any other silicic locus outside the southern part of the Cascade zone. The Mono-Long Valley locus is made up of perhaps four semiindependent systems complicated by their position on a great offset in the fault system that bounds the eastern Sierra Nevada. Because these systems may be hydraulically interconnected at depth, forecasting the kind of eruption that might occur here is tenuous. The minimum eruption would be a basaltic lava flow, whereas a maximum eruption could rival the Bishop Tuff of the Long Valley caldera-forming eruption of 700,000 yr ago. Many intermediate types of eruptions are possible. The intense earthquake activity, measurable abnormal ground deformation, and increased hot-spring activity of the past few years indicate that we can expect some kind of volcanic outbreak in the not too distant future. (This activity prompted the U.S. Geological Survey to issue a notice of potential volcanic hazard in May 1982.) The most recent eruptions in the locus took place only a few hundred years ago on the Inyo Craters chain.

Western Cordilleran Rift (WC)

This very complex rift system (Figure 4.5) contains most of the 0- to 5-m.y. volcanism found in the California zone north of the Garlock Fault and extends into the southern Cascade province. The long-continued magmatic activity and tectonic complexity of this region have allowed magma storage, crustal preheating and melting, and the development of several loci that have a high potential for future silicic volcanism. The most prominent of these silicic loci is the Mono-Inyo Dome system and Long Valley, discussed by Bailey *et al.* (1976).

Of the several subsidiary rifts that make up the Western Cordilleran rift as discussed here, one of the most prominent we term the *Inyo rift*. This term is tentatively assigned to a zone of east-west extension containing several volcanic loci and centering on the Owens Valley region. The dominant structural feature of this region is the Owens Valley graben, which has spectacular geomorphic expression and a general north-northwest trend and which bounds the eastern escarpment of the Sierra Nevada. Volcanism in this region is found both within the graben and outside it. In either place the dominant general arrangement of volcanic loci and vents within them is north-south, and we interpret this to mean control by east-west extension, as in the Rio Grande and Sevier rifts. If so, the loci in the high Sierra Nevada outside the obvious structures are nevertheless controlled by north-south fractures that are within the California zone and that extend southward from the Walker Lane. These same fractures determine the northern trends of Basin and Range elements that intersect the Sierra Nevada

from the Inyo rift to the Tahoe regions. It seems, therefore, that volcanism in this region is controlled by a series of short north-south rifts en echelon to the northwest. These rifts include the Inyo rift, the Mono rift, the West Walker-Carson rift(s), and the Tahoe rift. Important work on this region, published since our compilation, includes that of Bacon and Duffield (1980), Moore and Dodge (1980), Bacon *et al.* (1981), and Dodge and Moore (1981).

North of Tahoe the dominant rifting direction and vent alignment are northwest-southeast, and the extension direction and volcano migration direction are westerly to northwesterly (MacLeod *et al.*, 1976; L. J. P. Muffler, U.S. Geological Survey, written communication, 1981; McKee *et al.*, 1983) at least as far north as the Newberry region of Oregon (see discussion of Cascade zone and plates 1, 2, and 3).

Recent seismic activity throughout the length of this zone (Smith, 1978), especially in the Mono-Long Valley and Coso loci, suggests renewed extension along much of the Western Cordilleran rift.

Virgin Valley Zone (VV)

The Virgin Valley zone is predicated on the basis of a locus of 1- to 2-m.y.-old basalts in northwestern Nevada and southern Oregon, which suggests a tie between the northern part of the California zone and the Brothers zone in Oregon. This zone is anomalous in that the westward-migrating wave of volcanism passed through this region more than 10 m.y. ago. However, a K-Ar date of 1.2 m.y., together with many widely scattered young basalt vents (1 to 5 m.y.) in the Humboldt River zone and the eastern part of the Brothers zone (< 10,000 yr), suggest at least a sporadic resurgence of volcanism in northwestern Nevada and southeastern Oregon within the last few million years. We suspect that this entire region has potential for future volcanism.

Brothers Zone (B)

The Brothers zone, as distinct from the Orevada transgression (Luedke and Smith, 1982) of an earlier time increment, contains most of the 0- to 5-m.y. volcanic activity of Oregon exclusive of the Cascades region (Walker and Nolf, 1981, Figure 3). The zone extends eastward into Idaho to join the Snake River zone and westward to join the Cascades zone. At the Cascade junction (Newberry caldera area) there is much recent volcanism, including both basalt and rhyolite and lesser amounts of andesite and dacite. East of Glass Buttes, Oregon, the zone is dominantly, if not wholly, basaltic.

Several loci in this zone have had eruptions less than 10,000 yr ago (Figure 4.8). At its western end the Brothers zone appears to merge with the space-time terminus of the Orevada transgressive silicic volcanism. This junction may provide a partial explanation for the inflection, in isochrons, beginning about 5 m.y. ago, as shown by MacLeod *et al.* (1976).

The Newberry volcanic system erupted rhyolite about 1400 yr ago (MacLeod *et al.*, 1981) and must be considered to have high potential for future eruptions of either basalt, rhyolite, or intermediate compositions.

Snake River Zone (SR) and Humboldt River Zone (HR)

Within the eastern arm of the Snake River Plain, large volumes of alternating basalt and rhyolite, particularly rhyolitic outpourings and associated caldera formation followed by basaltic infilling, have been a volcanic mode for 15 m.y. Christiansen and McKee (1978) suggested that this volcanic pattern began somewhere in the region of the northern Nevada-Oregon-Idaho boundaries and progressed northeastward to its present position in the Yellowstone region. The Yellowstone system surely contains silicic magma today (see Chapters 6 and 7 of this volume) and has potential for future eruptions, possibly of catastrophic magnitude. Basaltic eruptions are a distinct possibility almost anywhere in the Snake River Plain region. Vent alignments throughout the Yellowstone and the eastern Snake River Plain (Craters of the Moon and eastward) reflect the northwest trend of the Rocky Mountain hinge zone.

The western arm of the Snake River Plain appears to be more closely related to the Brothers zone of Oregon in its tectonic and volcanic associations (Figures 4.2 and 4.5).

It is tempting to project the present Snake River zone southwestward of the position of its major 5- to 10-m.y.-old activity to join an area of volcanic resurgence in western Nevada, here termed the Humboldt River zone. This combined zone marks the position of some kind of profound tectonic zone seen, but not understood, on all small-scale topographic and geologic maps and on areal photos and satellite imagery. The very young volcanism (in part < 1.5 m.y.) of the Humboldt River zone, in a region of extensive volcanism 10 to 15 m.y. ago but with no volcanism 5 to 10 m.y. ago, strongly suggests a resurgence of activity in the last few million years and offers the prospect of possible future volcanism in the gap between the Humboldt River zone and the Snake River zone (the area of dashed zone boundaries in Figure 4.2) and further volcanism in the region designated by the Humboldt River zone.

The southwestern terminus of the combined Snake River-Humboldt River zones joins the California zone in the Tahoe-Virginia City region. This general region has been a locus of intermittent volcanism for more than 20 m.y., with both rhyolitic and basaltic volcanism less than 1.5 m.y. ago.

San Andreas-Related Loci

The Santa Clara locus (SC) and the Clear Lake locus (CL) are related to the San Andreas fault system. The Santa Clara locus is based on a single K-Ar date of 3.57 m.y. and two small areas of basaltic lava.

The Clear Lake locus is an extremely interesting and unique volcanic area. For more than 10 m.y., volcanism has been migrating northward across an active northwest-trending transverse fault system that has dominant right-lateral movement. The net result of these motions is progressive shredding of a series of volcanic fields with increasing age (see Luedke and Smith, 1981, inset C). The Clear Lake locus is the world's largest producer of electricity from geothermal steam; the youngest volcanic eruption within the locus was ~10,000 yr ago. We see no reason why this field will not continue to produce future eruptions of basalt, andesite, dacite, or rhyolite.

The origin of the magmas in this field and the controlling tectonic driving forces are controversial; for details, see McLaughlin and Donnelly-Nolan (1981).

Salton Sea Rift (SS)

The Salton Sea rift (as shown in Figures 4.2 and 4.5) includes the Salton Sea locus (a group of rhyolitic domes) and the Cerro Prieto volcano and associated geothermal field in Baja California, Mexico. These two loci are generally thought to be thermally (and magmatically) fueled by segments of the East Pacific Rise underlying the Gulf of California rift. The Salton Sea rhyolite domes are probably less than 10,000 yr old (Friedman and Obradovich, 1981) and are associated with a high-temperature geothermal system. This volcanic system should be considered potentially active.

Other Loci

The Eagle locus (E) in northwestern Colorado is of special interest because of its relative isolation in the Rocky Mountain hinge zone and the very young eruption about 4000 yr ago (Larson et al., 1975). This young locus is spatially associated with older loci that have had intermittent magmatism in the region for more than 20 m.y. The surface expression of this magmatism has always been weak, probably because of the great crustal thickness (50 ± km) of the region (Smith, 1978).

The Leucite Hills locus (LH) of southwestern Wyoming, a young volcanic field also isolated in the Rocky Mountain hinge zone, is of interest because of its extreme alkalic composition and approximately 1-m.y.-old volcanism. Three other widely scattered vent areas (Figure 4.2) in southwestern Colorado, northeastern Utah, and south-central Oregon are the only three areas not constrained by the zones and rifts of Figure 4.5 and are of unknown significance.

Cascade Zone (C)

The Cascade zone contains all of the volcanoes usually designated "active" and most of the volcanoes heretofore considered "potentially active" in the conterminous United States. Twelve major High Cascade andesitic stratovolcanoes or volcanic systems are probably potentially active. All of these are thought to have been active within the last few thousand years, and five have had eruptions of record even though some of the records are equivocal. Ten of the twelve represent thermal areas today.

Interspersed with these High Cascade volcanoes are many areas of recent volcanism in the form of andesitic and basaltic cinder cones, lava flows, and incipient stratovolcanoes. Most of this activity is found in the central and southern Cascades, but we see no valid reason why future volcanism could not occur anywhere in the chain; it is most likely to recur in those areas of the longest history of intermittent activity.

The widespread young basaltic volcanism of both the High Cascades and the flanks of the range is enigmatic and appears to us to be more closely related to the extensional volcanism so prevalent east of the Cascades (see Taylor, 1981; McKee et al., 1983). In and east of the region of the central and southern Cascades, basaltic and, locally, rhyolitic volcanism has been migrating westward for over 10 m.y. During the last 5 m.y. the basaltic volcanism has roughly coincided spatially with the so-called High Cascade andesites. The front of this wave of basaltic migration apparently continued westward during the last 5 m.y. to the line of High Cascade cones, and locally beyond it, but the loci of greatest intensity of volcanism are east of the main chain and from south to north center on the Lassen Peak area, Medicine Lake Highlands area, Newberry crater area, and Simcoe volcanic field near Mount Adams. It is significant that all of these areas have developed silicic foci.

The present chain of High Cascade andesitic stratovolcanoes has developed mostly over the last 1 m.y. (McBirney, 1978). The "potentially active" volcanoes are probably less than 700,000 yr old, and the youngest cycles of activity of these cones may have begun less than 100,000 yr ago.

Perhaps the area of most intensive recent volcanism within the main chain is the Three Sisters region, where silicic and basaltic volcanism have produced a complex unlike other High Cascade volcanoes but one that has a high potential for eruptions of either silicic or basic composition. On the south flank of South Sister, rhyolitic volcanism as recent as 2000 yr ago suggests a high probability for future caldera formation. With eruptions only about 70 yr ago, Lassen Peak may be viewed similarly.

Mogollon Rim Lateral Volcanic Transgression (MR)

Along the Mogollon Rim in Arizona there are three large volcanic loci that are strongly migratory in a northeasterly and easterly direction onto the Colorado Plateau. These loci are as follows:

1. Western Grand Canyon locus located in the Sevier rift near its southern terminus and at the junction with the St. George zone. Best and Brimhall (1974) have described the eastward migration of hawaiite vents in this area (Figure 4.10).

2. San Francisco Mountains locus located on the southwestern edge of the Colorado Plateau midway between the St. George zone and the Jemez zone (Figure 4.5) shows a dramatic northeastward and eastward migration of silicic foci over the last 4 m.y. (Figure 4.9). This migration from consideration of earlier time increments (plate 1 and Figure 4.3; plate 2 and Figure 4.4) appears to have been more or less continuous for more than 10 m.y. During the last 1.5 m.y. motion may have been more to the east than to the northeast. The San Francisco Mountains locus is probably associated with incipient rifting in the Verde Valley, which is shown in Figure 4.5.

3. Springerville locus is located in the Jemez zone near the eastern end of the Mogollon Rim. This locus is very large, probably complex, and not well known, but map patterns clearly show northward, northeastward, and perhaps eastward migration of vent areas over the last 10 m.y. (plates 1, 2, and 3).

These three loci form a northwest-trending zone that if extended southeast to the Rio Grande rift includes the Truth or Consequences locus and if extended northwest into south-central Nevada includes the Pancake locus. Both of these loci

appear to show northeastward, and the Pancake perhaps also eastward, migration during the last 5 m.y. Collectively the loci in this zone suggest a lateral transgressive migration of approximately 50 km across a front nearly 1000 km broad in less than 5 m.y.; this is a translation rate of ~1 cm/yr. It is doubtful that the entire length of this zone is a tectonic entity; rather it appears to be a fortuitous artifact of five volcanic loci that lie approximately in a "zone" or line subparallel to the Las Vegas shear zone, the Mogollon Rim, and the Texas Lineament. All of these loci are complex enough to record similar components of volcanic migration that also appear to be recorded over a much larger area. The three central loci of this zone may have some specific tectonic meaning with respect to the evolution of the Mogollon Rim. Four of the five loci discussed here record eruptions less than 25,000 yr old (Figure 4.9).

VENT AND COMPOSITION DISTRIBUTION IN TIME

As originally conceived, the maps that form the major part of our data base were to have shown only vent distributions, because the vent distribution best defines the thermal source, for evaluation of geothermal analyses. However, vent locations are often enigmatic, especially with increasing age and complexity of volcanic systems. Also, an isolated lava flow, even though the source is not known, may be counted as a vent for volcanological analysis. Considerable subjectivity was exercised in assigning ages and locations to vents. Thus, we cannot claim great accuracy in the vent analysis shown in Table 4.2. The vent count and analysis are subject to correction and refinement as more and better data become available, but we doubt that future refinement will necessitate major changes in our general conclusions.

We count, exclusive of the Cascade zone, about 5000 vents as having formed in the last 5 m.y., to give an average rate of 1 new vent per 1000 yr; 5000 is certainly a minimum number, but to find it low by a factor of 2 would be surprising. Loss of vents by erosion and/or burial prohibits accurate assessment of vent populations in stratovolcanoes and other complex volcanic systems. However, by excluding the Cascade zone from the vent count, approximately 90 percent of the counted vents represent monogenetic volcanoes. Although most of these vents are related to multiple vent systems, even cursory examination of our maps suggests minimal loss by either burial or erosion during the last 5 m.y. Greatest obvious loss has been in the eastern Snake River Plain region, Idaho. Notable loss also has occurred, because of the complexity of the volcanic systems at the Jemez locus, New Mexico; Newberry locus, Oregon; and the San Francisco locus, Arizona. A few additional areas might be cited, but the losses must be trivial in the total context of the rate estimates of Figure 4.11. Most of the losses in the areas cited were by burial in collapsed calderas, by stacking of units in composite volcanoes, or by cover with ash flows. If, ultimately, we decide to include much of the Cascade basaltic volcanism as being of extensional tectonic origin, the number will be much larger. However, it would require 50,000 vents total (an impossible number) to give an average rate of 1 vent

TABLE 4.2   Total Vents According to Rock Type and Time Periods During the Last 5 m.y., Exclusive of the Cascade Zone

Age (m.y.)	Basanite	Basalt	Andesite	Dacite	Rhyolite	Total
5	372	3984	142	156	313	4967
1	301	1517	69	51	227	2165
1.0-0.1	284	845	36	49	100	1314
0.1-0.01	9	537	27	2	101	676
0.01-0.001	4	130	6	0	24	24
0.001-0.0001	4	5	0	0	2	11

SOURCE: Data from Luedke and Smith (1978a, 1978b, 1981, 1982, 1983).

per 100 yr for 5 m.y. Figure 4.11 shows that the average rate had accelerated from 1 vent per 1000 yr to about 1 vent per 100 yr by 100,000 yr ago and then remained relatively constant until about 100 yr ago. The numbers for vents less than 10,000 yr old may be too small to be significant, but they too are minimum numbers, and the average rate of vent formation may well exceed 1 vent per 100 yr today. The average rate is probably not meaningful in a real sense because of the episodic nature of activity in most loci. However, it may be a real indicator of increasing volcanic activity over the last 5 m.y. and specifically over the last 100,000 yr. It is possible that averaging over the 1- to 5-m.y. time period obscures episodic peaks of volcanism with durations of 100,000 yr to 1 m.y. The data base does not allow a more detailed analysis at the present time.

Confirmative evidence for increase in vent formation is found in the number of loci that have been active during various time intervals (Figure 4.11). We count, exclusive of the Cascades, 112 loci (see Figure 4.2 and Table 4.1) active in the last 5 m.y. (the Snake River zone counted as 3). Of these 112 loci, 39 were active in the last 1 m.y., 31 were active in the last 100,000 yr, 24 were active in the last 10,000 yr, and 10 were active in the last 1000 yr. Of the 112 loci active in the last 5 m.y., about 45 have a history predating 5 m.y. and perhaps 10 have a history

predating 10 m.y. Perhaps 5 of these have been active in the last 100,000 yr.

When we consider that the eruptive periodicities of individual loci may range from 1000 yr to 1 m.y. (depending on their evolutionary state) and that episodicity may extend eruption intervals to perhaps 10 m.y. or more, it is presumptuous to think of these loci as extinct as long as the plate motions continue in their present mode. Also, we see no evidence for major hiatuses in volcanism over the last 15 m.y., when the entire map area is considered. However, local hiatuses exist within loci and within zones.

Figure 4.12 shows the distribution with time of the 5 major rock types, delineated on plate 3, as determined from vent counts. These cumulative curves simply show relative abundances of rock types. The overwhelming predominance of basalt is of course also seen by inspection of the maps (plates 1, 2, and 3) and is the basis for the conclusions formulated by Christiansen and Lipman (1972) and Lipman *et al.* (1972) for predominantly "andesitic" volcanism prior to about 16 m.y. ago and predominantly basaltic volcanism since that time.

These authors attributed the andesitic volcanism to subduction mechanics and the basaltic volcanism to crustal extension following cessation of subduction and formation of the San An-

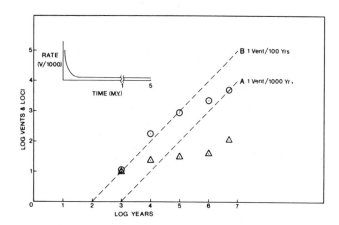

FIGURE 4.11 Cumulative curves showing volcanic vents and loci active during specific time periods over the last 5 m.y., exclusive of the Cascade zone. The data show an increase in the rate of vent formation within the last 1 m.y., particularly over the last 100,000 yr. The circles indicate the vents; the triangles indicate the loci.

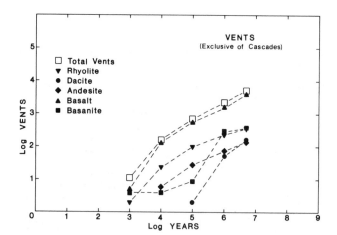

FIGURE 4.12 Cumulative curves showing temporal distribution of vents, by rock composition, over last 5 m.y., exclusive of the Cascades zone. The curve for total vents (squares) is the same as that for Figure 4.11. Vent counts and compositions are from Table 4.1.

dreas fault system. Irrespective of the modus operandi, there was a profound change in the "dominant" style of volcanism about 20 to 15 m.y. ago, the times differing slightly from region to region. We recognize that this change was related to a shift in tectonic mode, but the effect of the change was to accelerate magma rise from the mantle through the crust, thereby inhibiting thermal (magma) storage in the crust. Thus, the tectonic shift effectively increased the ratio of mantle melts to crustal melts erupted at the surface. Whether the rate of formation of mantle magma also changed is problematical and will be known only when definitive data on age, volume, and composition of all Tertiary volcanic rocks become available.

The high incidence of basanites in the last 5 m.y. is noteworthy in that it must signify a progressive change in the depth or conditions of magma generation over much of the southwestern United States in the same region where waxing volcanism is most pronounced (see Chapter 1 of this volume). We have found only trivial expression of basanite on the 5- to 10-m.y. map (plate 2) and none in the 10- to 15-m.y. increment. This may reflect, in part, a data bias that indicates a general preference for the study of younger rocks, but this cannot be the entire explanation. The deflection in the basanite curve in Figure 4.12 at 10^4 and 10^5 yr is probably not a valid representation. Rather it reflects a data gap suggesting that recognition and mapping of basanites lags behind collection and analysis of basanites. For example, we know of basanites not shown on our maps because no distinction had been made on the source maps.

Although the basalts and basanites together offer the greatest probability for future eruptions, such eruptions should be largely confined to cinder cones or lava flows. A cinder-cone eruption almost anywhere in the United States today would almost surely have an adverse effect on our population in some way, largely from ash falls. Maar-type eruptions, as have occurred in such loci as the Potrillo and Zuni-Bandera volcanic fields in New Mexico and the Bernardino field in Arizona, could be hazardous, but their probability is much lower than for more typical basaltic eruptions. Our maps record about 100 maars in the 0- to 5-m.y. interval, only about 2 percent of total vents for the same period.

Andesitic and dacitic eruptions (exclusive of the Cascades) would be most likely to occur today from complex multicompositional loci, such as the Long Valley-Mammoth Mountain system in California, the San Francisco locus in Arizona, or the Jemez locus in New Mexico. The Medicine Lake and Clear Lake systems in California and the Newberry system in Oregon also have high potential for eruptions. Such eruptions, of dacite or high-silica andesite, might be highly explosive (e.g., Mount St. Helens). Low-silica andesite can be expected to behave more like basalt.

In general, the rhyolitic systems, which might also involve other compositions, should be deemed to have the highest explosive potential. We count about 45 rhyolitic systems active in the last 5 m.y. exclusive of the Cascades and about 25 silicic (rhyolite and dacite) systems active in the Cascades in the last 1 m.y. (see Figure 4.13). Smith and Shaw (1975, 1978) assumed that most of these silicic systems derive from high-level magma chambers, i.e., within 10 km of the surface. Most of these

systems were evaluated in a preliminary way for their geothermal energy potential by means of a Smith-Shaw diagram (see Figure 4.14). New data and more detailed studies have modified the preliminary estimates in some areas but have caused no changes approaching an order of magnitude. The use and meaning of the diagram are explained by Smith and Shaw (1975, 1978).

Although, to date, the diagram has been used only for geothermal evaluations, it was originally designed to test the probability of existence of magma in high-level silicic magma chambers and, hence, their potential for future eruptions. Whether a system can be accurately plotted on the diagram depends on the accuracy of certain types of data discussed by Smith and Shaw (1975) that can be used to estimate the volume of the chamber and the age of the last eruption. Figure 4.14 plots those systems for which we feel the data are accurate enough to be significant. This includes all the known large-volume silicic systems active in the last 5 m.y. and many smaller ones. In all probability *most* of the other smaller silicic systems, for which the data are inadequate for plotting in Figure 4.14, are extinct in the sense that they no longer exist as part of a silicic magmatic continuum. This does not imply that silicic magma cannot be regenerated in a locus or that future basaltic eruptions cannot take place.

The periodicity of these silicic systems is thought to be, in part, a function of their volume (Smith, 1979)(see Figure 4.1), which in turn is thought to be a function of the volume-rate and spacing of a basaltic thermal input. To develop and maintain a large volume of silicic magma for a long period of time the basaltic component must be stored and its thermal energy either given off as silicic eruptives or lost to the adjacent wall rock. In this model the volume erupted is a function of the time between eruptions. In an order-of-magnitude sense we think the existing data are strongly supportive (Smith, 1979, Figure 12), especially for caldera-forming eruptions. However, these large-volume systems may leak, and Bacon (1982) concluded that in such systems as the Coso rhyolite dome field leaks may defuse the system from caldera-type eruptions and the leak periodicity, i.e., the interval between eruptions, depends on the volume of the last eruption. These models are not mutually exclusive and both may operate. In any case a positive relationship exists between the size of the system and the frequency and volumes of its eruptions, such that the entire life span of a small system may be shorter than the interval between two rhyolite domes of a large system. This scale problem, common to the rhyolitic systems of extensional tectonic regions, makes long-term evaluation and prediction risky given our present state of knowledge. For additional insight into the relationship between mafic and silicic magmas, Hildreth's review (1981) is required reading (see also Shaw, 1980).

THE SIGNIFICANCE OF MIGRATING VOLCANIC LOCI

We have briefly discussed migrating volcanic loci in connection with specific zones. To summarize the available data, the arrows in Figure 4.15 show migration directions as we understand

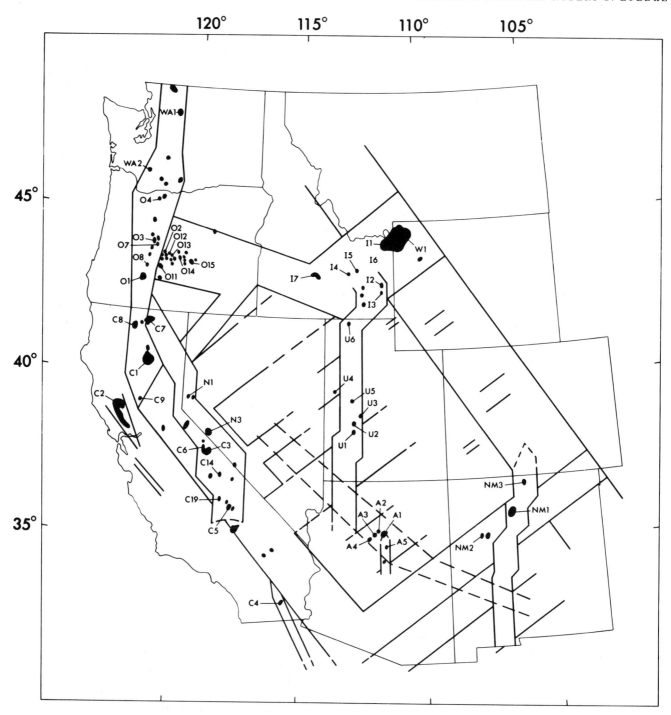

FIGURE 4.13   Distribution of rhyolitic volcanism, 0 to 5 m.y. Numbered loci are the systems evaluated in Figure 4.14.

them for the last 5 m.y. Our interpretation of migration directions is determined from vent or vent-cluster distributions and available age data or geologic analysis. Data for some regions show waxing or resurgent volcanism within the last 5 m.y. We think the waxing phase encompasses the present time.

Given the motions described for the Mogollon Rim and Snake River zones, all of which appear real and have some docu-

mentary basis, we can look at the loci of other zones and, even in the absence of critical geochronological data, see loci, or groups of loci, that have vent distributions, suggesting northeastward or eastward migrations. These areas represent data gaps where further study is needed to confirm or deny our suggestions. Some of these areas are shown in Figure 4.15 by the open arrowheads.

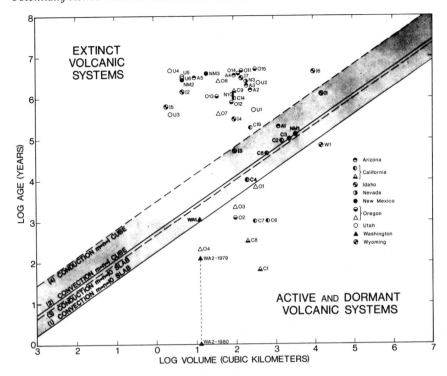

FIGURE 4.14 Smith-Shaw (1975) diagram showing potentially active and extinct silicic volcanic systems less than 5 m.y. old. Numbers are from Figure 4.13; see text for details.

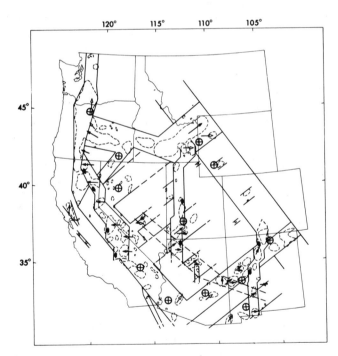

FIGURE 4.15 Migration directions of volcanic loci during the last 5 m.y. Solid arrowheads indicate the direction of migration of volcano-tectonic rift boundaries as deduced from vent and loci alignments. Solid arrows with tails indicate migration direction of volcanism as deduced from geochronological or geologic data. Open arrows with tails indicate permissable migration directions of volcanism, but confirmative geochronologic data are not available. Circles with crosses indicate general areas of increasing (waxing) intensity of volcanism over the last few million years.

The sum of the data suggests that there are northeastward migrations of loci in Idaho, Montana, Wyoming, Colorado, Utah, New Mexico, Arizona, and eastern Nevada. We think the most reasonable interpretation of these migrations, over the last 5 m.y., is that they result from southwestward motion of the North American Plate over magma sources in the mantle either "fixed" within the time frame under consideration or migrating westward but at a slower velocity than the overlying crust. We find it difficult to equate each of the many loci of such a large region with its own deep-mantle plume. Rather we think that the individual distribution of loci reflects the specific idiosyncrasies of the stress regime of the lithospheric plate in motion, overriding a mantle capable of generating magma anywhere that the stress regime permits magma rise. In regions of extreme crustal extension or thinning, deeper-mantle upwelling may take place. Isolated "spots" are not precluded, but a "hot spot" (mantle plume) model, as usually depicted, is a vast oversimplification for most of the western United States. It appears to us that the many loci are hot spots in the lower crust over reduced-pressure regions in the mantle.

In California, westernmost Nevada, and Oregon, volcano migration directions are generally westward or northwestward (MacLeod *et al.*, 1976) and must involve a different cause for the motion than for the areas discussed above. In this region the mantle flow and migration of the magma-forming domains have acted on the upper crust to cause rifting and drifting of Basin and Range type. In the Cascade region the subducting Juan de Fuca Plate has probably acted to constrain the westward-migrating magma sources, possibly redirecting them northward at least as far as the region between Newberry Volcano and Mount Adams, where extensive recent basaltic volcanism appears anomalous.

It is tempting to suggest that these westerly migrations are the result of westward-directed mantle flow in front of the westward-advancing "deep keel" of the North American Plate. In the Pacific Northwest of California and Oregon the migration during the last 15 m.y. has been a counterclockwise windshield-wiper-like sweep pivotal around the north end of the Sierra Nevada. Today the major zone of potentially active volcanism is virtually coincident with the High Cascade volcanoes, which we would relate to subduction mechanics.

South of the northern terminus of the Sierra Nevada, dextral tectonic motions associated with the San Andreas fault zone and the subducted Mendocino fracture zone appear to have placed the entire region between the San Andreas fault and the Walker Lane into a complex shear couple as far south as the Garlock fault. The northward-trending fractures in this couple form the southern half of the Western Cordilleran rift zone (as shown in Figure 4.5) and appear to localize volcanism in the California zone between the Garlock fault and the Tahoe-Sierra Valley rift. East-west extension in this region allows magma storage and the development of silicic nuclei, some of which, like the Mono-Long Valley complex, have high potential for future explosive eruptions.

Available data for the area south of the Garlock fault do not indicate definitive volcanic migration motions, but the pattern is consistent with the general westward migration of the California zone, and very recent volcanism in this region suggests a continuation of the waxing phase evident in the 0- to 5-m.y. pattern of loci.

Data from southwestern Arizona are limited, but "permissible" vent patterns in the Centennial locus suggest that this area may now be controlled by the westward motion of the North American Plate rather than by westward drifting of the magma source. If so, the westward and eastward migrations probably owe their directions simply to different relative velocities between mantle flow and overriding crust, both of which are moving in the same general direction, in this case to the west. However, these relative motions have profoundly different effects on the crust, which on the one hand (eastward migration of volcanism) is moving faster than the mantle magma source and on the other hand (westward motion) is carried, attenuated, and disrupted by the mantle flow giving rise to Basin and Range-type tectonics. Ultimately this attenuated crust may heal and again behave wholly or partly in conjunction with the more stable upper crust to the east, and eastward migration of volcanism may return to the area as in the Pancake locus, which is located in the strongly attenuated crust of an earlier time.

We recognize that the obvious explanation for the divergence of migration directions of volcanism is an axis of subcontinental spreading from near the Gulf of California northward through the Basin and Range Province—in a broad sense similar to that envisaged by Suppe *et al.* (1975). However, without attempting their detailed analysis, we tentatively suggest that all motions may be explained by different relative velocities between the westward-migrating North American Plate and its subjacent mantle. Doe *et al.* (1982) suggested that most Yellowstone basalts were derived from a lithospheric "keel" rather than from a deeper primitive mantle. If this interpretation applies to other migrating basaltic systems, then either laminar flow or horizontal shearing within the lithosphere is required. Furthermore, basaltic systems originating from different depths could have different migration rates at the surface. This probability offers exciting opportunities for future studies and could well provide the key to directions and relative velocities of mantle flow and relative depths of origin of magmas.

SUMMARY

Pulses of motion, alternately related to the opposed drifts of the North American and Pacific plates and their subjacent mantle, together with the additional complexity implied by the San Andreas transform dynamics and the incipient subduction of fragments of the East Pacific rise, have combined to place both the Pacific margin and the western intermontane region into an extremely complex and everchanging stress state.

Pulses of motion, especially those accompanied by slight directional changes, probably trigger magma formation and magma rise from the mantle into extensional fractures in the crust. The actual mechanics of magma generation, movement, and eruption are not well understood, but the existing fragmental data indicate that the process, as it leads to the formation and evolution of volcanic systems or loci, is uneven in time, space, and duration. Some systems are represented by a simple eruptive event, whereas others have recorded hundreds of eruptive events over periods of time as long as 10 to 15 m.y., although most loci are probably much shorter lived.

The elements of the 0- to 5-m.y. distribution pattern of loci began to emerge as early as about 10 m.y. ago. Other than the migration patterns of certain loci, and the general waning intensity of volcanism (exclusive of the Cascades) in the northwest and the notable waxing intensity in the east and south, the areal pattern has not changed in the past 5 m.y. Distribution patterns of loci active within the last 1.0, 0.1 and 0.01 m.y. (Figures 4.6, 4.7, and 4.8) clearly show that the total areal distribution of potentially active loci is the same today as it was 5 to 10 m.y. ago. We reason that if the areal distribution pattern has endured for such time and had not changed as of 10,000 yr ago it has not changed today. About 70 loci have appeared or been reactivated in a new position within their respective volcanic zones within the last 5 m.y., and perhaps 10 new loci have appeared within the last 1 m.y. About 30 loci have been active in the last 10,000 yr.

Data are inadequate for meaningful long-range predictive analysis. However, as long as the present extensional mode of tectonic motion continues, volcanism will continue to follow the present patterns. Silicic systems that have already generated high-level magma chambers will release energy when they crystallize or if they receive new thermal injections of basalt. The most advanced silicic systems are Yellowstone, Mono-Inyo-Long Valley, Coso, Medicine Lake, and Newberry; perhaps the Valles, San Francisco Peaks, Clear Lake, Salton Sea, Mount Lassen, and Mineral Mountains also should be listed. Within the Cascades the most advanced silicic systems are Lassen and South Sister, with Crater Lake and Shasta probably next in line. We see no valid reason why basaltic eruptions could not occur from any of the loci shown in Figure 4.2, but of course the probability will vary greatly from locus to locus and from zone to zone. Also, we see no reason why new loci (or, for that matter, new zones) could not develop. In terms of our lifetimes these possibilities may seem inconsequential, but for land-use planning beyond 100 yr, we need to know a great deal more about these volcanic loci than we do today—including their past periodicities, their states of chemical and physical evolution, their probable life spans, and whether they are randomly born and evolved or are products of systematic tectonic design. Systematic tectonic design will be revealed only through cumulative detailed studies of individual loci.

ACKNOWLEDGMENTS

Among those who have aided us with the writing of this paper we wish to thank Karen Gray for the vent count and classification that led to Table 4.2 and Figures 4.11 and 4.12 and R. G. Robinson for his helpful suggestions. We also thank F. R. Boyd, G. H. Heiken, B. C. Hearn, Jr., and R. I. Tilling for their detailed critical reviews, which considerably sharpened the paper.

REFERENCES

Bacon, C. R. (1982). Time-predictable bimodal volcanism in the Coso Range, California, *Geology 10*, 65-69.

Bacon, C. R., and W. A. Duffield, eds. (1980). Geochemical investigations in the Coso Range, California, *J. Geophys. Res. 85*, 2379-2516.

Bacon, C. R., R. Macdonald, R. L. Smith, and P. A. Baedecker (1981). Pleistocene high-silica rhyolites of the Coso volcanic field, Inyo County, California, *J. Geophys. Res. 86*, 10223-10241.

Bailey, R. A., G. B. Dalrymple, and M. A. Lanphere (1976). Volcanism, structure, and geochronology of Long Valley caldera, Mono County, California, *J. Geophys. Res. 81*, 725-744.

Best, M. G., and W. H. Brimhall (1974). Late Cenozoic alkalic basaltic magmas in the western Colorado Plateau and the Basin and Range transition zone, U.S.A., and their bearing on mantle dynamics, *Geol. Soc. Am. Bull. 85*, 1677-1690.

Chapin, C. E., A. R. Sanford, and L. D. Brown (1980). Past and present magmatic and hydrothermal systems at Socorro, New Mexico (abstr.), *EOS 61*, 1150.

Chapin, C. E. *et al.* (1978). Exploration framework of the Socorro geothermal area, New Mexico, *New Mexico Geol. Soc. Spec. Publ. 7*, 114-129.

Christiansen, R. L., and P. W. Lipman (1972). Cenozoic volcanism and plate-tectonic evolution of the western United States, II, Late Cenozoic, *Phil. Trans. R. Soc. London 271*, 249-284.

Christiansen, R. L., and E. H. McKee (1978). Late Cenozoic volcanic and tectonic evolution of the Great Basin and Columbia intermontane regions, in *Cenozoic Tectonics and Regional Geophysics of the Western Cordillera*, R. B. Smith and G. P. Eaton, eds., Geol. Soc. Am. Mem. 152, pp. 283-311.

Dodge, F. C. W., and J. G. Moore (1981). Late Cenozoic volcanic rocks of the southern Sierra Nevada, California: II. Geochemistry, *Geol. Soc. Am. Bull. 92*, 1670-1761.

Doe, B. R., W. P. Leeman, R. L. Christiansen, and C. E. Hedge (1982). Lead and strontium isotopes and related trace elements as genetic tracers in the Upper Cenozoic rhyolite-basalt association of the Yellowstone Plateau volcanic field, *J. Geophys. Res. 87*, 4785-4806.

Friedman, I., and J. Obradovich (1981). Obsidian hydration dating of volcanic events, *Quat. Res. 16*, 37-47.

Hildreth, W. (1981). Gradients in silicic magma chambers: Implications for lithospheric magmatism, *J. Geophys. Res. 86*, 10153-10192.

International Association of Volcanology (1951-1975). *Catalogue of the Active Volcanoes of the World Including Solfatara Fields, Parts I to XXII.*

Kennett, J. P., A. R. McBirney, and R. C. Thunell (1977). Episodes of Cenozoic volcanism in the circum-Pacific region, *J. Volcanol. Geotherm. Res. 2*, 145-163.

Larson, E. E., M. Ozima, and W. C. Bradley (1975). Late Cenozoic basic volcanism in northwestern Colorado and its implications concerning tectonism and the origin of the Colorado River system, in *Cenozoic History of the Southern Rocky Mountains*, B. Curtis, ed., Geol. Soc. Am. Mem. 144, pp. 155-178.

Lipman, P. W., H. J. Prostka, and R. L. Christiansen (1972). Cenozoic volcanism and plate-tectonic evolution of the western United States, I. Early and Middle Cenzoic, *Phil. Trans. R. Soc. London 271*, 217-248.

Luedke, R. G., and R. L. Smith (1978a). Map showing distribution, composition, and age of Late Cenozoic volcanic centers in Arizona and New Mexico, *U.S. Geol. Surv. Misc. Invest. Ser. Map I-1091-A.*

Luedke, R. G., and R. L. Smith (1978b). Map showing distribution, composition, and age of Late Cenozoic volcanic centers in Colorado, Utah, and southwestern Wyoming, *U.S. Geol. Surv. Misc. Invest. Ser. Map I-1091-B.*

Luedke, R. G., and R. L. Smith (1981). Map showing distribution, composition, and age of Late Cenozoic volcanic centers in California and Nevada, *U.S. Geol. Surv. Misc. Invest. Ser. Map I-1091-C.*

Luedke, R. G., and R. L. Smith (1982). Map showing distribution, composition, and age of Late Cenozoic volcanic centers in Oregon and Washington, *U.S. Geol. Surv. Misc. Invest. Ser. Map I-1091-D.*

Luedke, R. G., and R. L. Smith (1983). Map showing distribution, composition and age of Late Cenozoic volcanic centers in Idaho, western Montana, western South Dakota, and northwestern Wyoming, *U.S. Geol. Surv. Misc. Invest. Ser. Map I-1091-E.*

MacLeod, N. S., G. W. Walker, and E. H. McKee (1976). Geothermal significance of eastward increase in age of upper Cenozoic rhyolitic domes in southeastern Oregon, in *United Nations Symposium on the Development and Use of Geothermal Resources*, U.S. Government Printing Office, Washington, D.C., pp. 465-474.

MacLeod, N. S., D. R. Sherrod, L. A. Chitwood, and E. H. McKee (1981). Newberry volcano, Oregon, in *Guides to Some Volcanic Terranes in Washington, Idaho, Oregon, and Northern California*, D. A. Johnston and J. M. Donnelly-Nolan, eds., U.S. Geol. Surv. Circ. 838, pp. 85-91.

McBirney, A. R. (1978). Volcanic evolution of the Cascade Range, *Ann. Rev. Earth Planet. Sci. 6*, 437-456.

McKee, E. H., R. L. Smith, and H. R. Shaw (1974). Preliminary geothermal exploration San Francisco volcanic field, northern Arizona, *Geol. Soc. Am. Abstr. Programs 6*, 458.

McKee, E. H., W. A. Duffield, and R. J. Stern (1983). Late Miocene and early Pliocene basaltic rocks and their implications for crustal structure, northeastern California and south-central Oregon, *Geol. Soc. Am. Bull. 93*, 292-304.

McLaughlin, R. J., and J. M. Donnelly-Nolan, eds. (1981). *Research in the Geysers-Clear Lake Area, Northern California*, U.S. Geol. Surv. Prof. Pap. 1141, 259 pp.

Moore, J. G., and F. C. W. Dodge (1980). Late Cenozoic volcanic rocks of the southern Sierra Nevada, California: I. Geology and petrology, *Geol. Soc. Am. Bull. 91*, 515-518.

Olsen, K. H., and C. E. Chapin, eds. (1978). *Program and Abstracts, 1978 International Symposium on the Rio Grande Rift*, Los Alamos Scientific Laboratory Report LA-7487-C, 105 pp.

Riecker, R. E., ed., (1979). *Rio Grande Rift—Tectonics and Magmatism*, Am. Geophys. Union, Washington, D.C., 438 pp.

Shaw, H. R. (1980). The fracture mechanisms of magma transport from the mantle to the surface, in *Physics of Magmatic Processes*, R. B. Hargraves, ed., Princeton U. Press, Princeton, N.J., pp. 201-264.

Simkin, T., L. Siebert, L. McClelland, D. Bridge, C. Newhall, and J. H. Latter (1981). *Volcanoes of the World: A Regional Directory, Gazateer, and Chronology of Volcanism During the Last 10,000 Years*, Hutchinson Ross, Stroudsburg, Pa., 240 pp.

Smith, R. B. (1978). Seismicity, crustal structure, and intraplate tectonics of the interior of the western Cordillera, in *Cenozoic Tectonics and Regional Geophysics of the Western Cordillera*, R. B. Smith and G. P. Eaton, eds., Geol. Soc. Am. Mem. 152, pp. 111-114.

Smith, R. L. (1979). Ash-flow magmatism, in *Ash-flow Tuffs*, C. E. Chapin and W. E. Elston, eds., Geol. Soc. Am. Spec. Pap. 180, pp. 5-27.

Smith, R. L., and R. A. Bailey (1968). Stratigraphy, structure, and volcanic evolution of the Jemez Mountains, New Mexico, in Cenozoic volcanism in the southern Rocky Mountains, *Colo. Sch. Mines Q.* 63, 259-260.

Smith, R. L., and H. R. Shaw (1975). Igneous-related geothermal systems, in *Assessment of Geothermal Resources of the United States—1975*, D. E. White and D. L. Williams, eds., U.S. Geol. Surv. Circ. 726, pp. 58-83.

Smith, R. L., and H. R. Shaw (1978). Igneous-related geothermal systems, in *Assessment of Geothermal Resources of the United States—1978*, L. J. P. Muffler, ed., U.S. Geol. Surv. Circ. 790, pp. 12-17.

Smith, R. L., R. A. Bailey, and C. S. Ross (1970). Geologic map of the Jemez Mountains, New Mexico, *U.S. Geol. Surv. Misc. Invest. Map I-571*.

Smith, R. L., R. A. Bailey, S. L. Russell (1978). The volcanic evolution of the Jemez Mountains and its relationship to the Rio Grande rift, in *1978 International Symposium on the Rio Grande Rift*, K. H. Olsen and C. E. Chapin, eds., Los Alamos Scientific Laboratory Report LA-7487-C, pp. 91-92.

Suppe, J., C. Powell, R. Berry (1975). Regional topography, seismicity, Quaternary volcanism, and the present-day tectonics of the western United States, *Am. J. Sci. 275-A*, 397-436.

Taylor, E. M. (1981). Central High Cascade roadside geology, Bend, Sisters, McKenzie Pass, and Santian Pass, Oregon, in *Guides to Some Volcanic Terranes in Washington, Idaho, Oregon, and Northern California*, D. A. Johnston and J. M. Donnelly-Nolan, eds., U.S. Geol. Surv. Circ. 838, pp. 55-58.

Walker, G. W., and B. Nolf (1981). High Lava Plains, Brothers fault zone to Harney Basin, Oregon, in *Guides to Some Volcanic Terranes in Washington, Idaho, Oregon and Northern California*, D. A. Johnston and J. M. Donnelly-Nolan, U.S. Geol. Surv. Circ. 838, pp. 105-111.

Williams, H., and A. R. McBirney (1979). *Volcanology*, Freeman, Cooper & Co., San Francisco, Calif., pp. 237-239.

Mechanics and Energetics of Magma Formation and Ascension

5

BRUCE D. MARSH
The Johns Hopkins University

ABSTRACT

Magma is invariably generated through partial melting of existing rock. The most effective and realistic means of melting in the mantle is by convection of rock above its solidus and in continental crust by melting in the wallrock of intrusions from the mantle.

Extraction of melt from a partially melted assemblage probably occurs principally over small areas as a result of diapirism and over large areas by intercrystalline flow of compressed melt into fractures. Initial diapirism carries the entire source mass, allowing melting to reach the critical point (25 to 50 percent) where repacking of the solids allows extraction during, not before, ascent. Smaller amounts of melt can be extracted only if the network of solids is deformed to fill the potential voids left by the melt or, if over a large area, the initially compressed melt expands into fractures. The coalescence of droplets of melt seems possible only in special flows with special configurations of the melting phases.

Diapirism, stoping, and dike propagation are the principal means of magma ascent. Each process is constrained by the fact that magma must be transported fast enough to reach the surface while partially molten. Heat transfer, coupled with studies of dynamics, has yielded several rules; for example, successful diapirism is multiple-diapirism, stoping magmas easily become congested with blocks, and dikes must travel about 10,000 times faster than diapirs.

Generation of significant amounts of granitic magma from intrusion of basalt into continental crust is difficult, unless the volume of basalt is large or unless through repeated ascent of basalt it maintains its passageway at a high temperature. The latter event can have a profound effect on the crust and can trigger regional crustal instabilities and reorganization. Melts produced in the wallrock of passageways can erupt from a common vent area, producing a zoned pluton or lava series having a spectrum of both mantle and crustal sources.

Whether a magma becomes a pluton or a lava is intimately tied to its composition and cooling history. All lavas have less than about 60 percent phenocrysts, which implies that beyond this critical crystallinity they are too viscous to erupt; this critical amount decreases with increasing silica content of the magma. Beyond this critical point the magma forms a pluton. In addition, because crystal production is nonlinear with decreasing temperature, the most probable temperature at which to find magma is when it is about half crystallized. Because granitic magmas reach their critical crystallinity with less cooling than basalts, they are more likely to form plutons, whereas basaltic magmas (with all else equal) are more likely to form lavas.

67

INTRODUCTION

The physics and chemistry of the origin and evolution of magma are central to understanding volcanism and igneous rocks in general. The movement of magma to the outer regions of the Earth is a form of penetrative convection and heat transfer that is important not only in forming both oceanic and continental crust but also in providing energy to remelt and allow reorganization of continental crust. The variety of igneous rocks found around volcanoes and plutons has led petrologists to concentrate on the chemical origins of these suites. Beginning with the desire to know the chemical composition of rocks and minerals, the physical chemistry of igneous rocks has been developed to the extent that it, in itself, is now an important branch of inorganic chemistry. But being exceedingly difficult, if not actually impossible, to observe magma in its various stages, a physical theory of magmatism has only now begun to be needed and developed. Although the chemistry of silicate melts is a mature science, physical igneous petrology is in its infancy. In this paper I shall attempt to sketch the physical evolution of magma from its inception as a partial melt to its final solidification as a pluton or lava.

HISTORY

Physical igneous petrology had its beginning with Reginald Aldworth Daly (1871-1957), who, recognizing the demonstrably undisturbed nature of the wallrock of certain plutons, proposed stoping as a principal means of magma ascent. Unknown to Daly, this mechanism, whereby blocks of roof-rock settle through the magma allowing advancement, had been suggested earlier by Goodchild (1894) and Lawson (1896). But Daly (1933) developed the idea to its present state of appreciation. In convincing others of stoping, Daly sought to understand how the roof-rocks are dislodged, how they settle through the magma, and how they interact chemically with the magma itself. He studied thermal stresses as the agent in fracturing the rock and applied Stokes's law to the settling of blocks through the magma. He considered the magmatic thermal history and mass transfer by diffusion to evaluate assimilation of stoped blocks. And he always kept a hand on the field relations and chemical constraints of petrogenesis. Daly saw the inseparability of physical and chemical processes.

Even to get rough answers, Daly had to obtain the physical properties of magma, i.e., density, viscosity, thermal and chemical diffusivity, elastic constants, and thermal expansion. His textbooks are models of what physical igneous petrology is beginning to be. But because the mechanics of igneous intrusion was inseparable from the origin of granite itself and because of the great strides being made in igneous physical chemistry, the subject was essentially stillborn. There has always been an interest in the subject, but it lacked a following of its own. The eventual verdict on stoping was summarized by Buddington (1960) in his review of granitic emplacement (p. 735): "After half a century . . . the stoping hypothesis still lacks verification in the sense that we do not have desirable supporting evidence of sunken blocks in the plutons of deeper

levels." Regardless of the fate of stoping, it became increasingly clear that a physical theory of magmatism was needed to answer questions that chemistry could not.

Important quantitative contributions to specific aspects of the subject have been made by Gilbert (1877) on the mechanics of near-surface intrusion, by Grout (1945) and Ramberg (1967) on scale modeling of diapirism, and by Shimazu (1959) on a range of subjects from differentiation to zone melting. The present impetus of the subject, however, owes its origin to H. R. Shaw for his contributions on the in situ physical properties of magma and for his introduction of the principles of continuum mechanics to igneous petrology (Shaw, 1965, 1969, 1973, 1980). Although, as in the past, igneous petrology has followers who primarily study chemical problems and also those who study solely physical problems, it is becoming ever more clear that it is the middle ground of simultaneous consideration of chemical, field, and physical problems that promises to yield the greatest rewards.

ORIGIN OF MAGMA

Magma is any mixture of solids and/or liquid and without exception is produced by partial melting of existing rocks. The Earth's outer 3000 km of mantle and crust are known seismically to be essentially solid. The seismic low-velocity zone, at a depth of about 100 to 350 km, may represent incipient melting, but in general the Earth must be locally heated, relative to its solidus (beginning of melting), to produce magma. The many processes suggested as producing heating (see Yoder, 1976, p. 60) can be summarized by considering an arbitrary volume (R) or space within the Earth whose average temperature (T) is monitored with time. Conservation of energy for the region requires that

$$\int_R \left[\frac{\partial T}{\partial t} + V_i \frac{\partial T}{\partial X_i} - K \frac{\partial^2 T}{\partial X_i^2} - e/\rho C_p \right] dR = 0, \quad (5.1)$$

where the subscript i indicates summation over the three dimensions ($i = x, y, z$), t is time, V_i is velocity, X_i is spatial coordinate, K is thermal diffusivity, e is any source (sink) of heat, ρ is density, and C_p is specific heat. The first term here ($\partial T/\partial t$) is merely a monitor (thermometer) of temperature in a region, R. The second term ($V_i \, \partial T/\partial X_i$) describes heat carried into or from R by fluid flow (convection). The third term ($K \partial^2 T/\partial X_i^2$) describes heat carried into or away from R by diffusion (conduction), and the last term is heat gained or lost by sources (e.g., radioactivity) or sinks (e.g., heat of fusion).

Eq. (5.1) describes the temperature in a region, R, fixed in space. Material and heat may move through this region, but the space being considered remains fixed with respect to x, y, and z. Beginning with a space containing material below its solidus, the minimum criterion to cause melting is that $T/\partial t > 0$, or

$$\int_R \left[V_i \frac{\partial T}{\partial X_i} - K \frac{\partial^2 T}{\partial X_i^2} - e/\rho C_p \right] dR < 0. \quad (5.2)$$

To estimate the relative importance of each term it proves

convenient to make Eq. (5.2) dimensionless by choosing new variables: $V_i' = V_i/V_0$, $\partial/\partial X_i' = L\partial/\partial X_i$, and $T' = T/\Delta T$, where V_0 is a typical flow velocity, L is a length that characterizes R, and T is the average temperature difference between R and its surroundings. Because the heat source e can take on various forms, we let it stand. Since Eqs. (5.1) and (5.2) hold for any region R, these integrals can be true only if the integrands themselves satisfy the conditions on the right. Dropping the integrals (but keeping in mind that L represents the size of R), substituting the new variables, and collecting terms yields

$$Pe\, V_i' \left(\frac{\partial T'}{\partial X_i'}\right) - \frac{\partial^2 T'}{\partial X_i'^2} - \frac{eL^2}{Kc\Delta T} < 0, \qquad (5.3)$$

where $Pe = V_0 L/K$ is the so-called Peclet number, and Kc is thermal conductivity ($= \rho C_p K$). The scales (L, ΔT, V_0) are chosen to make the primed quantities (T', X', V') each have a numerical size of order one, and thus derivatives of these quantities are also of order one. The two dimensionless terms Pe and $eL^2/Kc\Delta T$ completely characterize any solutions to Eq. (5.3). Any means of describing magma production within this region R is described by combinations of these two parameters. Below we consider several possibilities.

Radioactive Heating

Consider the source region R to consist of motionless radioactive rock $e \neq 0$, $V_0 = 0$, and Eq. (5.3) reduces to

$$\frac{\partial^2 T'}{\partial X_i'^2} + \frac{eL^2}{Kc\Delta T} > 0. \qquad (5.4)$$

If the region is warmer than its surroundings, the first term, describing the heat flux about R, is negative; heat is being conducted away from R. Because of scaling this term is of order 1. To cause an overall heating the second (heat source) term must more than compensate for this loss to conduction. Therefore, a necessary condition to produce magma is

$$H \equiv \frac{eL^2}{Kc\Delta T} \gg 1. \qquad (5.5)$$

For typical peridotitic rock $e \simeq 1.7 \times 10^{-7}$ erg/cm³-sec and $Kc \simeq 4.2 \times 10^5$ erg/cm-sec-deg; for granite rock $e \simeq 8.5 \times 10^{-7}$ erg/cm³-sec and $Kc \simeq 3.3 \times 10^5$ erg/cm-sec-deg. These values set a relationship between L and T for anomalous heating to occur, namely

$$L \text{ (km)} \gg 16\, \Delta T^{1/2} \text{ for peridotite,} \qquad (5.6)$$

and

$$L \text{ (km)} \gg 6\, \Delta T^{1/2} \text{ for granite.} \qquad (5.7)$$

These results mean that if a region is being heated by radioactivity at the same time that the heat is being conducted away, for any heat to accumulate, the length scale must be sufficiently large to make the heat flux away smaller than the rate of heat production. For a temperature rise of, for example, 100°C, a spherical region must have a radius of about 160 km in peridotite and 13 km in granodiorite for heat to accumulate. As the rock continues to heat with constant heat production,

an equilibrium will eventually be reached between heat production and conduction. The values of e used above are considered typical. For anomalous values twice as large, the respective length scales decrease by only $(2)^{1/2}$.

Therefore, motionless, anomalously radioactive regions of the Earth, such as might be produced by metasomatic concentration of phlogopite, can produce magma if the regions are sufficiently large to minimize conductive heat loss. This does not indicate the time it takes to heat, because this depends on the exact rate of heating relative to conductive losses.

The time factor in raising the temperature by radiogenic heating while simultaneously losing heat to conduction can be gauged by a simple model. Consider a layer of half-thickness L containing heat sources of strength e distributed as A^{-ax}, where A and a are constants and x is the spatial coordinate normal to the layer. Placing the origin at the middle of the layer, the heating source will produce a symmetrical thermal disturbance in the absence of preexisting strong temperature gradients. From symmetry considerations the problem need be solved only in the $x > 0$ half of the layer with an insulating boundary condition ($dT/dx = 0$) at the origin and the full solution found by reflection across the midplane. The solution is given by Carslaw and Jaeger (1959, p. 80).

Letting the heat sources diminish by the factor $e^{-1} = 1/2.71$ over the distance L (i.e., $aL = 1$), the highest temperature (T) (at the midpoint $x = 0$) relative to the initial temperature (T_0) is given by

$$\frac{T}{T_0} = 1 + \frac{AL^2}{T_0 Kc}\left[\frac{2f^{1/2}}{\sqrt{\pi}} + e^f erfc(f^{1/2}) - 1\right], \qquad (5.8)$$

where $f \equiv Kt/L^2$ and $erfc$ is the complementary error function. If Carslaw and Jaeger's Table I of their Appendix II is used to plot out the function of f in brackets, we find that

$$\frac{T}{T_0} = 1 + C\frac{AL^2}{T_0 Kc}f, \qquad (5.9)$$

where C is a constant near unity for $0 \leq f \leq .04$ and diminishes to about $1/4$ when $f \simeq 10$. For our purposes, assume that C is always of order unity. Using the definition of f, Eq. (5.9) becomes

$$\frac{T}{T_0} \simeq 1 + \frac{A}{T_0 \rho C_p}\,t. \qquad (5.10)$$

This is now an explicit relation for the time to raise the temperature at the midplane by an amount T/T_0 with a heat source of strength A. Perhaps somewhat surprising, but conveniently, this result does not depend on the thickness of the layer.

The time to raise the initial temperature by enough to cause melting, e.g., by 10 percent ($T/T_0 = 1.1$), is given by

$$t \simeq \frac{T_0 \rho C_p}{A} \times 10^{-1}. \qquad (5.11)$$

For peridotitic mantle rock with $A = 1.7 \times 10^{-7}$ erg/cm³-sec, $T_0 = 1250°C$, $\rho = 3.3$ g/cm³, and $C_p = 1.25 \times 10^7$ erg/g-deg, the heating time is 1000 m.y. Because convection occurs at the rate of 10^{-7} cm/sec in the mantle and in 1000 m.y. would transport material 30,000 km, the peridotitic region could be greatly disturbed before magma production. Also the time for heating is of the order of the half-life of long-lived radioisotopes, so the heat source could not be considered constant. If this mantle region, however, was anomalously enriched in radioisotopes, e.g., by a factor of 10, the heating time is reduced by the same amount. It may not be unreasonable to conceive of heating events that take place on the order of 100 m.y., which may give rise to, for example, the chemical plumes hypothesized by Anderson (1975).

For granodioritic rock with $A = 2 \times 10^{-5}$ erg/cm³-sec, $\rho = 2.7$g/cm³, $C_p = 1.25 \times 10^7$ erg/g-deg, and $T_0 = 1000°$ C, the time to increase the temperature by 10 percent is 53 m.y. Therefore, within the continental crust (granodioritic) it appears to be reasonable from this heating time that anomalous heating occurs. (But a more accurate solution with an initial temperature gradient should be considered.)

Such anomalous heating would initiate a gravitational instability that, depending on its means of ascent, may still not be successful in producing magma. The thermal history of such an ascent is treated in a later section.

Viscous Dissipative Heating

The rate of heating resulting from friction in viscous shear flow is given by the product of the shear stress (τ) and the strain rate, which in a Newtonian fluid of viscosity (μ) is of order $\mu V_0^2/L^2$. Compared once again to conductive losses, the condition in Eq. (5.5) becomes the so-called Brinkman number (Br), a measure of the relative importance of viscous heating to conduction losses,

$$Br \equiv \frac{V^2}{Kc\Delta T} >> 1. \qquad (5.12)$$

This result is independent of any length scale as long as the shear flow and resulting thermal anomaly have the same length scales.

For an upper mantle viscosity of 10^{21} poise and with $Kc \simeq 4 \times 10^5$ erg/cm-sec-deg,

$$V >> 6.56 \times 10^{-2} \Delta T^{1/2} \text{ (cm/yr).} \qquad (5.13)$$

This can be an effective method of heating. For conduction losses controlled by $T \simeq 500°$ C, the velocity must be significantly larger than about 1.5 cm/yr to elevate the temperature to or beyond the solidus. This condition can probably be met in many places within the upper mantle, especially near Benioff zones. This heating can also be effective to cause melting in the crust along fault zones. But because crustal rocks are generally far from their solidus, the temperature rise must be large (~500° C), and this can be achieved only over a limited region for a short time because fault zones are narrow and conductive losses are large. Thus, to actually be effective in heating, the original temperature gradient must be relatively small, but perhaps, more important, there must be a strong source of shear. Although a model of steady shear is easy to pose, concentrated shear within the mantle is difficult to sustain for long times because, through conservation of mass, the flow rate is severely limited by the regional dynamics of the surrounding regions. In essence, as originally noted by Tozer (1967), mantle

viscous processes are self-regulating; there is a maximum shear stress that can be exerted on a silicate. So except in unique areas, such as the corner flow beneath island arcs or the wall-rock of diapirs, viscous heating is probably of limited importance within the mantle.

In non-Newtonian, power-law (shear thinning) flows, these effects can be accentuated but only over more restrictive zones. Such models have been investigated by Marsh (1976), Yuen et al. (1978), and Spohn (1980, 1981).

Flow of Hotter Rock

A region within the Earth fixed in space can also become hotter as a result of the flow of hotter material into the region. The amount of heating depends on the magnitude of convective relative to conductive heat transfer, which is measured by Pe. For a unidirectional flow containing no heat sources ($e = 0$) that is hottest upstream from the region R, $V_i' \partial T'/\partial X_i'$ 0, and $\partial^2 T'/\partial X_i'^2 < 0$. Thus, for R to become hotter,

$$Pe \equiv \frac{VL}{K} >> 1, \qquad (5.14)$$

and for $K = 10^{-2}$ cm²/sec,

$$V >> 10^{-2}/L \text{ (cm/sec).} \qquad (5.15)$$

For a modest length scale of even 1 km, the convective velocity must be larger than a few centimeters per year to override conductive losses. For larger regions this condition is easily satisfied by even the slow convective velocities associated with the crustal isostasy. This shows the strong effect of convection in transporting heat.

It is now possible to evaluate the overall condition defined by Eq. (5.3). For any heat source ($e \neq 0$) the last term ($eL^2/Kc\Delta T$) is positive, and for fluid (that is hotter upstream) flowing into the region the first term ($PeV_i'\partial T'/\partial X_i'$) is negative (i.e., for fluid flowing in the positive X direction $V_i > 0$, while $\partial T'/\partial X_i' < 0$). The conduction term is always negative for a region hotter than normal. So to satisfy Eq. (5.3)

$$\left| Pe V_i' \left(\frac{\partial T'}{\partial X_i'} \right) \right| + \left| \frac{eL^2}{Kc\Delta T} \right| - \left| \frac{\partial^2 T'}{\partial X_i'^2} \right| > 0. \qquad (5.16)$$

The size of the last (conduction) term is of order 1; thus, if the sum of the other two terms is much larger than 1, magma will form in the region R. But we have also seen that the second (heat source) term will probably never be very large (i.e., > ~100). And if the fluid flowing through R is unheated relative to its surroundings, the flow can easily carry away this (radioactive) heat if

$$Pe >> \frac{eL^2}{Kc\Delta T}, \qquad (5.17)$$

which from the earlier examples could easily be so. That is, for $V_0 = 3$ cm/yr, $L \simeq 10$ km, and $K = 10^{-2}$ cm²/sec, $Pe = 10$, while for $\Delta T = 100°$ C, $e = 3.5 \times 10^{-7}$ erg/cm³-sec (a peridotite twice as radioactive as usual), and $Kc \simeq 4.2 \times 10^5$ erg/cm-sec-deg, the heat source parameter (H) is 10^{-2}. So $Pe/H \simeq 10^3$ and convection could easily keep the region swept free of any anomalous heat.

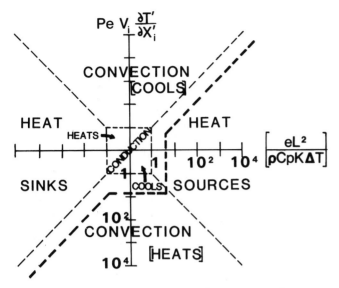

FIGURE 5.1 Magma production—a solution map to the energy equation delineating the regions where various processes dominate heat transfer. The vertical axis represents convective heat transfer— flows cooling on the positive portion and flows heating on the negative portion. The horizontal axis represents heat sources (positive portion) and heat sinks (negative portion). Only near the origin is conduction important. Along the dashed boundaries both convection and heat sources (sinks) are important. Magma is produced in any process that falls on the lower right-hand side of the diagram to the right of the heavy dashed boundary.

The importance of convection in transporting heat within the Earth cannot be underestimated. The most practical and probably only means of generating magma within the region R is by convection. Anomalously hot rock is carried into a region where its solidus is exceeded, or this unusually hot rock heats surrounding rock to its solidus, thereby producing magma. In the first instance it is the nearly adiabatic decompression of material that causes melting; in the other instance it is penetrative convection that brings unusually hot material (perhaps even magma itself) into contact with cooler material that can be melted. In both instances the necessary condition is that $Pe \gg 1$. The recognition that all magma is, strictly speaking, probably a result of convection was by Verhoogen (1954). By this analysis, nevertheless, there does stand the chance that large, unusually radioactive regions that are carried undisrupted for long periods within a larger convecting body may become hot enough to stimulate diapirism (i.e., penetrative convection) and eventually produce magma. The possibilities are shown diagrammatically in Figure 5.1.

EXTRACTION OF MAGMA

It is generally agreed that few, if any, magmas arrive as lavas being whole samples of their original source rock. Somewhere a fluid (perhaps containing some crystals) has been extracted from the source rock. The chief means of extraction is probably diapirism or flow of compressible melt. The size of the source

region and its dynamical characteristics are critically important in determining the amount of melt that can be extracted. In short, small systems (e.g., diapirs) must undergo extensive melting for separation of liquid and solids, and very large systems (e.g., beneath ocean ridges) may separate small amounts of melt. These and other mechanisms of extraction are discussed below.

Diapirism

Diapirism begins as soon as there is a density inversion. The rate of rise is determined essentially by the size of the body and the viscosity of the *surrounding* material. If the rise is fast enough (i.e., $Pe > 10$), cooling by conduction to the wallrock will be minimal, and the body will undergo increased melting as the temperature increases relative to its solidus. Although for any specific body the heat transfer may be difficult to know, for certain standard-state conditions cooling curves can be calculated (Marsh, 1978, 1982). A set of these cooling curves for an arbitrary body is shown as Figure 5.2.

A successful diapir can be arbitrarily defined as one that follows a cooling curve that allows it to get within about 10 km of the surface and still have a temperature above its solidus. During this ascent, melting increases to the extent that the

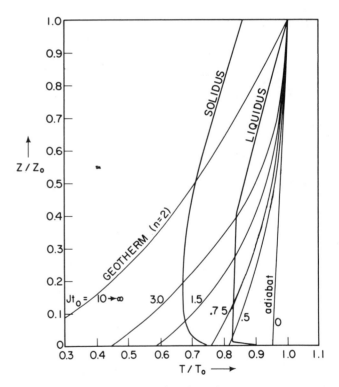

FIGURE 5.2 Cooling curves describing the average temperature (horizontal axis) in a magma ascending (depth, vertical axis) at a constant velocity by diapirism, stoping, or as a dike. The shape of the family of curves comes from the geotherm shown. As the ascent time decreases, the cooling curve approaches the adiabat. A representative solidus and liquidus for a magma containing about 1 wt. % water also is shown, and the magma is initially free of crystals (after Marsh and Kantha, 1978).

solids can be repacked more tightly, yielding a liquid. The mechanics of the extraction process depend critically on how much melt must be produced before the solids can be repacked to free a liquid.

As is well known from sedimentology and the processing of particulate matter, a collection of spheres of equal size can have a porosity of anywhere from 26 percent by volume at closest packing to 52.4 percent at poorest packing. If melting rocks have an assemblage of solids that show similar packing characteristics, somewhere between about 25 and 50 percent melting must take place before melt can be extracted during diapirism. That is, only at degrees of melting greater than 25 to 50 percent can the solids be repacked to free some liquid. The packing characteristics of igneous minerals are unknown, but there is some evidence from volcanic rocks as to what the critical porosity may be.

Lavas with more than about 55 percent phenocrysts are rare (Ewart, 1976; Marsh, 1981). Evidently, beyond this degree of crystallinity the magma reaches a locking point where it becomes too viscous to erupt. And this does not seem to depend on the type of volcanism, for basalts through rhyolites show this characteristic. There is, nevertheless, a reduction in this critical crystallinity with increasing silica content of the lava itself. In alkali basalts the crystallinity may become slightly greater than 55 percent, and in rhyolite lavas it may be as low as 15 to 20 percent (Hildreth, 1981; Marsh, 1981). This trade-off in crystallinity and silica content reflects these effects on the viscosity of the magma. It increases both with increasing silica and crystallinity. This evidence suggests that melt extraction can only take place at rather large degrees of partial melting—at perhaps a minimum of 45 percent melt. If so, the diapir may rise a significant distance from the depth of its original source before a magma is eventually extracted. And it is important to realize that the chemical imprint of the source on this magma will be controlled by a set of pressure-temperature conditions significantly different from those of the original unmelted source rock. This means of extraction is likely to be important in island arcs.

Filter Pressing

Filter pressing is a classical mechanism whereby a partially melted rock can have its melt extracted by squeezing the solids into tighter packing and lower porosity. At any degree of crystallinity above the one of closest packing and with rigid crystals, any attempt to squeeze the assemblage will bring about dilation. This was first noted by Bagnold (1954) in his study of sand, and it has been noticed experimentally during deformation of partially molten granitic rock by van der Molen and Patterson (1979). Squeezing to the extent of crushing may only produce a geometrically similar assemblage with little change in packing characteristics. (Squeezing nonrigid but highly viscous crystals places a critical limit on the time for pressing, which is evaluated below.) A great deal needs to be learned, but the packing characteristics of common assemblages of igneous minerals already promise to yield information critical to understanding melt extraction and the rheology of magma (Marsh, 1981).

A somewhat related mechanism of melt accumulation is the coalescence of droplets of melt in a shear flow (Weertmen, 1968, 1972). The differential strain across a shear is thought to make one droplet collide with another, eventually growing into a sizable body of magma. Although the idea is appealing, the process seems inefficient unless the flow has special characteristics or unless the drops remain spherical. This apparent difficulty appears when considering a geologically reasonable, simple shear flow (incompressible flow in the x direction, velocity varying in the y direction, and streamlines parallel to x axis) where the drops are contained in the x-y plane. Because of their small concentration, size, and nonrigidity, the drops do not distort the streamlines. Since by definition fluid does not cross streamlines, any two points on neighboring droplets connected by the same streamline will *always* remain the same distance apart; thus, the droplets will not touch during the flow, prohibiting coalescence. Furthermore, if coalescence could occur, to maintain chemical equilibrium the surrounding solids must also remain in the correct volumetric proportions to the melt. Otherwise the melt will partially solidify, which may act to continuously stabilize the accumulation process. Because drop shape also depends on chemical equilibrium with the surrounding solids, the droplets will become stretched into streaks along with the solids during deformation. Although this can be offset through continual recrystallization, the overall effect during deformation is to reduce the collision cross section. If the shape could be held spherical by surface tension, which varies inversely with droplet size, increasing coalescence would also cause this mechanism to become ineffective. Nevertheless, silicate melt generally does not form spherical drops but is distributed along grain boundaries. This mechanism thus appears feasible only in special flows having pinched streamlines.

Compressible Melt Flow

Regardless of how the packing characteristics of minerals may limit melt extraction, there is unequivocal chemical evidence from rocks of ocean ridges that tholeiitic magma represents only 20 to 30 percent melting. How, then, is this magma extracted? The situation discussed earlier concerns a diapir that ascends in an essentially incompressible flow. The diapir moves by changing place with its roof-rock (see below for details). There is no regional dilation, and the total volume of diapir is relatively small. Beneath ocean ridges the convective flow separates into two limbs; the flow separates. The melting (and volume increase) in response to the adiabatic decompression attending this flow produces a regional volume increase and a deviatoric stress that is relieved partly by compressing the melt and by possibly initiating fractures, which can produce melt-filled dikes that can propagate to the near surface. Although the unmelted rock could probably withstand these deviatoric stresses without fracturing, Shaw (1980) has shown that, on melting, a pore overpressure develops that greatly reduces the strength; the Mohr circle slides toward the origin and intersects the fracture envelope. (Shaw's model does not, however, explicitly address the problem of melt extraction.)

Once a crack begins to form and propagate, the melt is

driven, as flow in porous media, by the need for the fluid to expand relative to the solids. In this way melt can be separated from solids without leaving a void. The key to this mechanism seems to be a partially melted source region large enough to "store" by compression a significant amount of volume. Although in many terranes these propagating dikes would be rapidly quenched because of their high rate of heat loss, the regional hotness of oceanic-ridge areas insulates them. The much smaller nature of diapirs, e.g., beneath island arcs, suggests that if dikes of this sort form in the ascending body they will be quenched as they attempt to propagate beyond the diapir itself. [The observation of thin (~20 cm) dikes in the field means of course that they are not far from their source and were injected quickly.] Some alkali basalts are thought, on chemical grounds, to represent small (<10 percent) degrees of melting. If so, this basalt may be an extreme example of ridge-type compressible extraction occurring deeper within the mantle, and it may be a direct indicator of the flow regime.

Incompressible Melt Flow

If a model of melt extraction is based on differential flow of solids and liquids in essentially a fluidized system (Sleep, 1974) or on the principle of flow in a porous media alone (Walker *et al.*, 1978), the residual solids must be viscously deformed to fill the potential void left by the melt. That is, the solids cannot be packed any tighter without first being deformed, and, if they are so rigid that they cannot deform, melt cannot be withdrawn. The rate of extraction of melt is regulated by the rate of deformation of the solids. (The solids may deform partly by pressure solution.) The rate of deforming this assemblage is dictated by the viscosity of the solids, not the melt. And the deformation must take place in less time than that for the whole, anomalously hot region to cool and once again become solid. For the pressure gradient $\Delta\rho g$ caused by density difference ($\Delta\rho$) between melt and solid, the time to condense the mass by a distance a is, by dimensional analysis,

$$t_v \simeq \frac{\mu a}{\Delta\rho g L^2}, \qquad (5.18)$$

where L is the characteristic length of the region, μ is the viscosity of the solids, and g is gravity. If the melt fraction is X, to squeeze out all the melt $a = XL$.

The time (t_c) taken for this region of dimension L to cool and solidify is well known to be of the order $t_c \simeq L^2/K$, where K is the effective thermal diffusivity ($\simeq 10^{-2}$ cm²/sec). To extract melt the ratio of these times (t_v/t_c) must be such that

$$\frac{t_v}{t_c} \simeq \frac{\mu X K}{\Delta\rho g L^3} << 1. \qquad (5.19)$$

For $\mu \simeq 10^{22}$ poise, $X = 10^{-2}$ (i.e., 1 percent), and $\Delta\rho = 0.1$ g/cm³, this condition suggests that all bodies of a characteristic size much larger (e.g., 10 times) than

$$L >> \left[\frac{\mu X K}{\Delta\rho g}\right]^{1/3} = 20 \text{ km} \qquad (5.20)$$

will deform fast enough to keep the melt from solidifying.

If, however, the melt attempts to migrate or collect beyond the margins of the partially melted region, it will be quenched, because of its small volume. This complete lack of thermal inertia of a small fraction of melt (<~10 percent) is probably what promotes the stability of the asthenosphere.

MAGMA ASCENT

Several means of magma ascent have been mentioned—diapirism, porous media flow, and dike propagation. In addition, there is stoping, whereby blocks of roof-rock fall through the magma, and zone melting, where magma moves by melting roof-rock in concert with solidification at the floor. Because of the limitations on porous media flow mentioned earlier, intercrystalline flow is not a serious contender for transport of magma through the lithosphere or within the crust. The usefulness of zone melting is seriously limited by pervasive geochemical evidence suggesting that magma may transit the lithosphere of oceans and continents alike without complete chemical adulteration, which is a necessary condition of zone melting. So in what follows, transport of magma by dikes, stoping, and diapirism is considered viable. Each process is first evaluated in terms of heat transfer and then in terms of mechanics.

Heat Transfer

The most basic characteristic of magma that is valuable in understanding its ascent is that it arrives on the Earth's surface still partially molten. In fact, as mentioned already, nearly all magmas contain less than 55 volume percent crystals upon eruption. Any means of transport must bring the magma to the surface in less time than that for solidification. Because the relatively low viscosity magma can internally transport heat easily to its margins by thermal convection, it is the wallrock that controls magmatic heat transfer. The ability of the wallrock to transport heat away from the magmatic body is the rate-controlling step in the cooling of an ascending body.

The cooling of an ascending body of magma can be described on general (phenomenological) grounds without an accurate knowledge of the actual dynamics of ascent. Limits placed on the ascent velocity by cooling models are, however, essential in evaluating dynamic models of ascent. The total heat flux (Q_T) from an arbitrary body of ascending magma relative to the purely conductive heat flux (Q_{cd}) when it is not ascending is defined as the Nusselt number (Nu), a pure number:

$$Nu = \frac{Q_T}{Q_{cd}}. \qquad (5.21)$$

If Nu is known on independent grounds, the heat transfer from the body is just

$$Q_T = Nu Q_{cd} \qquad (5.22)$$

or

$$Q_T = Nu \frac{KcA}{L} (T - T_m(t)), \qquad (5.23)$$

where the conductive heat flux has been substituted for a body of average temperature T in a medium whose temperature far from the body is $T_m(t)$, whose area is A, and whose characteristic length scale is L (e.g., L = radius, for a sphere); Kc is the thermal conductivity of the wallrock.

This heat flux is supplied by the magma as it cools. By conservation of energy,

$$\frac{dE}{dt} = -Q_T, \qquad (5.24)$$

where E ($= \rho C_p V'T$) is the thermal energy (enthalpy) of the magma. For constant physical properties (ρ = density, C_p = specific heat) and volume (V'), Eq. (5.24) becomes

$$\frac{dT}{dt} + JT = JT_m(t), \qquad (5.25)$$

where

$$J = Nu \left(\frac{Kc}{\rho C_p}\right) \left(\frac{A}{LV'}\right). \qquad (5.26)$$

Eq. (5.25) is fundamental to understanding heat transfer from ascending magma regardless of its means of ascent (Marsh, 1978; Marsh and Kantha, 1978). This equation can be integrated by using the integrating factor $\exp(Jt)$:

$$T = J \exp(-Jt) \int \exp(Jt) \, T_m(t)dt, \qquad (5.27)$$

where the temperature of the wallrock far from the body has been left as a function of ascent time [$T_m(t)$]; for a constant velocity of ascent, t can always be converted to distance (depth, z) through $z = Vt$.

As an (unrealistic) example, consider $T_m(t)$ to be constant at T_m, then integrating Eq. (5.24) we find

$$T = T_m + C \exp(-Jt), \qquad (5.28)$$

where C is the constant of integration. Employing the initial condition that $T = T_0$ when $t = 0$, $C = T_0 - T_m$ and

$$\frac{T - T_m}{T_0 - T_m} = \exp[-Jt_0(t/t_0)]. \qquad (5.29)$$

Notice that the rate of cooling is completely controlled by the dimensionless parameter Jt_0, where t_0 is the total ascent time. More complicated functions for $T_m(t)$, such as might describe a geotherm, yield solutions that are also controlled by this parameter (Jt_0).

A general solution for an arbitrary geotherm has been given by Marsh and Kantha (1978), and a set of cooling curves calculated from this solution is shown in Figure 5.2. For values of $Jt_0 \gtrsim 1$ the magma strikes its solidus. That is, for a successful ascent

$$Nu \left(\frac{Kc}{\rho C_p}\right) \left(\frac{A}{LV'}\right) t_0 \simeq 1. \qquad (5.30)$$

The group $Kc/\rho C_p$ is an effective thermal diffusivity containing Kc of the wallrock and ρ and C_p of the magma; this group is of order 10^{-2} cm²/sec. The group A/LV' is the ratio of the surface area of the body to the volume and length scale (e.g., radius) of the body.

Dike To see the strong effect of this group (A/LV') on cooling, consider a generalized body represented by a triaxial ellipsoid with axes a, b, and c, where its volume is given by $(4/3) \pi abc$. A dike can be approximated by this body when $a = c$, $L = b$, and $A \simeq 2a^2$, we find $A/LV' = (3/2b^2)$, where b is the half width of the dike. Compare this with a sphere where $A/LV' = 3/(a_s)^2$, where a_s is the sphere's radius and $L = a_s$. For equivalent volumes of magma it is simple to show that $a_s/b = R^{2/3}$, where R ($= a/b$) is the aspect ratio of the dike. Then

$$\frac{(A/LV')_{dike}}{(A/LV')_{sphere}} = \frac{1}{2}\left(\frac{a_s^2}{b^2}\right) = \frac{1}{2} R^{4/3}. \qquad (5.31)$$

Dikes commonly have aspect ratios in the range $R \simeq 10^3$. So with all else being equal, to reach the surface at the same average temperature as an equivalent spherical volume, a dike must travel about 10^4 times faster than the sphere.

As dikes leave their source and cut cooler wallrock a solidified rind may develop at the contact. This rind can grow with time and stop the dike. This process has been investigated by Delaney and Pollard (1982), who showed that a dike 2 m thick that is traveling at 1 m/sec will solidify in the upper crust within a few hours if it flows more than a few kilometers from its source. At least in the upper crust, this places a useful constraint on the spatial relation of dikes to their magma chambers.

To obtain an actual ascent time for each of the cooling curves in Figure 5.2, a value of Nu must be known. It can be shown on dimensional grounds, and it has been verified by experiment, that $Nu = f(Pe)$. These relationships have been discussed by Marsh (1982). Some of these results are given below.

Diapir In diapiric ascent of a spherical body of radius 0.8 km the ascent velocity is $\simeq 10^{-2}$ cm/sec for $Jt_0 = 0.25$, $\simeq 10^{-3}$ cm/sec for $Jt_0 = 0.75$, and $\simeq 10^{-4}$ cm/sec for $Jt_0 = 3$. For a body of radius 6 km, $V \simeq 10^{-4}$ cm/sec for $Jt_0 = 0.25$, $\simeq 10^{-5}$ cm/sec for $Jt_0 = 0.75$, and $\simeq 10^{-6}$ cm/sec for $Jt_0 = 3$. As indicated above for a dike with the same values of Nu as for this diapir, each of these velocities must be increased by $R^{4/3}$ for equivalent volumes of magma. But, strictly speaking, because these dike velocities are much larger, the values of Nu used in Eq. (5.30) are larger than for the sphere, and these ascent velocities (i.e., $VR^{4/3}$) are minimum estimates.

The temperature distribution within a diapir depends on the convective motion of the magma. During the early stages of ascent, before the body has undergone enough melting to allow segregation of the melt, the convective velocity is probably small. This is so because in these deeper regions of the Earth the viscosity of the mush is large and also because horizontal temperature gradients within the body will be relatively small. There will surely be convection, but it will be much slower than when the body is near the surface. Once the melt segregates from the residual solids, the low-viscosity magma will convect more rapidly. Although the temperature field within the magma is difficult to calculate, an experiment of a very similar process in which the temperature can be measured was performed by Head and Hellums (1966). They measured the temperature in a heated drop of fluid suspended in another (laminar) flowing fluid. These results, which are shown in Figure 5.3, may indicate the general temperature distribution in

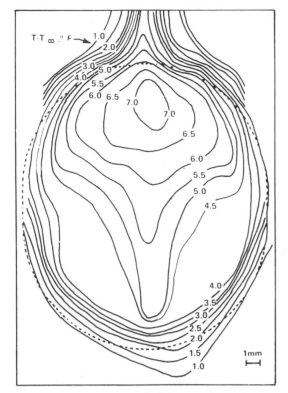

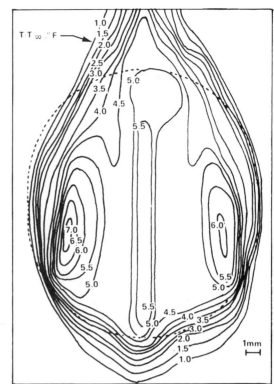

RETARDED CONVECTION   FULL CONVECTION

FIGURE 5.3 The distribution of temperature in a hot drop of fluid suspended in another flowing fluid. When viewed upside down, these temperatures should be analogous to those in an ascending diapir of magma. In one case the drop is not convecting (left), and in the other (right) it is convecting (after Head and Hellums, 1966).

an ascending magma (if viewed upside down). In one case the flow inside the droplet has been retarded, and the isotherms are similar to those for conduction in a sphere. In the other case the droplet is in full convection, and the isotherms show two hot vortexes straddling a cooler central portion. The temperature distributions are significantly different, and, if the body was suddenly tapped, the portion of fluid erupted might be quite different in each case.

Stoping The cooling effect of stoped blocks sinking through the magma depends on their size. To isolate the effect of stoping alone, consider a magma being cooled only by stoping with no lateral heat losses. Each cooling curve in Figure 5.2 can then be associated with a stoped block of a certain size. This size represents the average size of blocks traversing the magma. Although cooling increases as the block fraction increases, so does the speed at which the magma ascends. These effects surprisingly cancel one another, and the cooling is independent of the concentration of blocks (Marsh, 1982). [But therefore lateral heat losses are critically dependent on the concentration (i.e., ascent velocity) of blocks.] For a basaltic magma of viscosity 10^3 poise ascending 100 km through material whose density is 0.5 g/cm^3 greater, the magma will cool along the curve $Jt_0 = 1$ if the radius of the blocks is about 2 m. For larger values of Jt_0 the size decreases by about 50 percent, and for

smaller values of Jt_0 the block size must be larger. If the magma is granitic with a viscosity of about 10^7 poise, the block size must be about 20 m to follow the $Jt_0 = 1$ cooling curve. The block is larger because its residence time is longer, and this additional heat transfer must be offset by a reduction in the ratio of surface area to volume. An analytical treatment of this problem is given by Marsh (1982).

Mechanics

The problem in magma ascent is to displace roof-rock to the bottom of the body. Stoping, diapirism, and dike propagation each do this in different mechanical ways, and the constraints on each are different. These constraints are essential in evaluating the usefulness of each model; stoping and diapirism have been discussed by Marsh (1982) and hence are only briefly summarized here (see Figure 5.4).

Stoping The angularity of stoped blocks ensures that the trail of blocks left behind will be of considerable porosity. That is, on collecting on the floor of the magma chamber the blocks will take up more space than when they were roof-rock. This additional space is due to poor packing, and the pile of blocks will have a pore space that must be filled by magma. Consider a porosity of, for example, 50 percent. As the magma moves

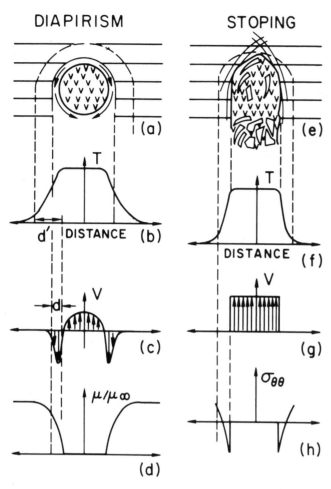

DIAPIRISM

STOPING

FIGURE 5.4 The general mechanics of the ascent of magma by (a) diapirism and (e) stoping, and their respective temperature fields (b, f), velocity fields (c, g), viscosity field (d), and thermal stress field (h) (from Marsh, 1982).

tain granitic xenoliths but also exhibit an unusual isotopic identity.

In sum, block choking apparently presents such a severe constraint on stoping that it is unlikely to be a useful means of magma ascent except very near the Earth's surface, which is only where Daly intended for it to apply. The possibility that stoping has occurred may be evident in the isotopic contamination of the magma.

Diapirism The body moves by heating the wallrock and roof-rock to the point that the wallrock and roof-rock flow around the body. The critical constraint on this process is that the diapir must contain enough heat to process an entire column of roof-rock from source to surface equal in radius to that of the body itself. An energy balance, including loss of potential energy and heat gained from latent heat of crystallization, shows that a single body can ascend only about half way to the surface (i.e., $\simeq$ 50 km) before solidifying. A second body following within $\simeq 10^5$ yr of the first body will be partially insulated by the heat of the earlier ascent, and it can approach within about 20 km of the surface (see Marsh, 1982). Once a pathway is forced to the surface, succeeding bodies can ascend the complete distance.

The major obstacle in calculating the velocity of ascent is the formulation of a law for the drag on the body moving in a fluid whose viscosity is exceedingly sensitive to temperature. This has been investigated by Marsh (1982) and Morris (1980, 1982), with the result that the ascent velocity seems to be limited to velocities of around 10^{-7} cm/sec. Higher velocities can be obtained only if the magma is able to melt the wallrock extensively. These velocities are significantly less than those required by the cooling curves in Figure 5.2 for a diapir that is still partially molten to reach the surface. Here, also, a multiple ascent is needed to bring magma eventually to the surface.

The ascent velocity is critically dependent on the width of the softened zone about the body (i.e., the thermal boundary layer). For these variable viscosity flows, it controls the velocity field, and hence the drag, about the body. It is thus expected that the ascent velocity will be only weakly dependent on the size of the body, unlike that for Stokes's law. But Morris (1980, 1982) found that there may be an optimum body size for a maximum ascent velocity. This is so because large bodies must pass a vast amount of softened roof-rock around the body through the thin thermal boundary layer, which requires an unrealistically large pressure gradient. At the other extreme, small bodies do not possess enough buoyancy to move very fast. Somewhere in between is an optimum body size for ascent. All larger and smaller bodies will solidify. Just what the dimension is of this optimum body is not yet clear.

Critical to determining the rate of diapiric ascent is information on the rheology of rock within its melting interval. But precious little experimental data near the solidus exist on this subject. Near the liquidus the increase in viscosity with increasing crystallinity has been noted by Shaw (1969), Shaw *et al.* (1968), and Murase and McBirney (1973), but the rapid increase of viscosity and the difficulty of measurement have thwarted extensive measurements. Sakuma (1953) used the deflection upon heating of a weighted beam of rock to measure

forward and displaces a volume of roof-rock equivalent to its own volume, the stoped roof-rocks will occupy a volume twice that of their original volume. The volume of magma fills the pore space, the body is completely congested, and ascent ceases.

A formula for the change in volume (V') of the magma as a function of the porosity (p) of the pile of stoped blocks can be derived (Marsh, 1982) as

$$\frac{V'}{V'_0} = (1 - 2p)^{z/z_1}, \qquad (5.32)$$

where V'_0 is the original volume corresponding to a height z_1, and V' is the reduced volume after stoping a distance z_1 above the original body. Even for the tightest packing of spheres (porosity of about 0.25), the volume of the block-free magma diminishes to about 6 percent of the original magmatic volume after ascending four body heights. Because stoped blocks present a large amount of surface area of the magma, the possibility of chemical contamination is great. Basaltic magma, having stoped its way through continental crust, should not only con-

visocity near and below the solidus of an Oshima basalt; these results are shown in Figure 5.5. Although completely crystalline rocks are known to have a high-temperature, but subsolidus, viscosity of about 10^{22} poise, this Oshima basalt is glassy and porphyritic and, consequently, has solidus viscosity of about 10^{15} poise. During the melting interval the visocity changes by a factor of 10^{10}, which is the most drastic change known for any physical property of rocks. Since drag is usually linearly related to viscosity, with all else equal upon partial fusion of wallrock, a diapir could increase its ascent velocity by many powers of 10. It is particulary important, then, to learn the shape of the relation of viscosity to temperature during melting. Does it decrease slowly near the solidus and then drop essentially discontinuously, as Sakuma's data suggest? Or does it drop with constant slope from solidus to liquidus? Knowledge of this relation is also criticial to understanding the mechanical behavior of magma during both melt extraction and crystal fractionation. Sakuma's data would suggest a critical point where upon further crystallization the magma reaches a locking point and passes from liquid to solid. I will return to this point later.

A great deal of first-hand information on the exact mechanics of ascent could also be learned by mapping the deformation (strain) in the wallrocks. This can only be done with excellent lateral and vertical exposure and preferably where the exposed wallrocks lie alongside or below the diapir. From the mechanics of fluids we can get a feeling for this pattern of deformation. The problem is to trace a particle's path in the wallrock as the diapir passes it. For a fluid (wallrock) having viscosity this problem is yet to be solved (let alone with variable viscosity). But even for viscosity-free fluid the results are highly instructive. The introduction of this problem has always been attributed to Darwin (1953), with a later more thorough and extensive treament by Lighthill (1956), but the much earlier work by Maxwell (1870) seems to have been overlooked. The results all agree. Figure 5.6 shows Lighthill's results for a spherical body in a infinite expanse of viscosity-free fluid.

The deformed horizontal lines can be thought of as originally flat marker horizons that have been disturbed by the body. After the body has passed by there will be a residual or permanent deformation and displacement or "drift" of the fluid in the direction of motion of the body. It is perhaps surprising that this nonviscous deformation resembles that often seen near the contact of plutons and stocks. To gain a more truthful impression of deformation within the Earth, this problem must be solved for a fluid whose viscosity is sensitive to temperature. For a constant viscosity fluid the deformation can be expected to extend to a far greater distance, probably more than 10 body radii. But for a temperature-sensitive viscosity, as possessed by rocks, the deformation zone thins, and it may be roughly similar to this viscosity-free result. Careful field studies of wallrock deformation coupled with an intimate understanding of the analogous fluid-mechanical (viscous or plastic) problem promises to give important insight into understanding the emplacement problem.

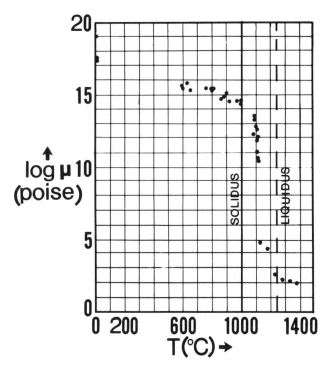

FIGURE 5.5 The apparent viscosity of a basaltic lava from subsolidus to liquidus temperatures. The viscosity changes by at least a factor of 10^{10} through the melting interval, and this change is surely the largest change in any physical property of a rock across the melting interval. Although these data from Sakuma (1953) are the only data of its kind, this information on plutonic rocks is critical to understanding the ascent of diapirs.

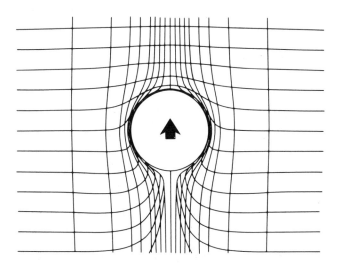

FIGURE 5.6 The deformation attending the movement of a sphere through an inviscid (i.e, no viscosity) fluid (after Lighthill, 1956). As the body advances, the originally horizontal marker lines are increasingly warped upward, leaving a trail of strongly deformed markers. Although this cannot be expected to describe deformation within the highly viscous Earth, because of the severe sensitivity of rock viscosity to temperature, deformation about a diapir may be broadly similar.

Dikes Although much is known about fracture propagation in solids (see the reviews by Anderson and Grew, 1977; Shaw, 1980; Spera, 1980), little is known about its constraints. Heat-transfer studies suggest that only in regionally warmed areas (e.g., ridge areas) can dikes carry magma effectively over significant distances. The maximum ascent velocity is limited by the rate of change of potential to kinetic energy to be less than a few kilometers per second. But this is not much of a limit. Delaney and Pollard (1981) carefully analyzed and described the near-surface behavior of magma in dike formation and development. This information promises to be useful in understanding magma transfer, particularly in the near-surface environment.

MAGMA ASCENT IN CONTINENTAL CRUST

In studies of continental silicic volcanism, the sporadic, but persistent, occurrence of contiguous and contemporaneous basaltic volcanism suggests that basaltic magma may have a role in silicic volcanism. The basalt is nearly always volumetrically subordinate to the silicic volcanics, but small amounts here and there suggest its continual presence. More often than not in these typically bimodal suites it is this basaltic magma that is called on to supply the heat to produce the silicic magma (see Smith, 1979). It is thus of some importance to understand the energetics of such a relationship.

Consider the case of a horizon in the continental crust with an initial temperature of T_1. If its solidus is T_{1S} and its liquidus is T_{1L}, to bring a volume V_1 of this rock to its liquidus an amount of energy

$$H_1 = V_1' \left(\rho_1 C_{p1} (T_{1L} - T_1) + \rho_1 L_1 \right) \qquad (5.33)$$

is needed, where ρ_1 is the density, C_{p1} is the specific heat (both taken to be approximately independent of temperature), and L_1 is the bulk latent heat of fusion. If a volume V_2' of basalt is injected into this horizon and solidifies, it gives up an amount of heat

$$H_2 = V_2' \left(\rho_2 C_{p2} (T_2 - T_{2S}) + \rho_2 L_2 \right). \qquad (5.34)$$

This heat from the basalt can fuse a volume of crust relative to the basalt volume of

$$\frac{V_1'}{V_2'} = \left(\frac{\rho_2 C_{p2}}{\rho_1 C_{p1}} \right) \left(\frac{(T_2 - T_{2S}) + L_2/C_{p2}}{(T_1 - T_{1L}] + L_1/C_{p1}} \right). \qquad (5.35)$$

For rocks where, roughly speaking, $\rho_2 C_{p2} \simeq \rho_1 C_{p1}$ and $L_1 \simeq L_2$, then

$$\frac{V_1'}{V_2'} = \frac{(T_2 - T_{2S})C_{p2}/L_2 + 1}{(T_1 - T_{1L})C_{p1}/L_1 + 1}. \qquad (5.36)$$

It is clear that if the basalt cools by exactly the amount that the crust needs to be heated to form a magma, $V_1' = V_2'$, a volume of crustal melt equivalent to the original basalt volume can be formed. If the basalt cools by, say, 200°C and its latent heat is similar to that of diopside (~150 cal/g), the crust must be heated to (at most) 1000°C with the same latent heat $V_1' = 0.47 \, V_2'$. If the crustal latent heat is, however, only half that of the basalt (~75 cal/g), $V_1' \simeq 0.56$. In short, under most con-

ditions the basalt contains enough energy in cooling to its solidus (or slightly below) to generate a similar (i.e., within a factor of about 2) volume of crustal melt. But this assumes a process with perfect efficiency where all the basaltic heat is used in producing crustal magma. The actual amount of crustal melt will generally be a great deal less. An optimistic efficiency for this process might be about 10 percent, with which the basalt can generate only a tenth as much crustal melt.

The fundamental difficulty in melting the crustal rock adjacent to a nonconvecting intrusion is that the highest temperature that can be attained is (for $\rho_1 C_{p1} \simeq \rho_2 C_{p2}$) about the average of the crustal and intrusive temperatures. Even taking into account latent heat of solidification increases the maximum temperature at the contact only enough to make it to about 65 percent of the sum of the temperature of the (initial) wallrock and intrusive. If basalt at 1200°C was injected against lower crustal rock at 750°C, the maximum contact temperature would be at least 975°, and this would occur in a narrow region along the contact. This inability to melt cool wallrocks with a single intrusion is abundantly clear in the common occurrence of the unfused wallrock of dikes. The extent of melting that can occur depends critically on the type of wallrock. For a high-pressure granulite with no hydrous phase, temperatures near 1100°C might be required to cause large degrees of melting. Near the surface a muscovite or biotite-bearing granitic rock could begin melting at fairly low temperatures (~ 700°C), but here the cool wallrock (~ 200°C) temperature could be heated only to about 700 to 900°C by an intrusive initially at 1200°C.

Any melting occurs in relatively thin regions along the contact. Patchett (1980) found this zone to have a width of about 10 to 20 percent of the thickness of the intrusion. This width can be increased, perhaps significantly, by convection within the intrusion. This effect maintains higher temperatures near the contact and may increase the melt zones to about 30 to 40 percent of the intrusive's width. This increase varies with depth, since the lower crust is much nearer to its solidus temperature than is the upper crust where the melt zones would be very thin.

So far, this discussion has concerned a single intrusion (dike or pluton), but melting can be greatly increased by heating from a plexus of dikes. A plexus or swarm of dikes allows overlap of their thermal fields, which slows heat loss and promotes higher temperatures near the contact. Extensive melting can also occur during stoping, where relatively small blocks become immersed in the intrusive. The time, t, for a spherical block of radius, L, to reach the temperature of the magma itself is $0.5L^2/K$, where K is thermal diffusivity. For a block with a radius of 1 m, this amounts to about 6 days; for $L = 5$ m, it takes about 5 months. During this heating the block is likely to fracture because of thermal stress and thereby reduce these times even more.

Another means by which to melt the crust is by maintaining conduits through it to volcanoes or to pond magma at the base of the crust (Hodge, 1974). As is well known, island arcs typically develop volcanic centers that remain fixed for relatively long periods of time (~ 10 to 20 m.y.). That is, volcanoes within a volcanic center may come and go on a much shorter time scale, but overall the volcanism remains in the same areas.

Over long periods of time the same regions of the crust are repeatedly heated by magma ascending to the surface. These volcanic centers are often fairly equally spaced. To judge their thermal effect on the crust, a simple model can be made whereby these conduits to the surface are maintained at a constant temperature.

Such a model is shown in Figure 5.7, which shows the increase in temperature at any depth around a magmatic conduit that is kept hot. [This is actually for a dike-like conduit, but one for a cylindrical conduit hardly changes things (Carslaw and Jaeger, 1959).] A thermal diffusivity of 10^{-2} cm/sec has been assumed, with no effect of latent heat of fusion. The effect of latent heat will be to increase the times, and it can be approximately included using Carslaw and Jaeger's method of decreasing the diffusivity by a factor of about 2.

This simple model indicates that after about 10 m.y. the entire crust between neighboring conduits will be at least slightly heated. This has several implications. Because crustal temperature increases with depth, deep in the crust this heating will produce a zone of partial melting that will shrink in volume with its approach to the surface. That is, whereas at depth the partially melted zone may span the distance between conduits, with decreasing depth it will be very local only around each conduit. The first effect of this is that magma from the more complete melting in the lower crust may ascend along the conduits and issue from the original volcanoes. A second effect is that this heating produces a low-density region in the crust that is gravitationally unstable and that may issue diapirs of its own at locations between the conduits. Since at any depth melting is most extensive near the conduits themselves, buoy-

ancy is concentrated here and large regions may begin to rise. This may induce a larger-scale motion in the whole crust itself.

The net result of this crustal reorganization is to produce a batholith consisting of plutons derived mostly from the underlying crust. The form of the batholith itself is a reflection of the original heating event that activated the reorganization toward a more gravitationally stable crust. Original geochemical variations within the crust can be sustained largely through this reorganization because each magnetic parcel is small relative to regional chemical variations. Good examples of this process are the unusually systematic spatial variations in $^{87}Sr/^{86}Sr$ and $^{18}O/^{16}O$ within the Southern California-Baja batholith (Taylor and Silver, 1978).

The composition of the magma that is eventually extracted from the crustal diapirs rising near the conduits depends on the composition of the source material and perhaps on the ascent distance. Consider a crust that is granodioritic in composition throughout and that is partially melted an equal amount at depths of 15 to 40 km (see Figure 5.8). As the deeper diapir begins to rise, it will attempt to move farther from its solidus (assuming it is undersaturated with water). This will produce more melt, and eventually enough melting might occur such that the solids can be repacked to free a melt. Although this melt may be of smaller volume than the original diapir, its density also is less and it may rise to very high levels in the crust. The shallow diapir will also attempt to rise away from its solidus and produce more melt as it rises. But the total potential rise distance is much less and may never produce enough melt for extraction of a free-melt phase. If melt is not extracted, the diapir may only be slightly less dense than its

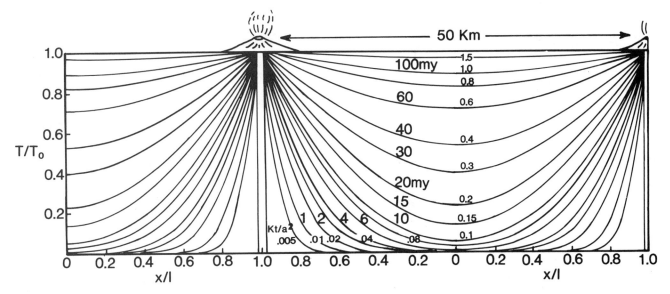

FIGURE 5.7 An oversimplified thermal model showing the effect of maintaining hot, equally spaced magmatic conduits through continental crust to volcanic centers. The temperature (T/T_0, where T_0 is the initial temperature) is given by the vertical axis, and the horizontal axis is the nondimensional distance. This model describes the evolution of a single isotherm (say, 1100°C) within the crust as if the crust initially had a uniform temperature. The nondimensional time (Kt/l^2) on each curve is also shown in units of millions of years (m.y.) for $K = 10^{-2}$ cm²/sec. This describes the temperature about the conduit at any depth. The effect of a real crustal thermal regime, which varies from top to bottom by, say, 800°C, will be to make heating increasingly slower with approach to the surface, and thus this is only a useful model in the lower crust. The important point is that in a relatively short time (~10 m.y.) the lower crust can be heated enough to be remobilized.

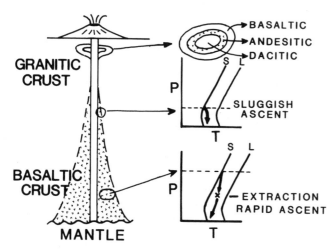

FIGURE 5.8 Magma production and extraction near a hot magmatic pathway leading to the formation of zoned plutons. The first magma to the surface is basaltic from a mantle source. Melting in the basaltic lower crust gives rise to a diapir ascending a considerable distance that allows extraction of an andesitic magma that continues to ascend and intrude the original basalt. Later partial melting in the upper granitic crust produces a sluggish (small-density contrast) diapir that may not ascend far enough to allow melt extraction, yet this body may intrude the others, forming a zoned pluton.

surroundings, its rise will be sluggish, and it may appear on the surface much later (if it does at all) than the melt from the deeper diapir. In fact, the shallow diapir may appear in the near-surface environment still containing recognizable remnants of its source rock (i.e., restite).

Since the composition of the crust surely becomes more basaltic with depth, the deeper diapir will yield a melt that is more basic than the shallow diapir. The sequence of intrusion in such an area would be basalt-andesite-dacite. Furthermore, if the bodies follow similar paths and become neutrally buoyant at a similar depth, later bodies will penetrate earlier ones and together produce a zoned pluton of material from completely different source areas (see Figure 5.8).

The above is evidence of why large bodies of silicic magma, which give rise to giant ash flows, are common to the upper continental crust while similar bodies of basaltic magma are not. Aside from the fact that basalt is denser than granitic magma, which retards buoyancy, basalts cannot normally be generated within the crust, whereas large, slow-cooling silicic bodies can be formed essentially in place provided the necessary heat is supplied. In sum, a basalt entering continental crust enters a hostile environment where its every move compromises either its composition or temperature, stiffling its longevity.

PLUTONS AND LAVAS

"The igneous rocks of the globe belong chiefly to two types: granite and basalt" (Daly, 1933, p. 41). And today even with the recognition of ash flows and seafloor basalt this statement

is still roughly true if the entire crust is considered. Daly also recognized that "To declare the meaning of the fact, that one of these dominant types is intrusive and the other extrusive, is to go a long way toward outlining petrogenesis in general." This question may have a rather simple answer if we ask what the probability is of observing a magma at any temperature or crystallinity. Aside from the preceding agreements, it can be shown that granitic magmas are more likely to form plutons than are basaltic magmas.

Consider a body of magma held in a magma chamber below a volcano. If the magma cools, for example, very quickly through a particular temperature interval, the probability of observing it (if we lived within the magma) in this interval is small. On the other hand, if the magma body spends a long time within this same temperature interval, the probability of observing it at these temperatures is greater. From this statement we can define the probability of observing the magma at a particular temperature as the *thermal probability* (P_T), and it is evidently inversely proportional to the rate of cooling. Thus

$$P_T = \left(\frac{dT}{dt}\right)^{-1}. \tag{5.37}$$

The total flux of heat (Q) from the magma is controlled by the ability of the wallrock to accept heat, and this heat is supplied by the magma from cooling and latent heat,

$$Q = V'\left[\rho C_P \frac{dT}{dt} + \rho L \frac{dX}{dt}\right], \tag{5.38}$$

where V' is the volume of magma, X is the volume fraction of crystals, and the other symbols are as defined earlier. Using the chain rule of differentiation on the last term and factoring,

$$\frac{Q}{V'} = \frac{dT}{dt}\left(\rho C_P + \rho L \frac{dX}{dT}\right) \tag{5.39}$$

and from Eq. (5.34)

$$P_T \simeq \frac{V'}{Q}\left(\rho C_P + \rho L \frac{dX}{dT}\right). \tag{5.40}$$

Since V', Q, and ρC_p do not vary significantly through the crystallization interval (this is not strictly a necessary assumption), this equation is the sum of two probabilities. The first, $V'\rho C_p/Q$, is essentially constant throughout the crystallization interval and is of no interest; we neglect it. The interesting probability is the term $(V'\rho L/Q)dX/dT$. It indicates that the probability of observation is controlled by the change in crystallinity with temperature. In fact, if we choose a temperature scale of $T = (T_L - T_S)T'$, where T_L and T_S are, respectively, the liquidus and solidus temperatures (at any pressure) and T' is a dimension-free temperature variable, then

$$P_T \simeq \left(\frac{\rho L V'}{Q(T_L - T_S)}\right)\frac{dX}{dT'}. \tag{5.41}$$

And since the first group of parameters is just a scale factor, it can be set to unity, and we define the *thermal probability* as

$$P_T \equiv \frac{dX}{dT'}. \tag{5.42}$$

An example of the use of this result is shown in Figure 5.9, where general X versus T' distributions typical to many igneous rocks are shown along with the associated bell-shaped P_T functions as computed from Eq. (5.42). Also shown are the standard deviations for each of these P_T functions. It is clear here that the most probable temperatures at which to find the body are those stradling the point when the magma is half crystallized. This is so because here the crystals are growing fastest (volumetrically) per unit change in temperature, the heat to the wallrock is supplied almost solely by latent heat of solidification, and the magma cools slowly. On the other hand, near T_S or T_L, X changes slowly with temperature, all the heat to the wallrock is supplied by cooling, and the body cools relatively quickly. It is improbable, then, to find lavas that are free of phenocrysts. If they are found, they must have had a short residence in the near-surface environment.

There is something incomplete, however, about the type of P_T function in Figure 5.9, i.e., that basaltic lavas are rarely seen with more than about 55 volume percent phenocrysts. Beyond this crystallinity it seems that the magma is too viscous

to be erupted. Since this critical crystallinity is near that for maximum packing of poorly arranged spheres, this cutoff in lavas may signify a locking point beyond which crystallinity of the magma is rheologically more solid than an eruptable fluid.

This suggests that the population of lavas actually observed represents the product of P_T and a rheological probability (P_R). We can define this as the eruption probability (P_E) as

$$P_E \equiv P_T P_R. \quad (5.43)$$

The general form of P_R can be estimated by noting that near the liquidus crystallinity is low and $P_R \simeq 1$, while as $X \to 0.55$, $P_R \to 0$. Evidently P_R has a step-like form being flat near the liquidus and tending sharply to zero near $X = \sim 0.55$.

To test these deductions we can use Eq. (5.43)

$$P_R = P_E/P_T, \quad (5.44)$$

where the eruption probability (P_E) is found by making a histogram of crystallinity of lavas of a like composition. The crystallinity is then converted to T' using the known X-T function for the rock. This is shown in Figure 5.10, where the general shape of P_R is found to be step-like.

A basaltic magma can be erupted at any crystallinity between $0 \leq X \leq 0.55$, but once it becomes more than about half crystallized it is too viscous to erupt and it passes on to become a pluton. Without an influx of heat the magma has no chance of becoming a lava. There is good observational evidence (Hildreth, 1981; Marsh, 1981) that the critical crystallinity decreases with increasing silica content of the magma. This is understandable because the interstitial liquid becomes increasingly viscous and increases the bulk viscosity, much as increasing crystallinity would. In granitic magmas this critical crystallinity might be at $X \simeq 0.20$.

A granitic magma can be erupted anywhere in the range $0 \leq X \leq 0.2$. This span of conditions is much narrower than for basalt, and because of this we see that granitic magmas have a much higher probability of becoming plutons. Basaltic magmas, on the other hand, have a much higher probability of being erupted as lava. This difference may partially explain Daly's bimodal distribution of basalts and granites.

There are other probabilities (e.g., density, water content) that can be multiplied with P_T and P_R to give a more complete P_E. As the number of factors increases, the conditions favorable for eruption and lava formation diminish. In other words, the possible choices of P, T, X_{H_2O}, and other factors, open to petrologists to explain the petrogenesis of lava, are significantly reduced from what we assumed previously. Other manifestations of the relations presented here are treated more fully by Marsh (1981).

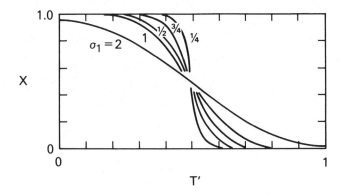

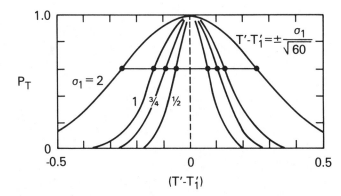

FIGURE 5.9 The top graph shows buildup of crystallinity (X) with decreasing temperature (T') that can be described analytically by the error function with an appropriate value of σ_1. For small σ_1, most crystallization occurs over a narrow temperature range. The probability of finding the magma at any temperature between solidus ($T' = 0$) and liquidus ($T' = 1$) is given by the slope of the curve at that temperature. These *thermal probabilities* (P_T) are given in the bottom graph. The most probable temperature at which to find the magma is when crystals are appearing most rapidly. The standard deviation about this most probable state also is shown.

CONCLUSION

We have sketched through physical processes our initial attempts to understand the life cycle of a magma. It is becoming more clear that much can be learned about this life cycle from a systematic, yet simple, mechanical analysis of the fundamental physical processes involved. Although the physical chemistry of silicates and their melts has been far advanced by

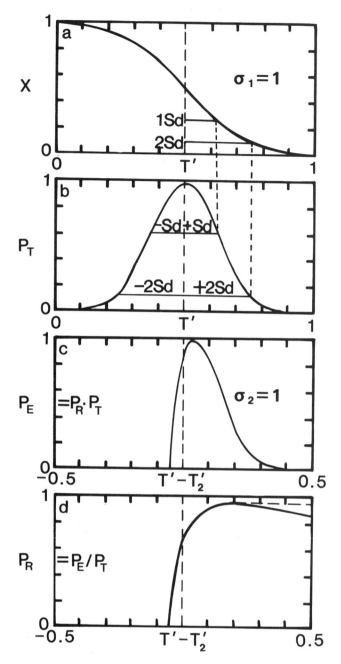

FIGURE 5.10   Relationship between the crystallinity (X), the thermal probability (P_T), the eruption probability ($P_E = P_R P_T$), and the rheological probability (P_R). (The crystallization for the $+1$ and the $+2$ standard deviations from the most probable P_T also are shown.) The rheological probability describes the viscous state of the magma with respect to its ability to be erupted. Near its liquidus there is no rheological constraint on eruption, but near a crystallinity of 0.55 (for a basaltic magma) the crystals approach closest packing, causing a drastic increase in viscosity and a precipitous decrease in P_R. The overall ability to be erupted (i.e., eruptions probability, P_E) is given by $P_R P_T$, and it should be measured by the observed histogram of lava crystallinity. If a magma cools to $X > 0.55$ it goes on to become a pluton. For granitic magmas the critical P_R is at lower degrees of crystallinity ($X \simeq 0.20$), and the probability is great that the body will form a pluton.

petrologists, physical igneous petrology is just now beginning to show a connected fundamental framework within which to study magmas. Many physical processes commonly invoked to explain petrogenesis may rest on the most feeble of physical principles. A thorough understanding of the many processes of paramount importance to petrogenesis will certainly take decades of study, but the advancement of petrology itself depends on this knowledge.

ACKNOWLEDGMENTS

James Brophy kindly helped in a number of ways. The work leading to this paper was supported by grant EAR-8212541 from the National Science Foundation to The Johns Hopkins University.

REFERENCES

Anderson, D. L. (1975). Chemical plumes in the mantle, *Geol. Soc. Am. Bull. 86*, 1593-1600.

Anderson, O. L., and P. C. Grew (1977). Stress corrosion theory of crack propagation with applications to geophysics, *Rev. Geophys. Space Phys. 15*, 77-104.

Bagnold, R. A. (1954). Experiments on a gravity-free dispersion of large solid spheres in a Newtonian fluid under shear, *Proc. R. Soc. London A225*, 49-63.

Bateman, P. C., and B. W. Chappel (1979). Crystallization, fractionation, and solidifaction of the Tuolumne intrusive series, Yosemite National Park, California, *Geol. Soc. Am. Bull. 90*, 465-482.

Buddington, A. F. (1959). Granite emplacement with special reference to North America, *Geol. Soc. Am. Bull. 70*, 671-747.

Carslaw, H. S., and J. C. Jaeger (1959). *Conduction of Heat in Solids*, 2d ed., Clarendon Press, Oxford, 510 pp.

Daly, R. A. (1933). *Igneous Rocks and the Depths of the Earth*, McGraw-Hill, New York, 598 pp.

Darwin, C. (1953). Note on hydrodynamics, *Proc. Cambridge Philos. Soc. 49*, 342-354.

Delaney, P. T., and D. D. Pollard (1981). *Deformation of Host Rocks and Flow of Magma During Growth of Minette Dikes and Breccia-bearing Intrusions Near Ship Rock, New Mexico*, U.S. Geol. Surv. Prof. Pap. 1202, 61 pp.

Delaney, P. T., and D. D. Pollard (1982). Solidification of basaltic magma during flow in a dike, *Am. J. Sci. 282*, 856-885.

Ewart, A. (1976). Mineralogy and chemistry of modern orogenic lavas—some statistics and implications, *Earth Planet. Sci. Lett. 31*, 417-432.

Gilbert, G. K. (1877). *Geology of the Henry Mountains*, U.S. Geograph. and Geol. Surv., 160 pp.

Goodchild, J. G. (1894). Augen-structure in relation to the origin of the eruptive rocks and gneiss, *Geol. Mag. 1*, 20-27.

Grout, F. F. (1945). Scale models of structures related to batholiths, *Am. J. Sci. 243-A*, 260-284.

Head, H. N., and J. D. Hellums (1966). Heat transport and temperature distributions in large single drops at low Reynolds numbers: A new experimental technique, *Am. Inst. Chem. Eng. J. 12*, 553-559.

Hildreth, W. (1981). Gradients in silicic magma chambers: Implications for lithospheric magmatism, *J. Geophys. Res. 86*, 10153-10192.

Hodge, D. S. (1974). Thermal model for origin of granitic batholiths, *Nature 251*, 297-299.

Lawson, A. C. (1896). The eruptive sequence, *Science 3*, 636-637.

Lighthill, M. J. (1956). Drift, *J. Fluid. Mech. 1*, 31-53.

Marsh, B. D. (1976). Mechanics of Benioff zone magmatism, in *The Geophysics of the Pacific Ocean Basin and Its Margins*, G. H. Sutton, M. H. Manghnani, and R. Moberly, eds., Am. Geophys. Union Geophys. Monogr. 19, pp. 337-350.

Marsh, B. D. (1978). On the cooling of ascending andesitic magma, *Philos. Trans. R. Soc. London A288*, 611-625.

Marsh, B. D. (1981). On the crystallinity, probability of occurrence, and rheology of lava and magma, *Contrib. Mineral. Petrol. 78*, 85-98.

Marsh, B. D. (1982). On the mechanics of igneous diapirism, stoping, and zone melting, *Am. J. Sci. 282*, 808-855.

Marsh, B. D., and L. H. Kantha (1978). On the heat and mass transfer from an ascending magma, *Earth Planet. Sci. Lett. 39*, 435-443.

Maxwell, J. C. (1870). On the displacement in a case of fluid motion, *Proc. London Math. Soc. 3*, 82-87.

Morris, S. (1980). An asymptotic method for determining the transport of heat and matter by creeping flows with strongly variable viscosity; Fluid dynamics problems motivated by island arc volcanism, Ph.D. dissertation, The Johns Hopkins U., Baltimore, Md., 154 pp.

Morris, S. J. (1982). The effects of a strongly temperature-dependent viscosity on flow past a hot sphere, *J. Fluid Mech. 124*, 1-26.

Murase, T., and A. R. McBirney (1973). Properties of some common igneous rocks and their melts at high temperatures, *Geol. Soc. Am. Bull. 84*, 3563-3592.

Patchett, P. J. (1980). Thermal effects of basalt on continental crust and crustal contamination of magma, *Nature 283*, 559-561.

Ramberg, H. (1967). *Gravity, Deformation and the Earth's Crust*, Academic Press, New York, 2143 pp.

Sakuma, S. (1953). Elastic and viscous properties of volcanic rocks and high temperatures. Part 3. Osima lava, *Earthquake Res. Inst. Bull. 31*, 291-303.

Shaw, H. R. (1965). Comments on viscosity, crystal settling and convection in granitic magmas, *Am. J. Sci 263*, 120-152.

Shaw, H. R. (1969). Rheology of basalt in the melting range, *J. Petrol. 10*, 510-535.

Shaw, H. R. (1973). Mantle convection and volcanic periodicity in the Pacific, evidence from Hawaii, *Geol. Soc. Am. Bull. 84*, 1505-1526.

Shaw, H. R. (1980). Fracture mechanisms of magma transport from the mantle to the surface, in *Physics of Magmatic Processes*, R. B. Hargraves, ed., Princeton U. Press, Princeton, N.J., pp. 201-264.

Shaw, H. R., T. L. Wright, D. L. Peck, and R. Okamura (1968). The viscosity of basaltic magma: An analysis of field measurements in Makaopuhi lava lake, Hawaii, *Am. J. Sci. 266*, 225-264.

Shimazu, Y. (1959). A thermodynamic aspect of Earth's interior-physical interpretation of magmatic differentiation process, *Nagoya Univ. J. Earth Sci. 7*, 1-34.

Sleep, N. H. (1974). Segregation of magma from a mostly crystalline mush, *Geol. Soc. Am. Bull. 85*, 1225-1232.

Smith, R. L. (1979). Ash-flow magmatism, in *Ash-Flow Tuffs*, C. E. Chapin and W. E. Elston, eds., Geol. Soc. Am. Spec. Pap. 180, pp. 5-27.

Spera, F. J. (1980). Aspects of magma transport, in *Physics of Magmatic Processes*, R. B. Hargraves, ed., Princeton U. Press, Princeton, N.J., pp. 265-323.

Spohn, T. (1980). Orogenic volcanism caused by thermal runaways? *Geophys. J. R. Astron. Soc. 62*, 403-419.

Spohn, T. (1981). On the origin of orogenic volcanism, *Geol. Runds. 70*, 154-165.

Taylor, H. P., and L. T. Silver (1978). Oxygen isotope relationships in pluton igneous rocks of the Peninsular Ranges batholith, southern and Baja California, *Fourth International Conference on Geochronology*.

Tozer, D. C. (1967). Towards a theory of thermal convection in the Earth's mantle, in *The Earth's Mantle*, T. F. Gaskell, ed., Academic Press, New York, pp. 325-353.

van der Molen, I., and M. S. Patterson (1979). Experimental deformation of partially-melted granite, *Contrib. Mineral. Petrol. 70*, 299-318.

Verhoogen, J. (1954). Petrological evidence on temperature distribution in the mantle of the Earth, *Trans. Am. Geophys. Union 35*, 85-92.

Walker, D., E. M. Stolper, and J. F. Hays (1978). A numerical treatment of melt-solid segregation: Size of eucrite parent body and stability of terrestrial low-velocity zone, *J. Geophys. Res. 83*, 6005-6013.

Weertman, J. (1968). Bubble coalescence in ice as a tool for the study of its deformation history, *J. Glaciol. 7*, 155-159.

Weertman, J. (1972). Coalescence of magma pockets into large pools in the upper mantle, *Geol. Soc. Am. Bull. 83*, 3531-3532.

Yoder, H. S., Jr. (1976). *Generation of Basaltic Magma*, National Academy of Sciences, Washington, D.C., 265 pp.

Yuen, D. A., L. Fleitou, G. Schubert, and C. Froidevaux (1978). Shear deformation zones along major transform faults and subducting slabs, *Geophys. J. R. Astron. Soc. 54*, 93-120.

Yellowstone Magmatic Evolution: Its Bearing on Understanding Large-Volume Explosive Volcanism

6

ROBERT L. CHRISTIANSEN
U.S. Geological Survey

ABSTRACT

The Yellowstone Plateau volcanic field of Wyoming, Idaho, and Montana evolved through three cycles during the past 2 m.y. Each cycle began and ended with protracted periods of episodic eruptions of rhyolitic lava but climaxed with an explosive caldera-forming pyroclastic eruption of 3×10^2 to 3×10^3 km^3 of rhyolitic magma, mainly as ash flows. These climactic eruptions occurred about 2.0, 1.3, and 0.6 m.y. B.P. In each cycle the voluminous pyroclastic eruptions, which lasted only hours or days, were preceded by the evolution of a high-level crustal magma chamber over 0.2 to 0.7 m.y., marked by the development of surficial ring-fracture systems and episodic eruptions of rhyolitic lavas aggregating 10^2 to 10^3 km^3. Chemical and isotopic compositions of the rhyolites evolved systematically during each cycle and, possibly, secularly from one cycle to the next. The preclimactic magmas were vertically compositionally zoned, including notable roofward enrichment in volatile constituents. About 10^2 km^3 of basalt has erupted marginally to the active rhyolitic source areas throughout development of the volcanic field, some of it directly through former rhyolitic sources about 10^6 yr after rhyolitic activity.

Fundamental elements of this evolutionary pattern are common to other well-studied voluminous pyroclastic fields, although specific aspects vary in different geologic frameworks. Common elements of the pattern indicate that a protracted crustal-magmatic evolution must precede explosive volcanism on the largest scale. The sustained process of energy transfer from the mantle by mafic magmas can lead to the formation of more silicic crustal magma chambers containing a large magmatic volume with major explosive potential. This sequence of events is recognized in various areas through characteristic time-space patterns of surficial volcanism and chemical and isotopic evolution of eruptive products. Shorter-term recognition of impending voluminous explosive volcanism may, however, be ambiguous.

Magma may now be present in the Yellowstone chamber. The Yellowstone Plateau volcanic field might even be in a new evolutionary cycle leading to a fourth voluminous pyroclastic eruption, although such an interpretation cannot yet be made confidently.

INTRODUCTION

Much of our current knowledge of explosive volcanism derives from direct study of dynamic processes in observed explosive eruptions or from analysis of their deposits. It is important, however, to also consider the evidence for superexplosive eruptions of magnitudes that have seldom, if ever, been recorded in written human history but for which there is abundant geologic evidence. Modern eruptions of these types would have devastating effects on large areas and would have significant global effects, including probable climatic changes. Although such large pyroclastic eruptions occur infrequently on the human time scale, the largest—such as that of the Yellowstone Plateau region 600,000 yr before present (B.P.)—occur with

84

a worldwide frequency of once every few hundred thousand years; similar, if not quite so large, eruptions, such as that of Mount Mazama-Crater Lake, probably occur every 1000 yr or less. Geologic studies of the source areas of these very large pyroclastic eruptions invariably indicate that they occur from complexly evolved voluminous magmatic systems, of which the great pyroclastic eruptions represent climactic evolutionary stages. It is useful, therefore, to consider the geologic, petrologic, and geophysical evidence for the preclimactic conditions that have preceded pyroclastic eruptions on the largest scales.

The Yellowstone Plateau volcanic field of northwestern Wyoming and nearby areas is one of the Earth's most voluminous accumulations of erupted rhyolitic magma. Much of this material was ejected in three brief eruptive events, each of which vented hundreds to thousands of cubic kilometers of pyroclastic deposits. The activity of this volcanic field began somewhat more than 2 million years (m.y.) B.P. Virtually all the volcanic rocks of the field are either rhyolites, containing more than 72 wt.% SiO_2 (mostly more than about 75 wt.%), or basalts, containing about 48 to 52 wt.% SiO_2; intermediate compositions are absent except for a few small-volume lavas that represent locally mixed rhyolitic and basaltic magmas.

Intensive study of the Quaternary volcanism, tectonism, and hydrothermal activity of the Yellowstone Plateau volcanic field reveals a major active system of energy transfer in the crust and upper mantle. Ongoing seismicity and faulting, vigorous hydrothermal convection, geologically young volcanism of exceptionally large volume and great explosivity but with long intereruptive reposes, and evidence for simultaneously active basaltic and rhyolitic magmas reveal a range of dynamic processes that are closely interrelated within a unified magmatic and tectonic system. Furthermore, because Yellowstone lies far from the more generally active margins of the lithospheric plates, it is possible here to develop a more comprehensive view of a coherent tectonomagmatic system than is commonly the case at plate margins, where lithospheric processes are linearly extensive and individual magmatic systems may be difficult to identify or may overlap complexly.

Boyd (1961) first laid out the broad geologic framework of the voluminous rhyolitic field of the Yellowstone Plateau and clearly established both the pyroclastic-flow origin of much of the erupted rhyolite and the immense size of some of its lava flows. More recent work shows that the volcanic activity of the field has occurred in three major cycles (Christiansen and Blank, 1972; Christiansen, 1979) and that the Yellowstone system is still to be considered active (Eaton *et al.*, 1975; R. B. Smith and Christiansen, 1980). This work is ongoing and includes studies of the stratigraphy, chronology, volcanic processes, tectonics, and petrologic evolution of the volcanic field.

FRAMEWORK OF YELLOWSTONE PLATEAU VOLCANISM

The Yellowstone Plateau (see Figure 6.1), a region of 6500 km² at an average elevation of about 2500 m, is flanked by mountains

that rise to nearly 4000 m. To the west and southwest the terrain drops gradually to the eastern Snake River Plain, a region of low relief at about 1000- to 1500-m elevation. The Snake River Plain is—like Yellowstone—a region of voluminous young volcanism, and the origins of the two regions appear to be closely linked. Island Park is a transitional area between the Plateau and the Plain but constitutes an integral part of the Yellowstone Plateau volcanic field.

Three regional late-Cenozoic tectonic trends intersect at Yellowstone: (1) the northeast-trending eastern Snake River Plain represents a great crustal downwarp or semigraben (Kirkham, 1931; Braile *et al.*, 1982); (2) approximately north-trending faults south of the Yellowstone Plateau define a block-faulted terrane of the Middle Rocky Mountains that is tectonically equivalent to the eastern marginal zone of the Great Basin system of regional tectonic extension (R. B. Smith and Sbar, 1974; Christiansen and McKee, 1978); (3) and east-west to northwest-trending faults from the west and northwest are reflected in the topography, tectonics, and seismicity of the Rocky Mountains north of the eastern Snake River Plain (Myers and Hamilton, 1964; R. B. Smith and Sbar, 1974). These three intersecting trends suggest a focus of extensional lithospheric stresses that has localized the Yellowstone Plateau volcanism. These tectonic systems are still active, as shown by young faults and earthquakes on the northern and southern parts of the region (Figure 6.1) and by late Pleistocene and Holocene basaltic volcanism in the eastern Snake River Plain.

Volcanism has been intermittently active in the Yellowstone-Island Park region for at least 2.2 m.y.; from a geologic perspective, there is no compelling reason to think that Yellowstone's volcanism has ended. The accumulated eruptive products of the youngest cycle of activity—predominantly rhyolitic and subordinately basaltic—form the Yellowstone Plateau (see Figure 6.2). In the past 2 m.y. volcanic rocks and deposits equivalent to more than 6500 km³ of magma have accumulated. Furthermore, magmas that have filled chambers of very large volume beneath the plateau and that have partly crystallized and solidified over time provide the source of heat for the well-known hydrothermal features of Yellowstone National Park. In a general model of continental silicic volcanism in the United States, R. L. Smith and Shaw (1975) indicated that intrusive volumes probably amount to about 10 times the erupted volumes. Thus, the Yellowstone system probably reflects an average magma production rate of more than 0.03 km³/yr, roughly comparable to that of the most active volcanic regions of the Earth, such as Hawaii, Iceland, and the mid-oceanic ridges. Nevertheless, the style and timing of magma transfer, storage, and eruption in these regions differ vastly. In Yellowstone, brief episodes of voluminous eruption that (as discussed below) probably lasted only a few hours, days, or months were separated by quiet intervals, some lasting hundreds of thousands of years (see Figure 6.3). This long-period episodic style of volcanism, which is characteristic both of silicic magmas in general and of magmas in many continental settings, contrasts with the more nearly continuous volcanism characteristic of large predominantly basaltic systems, particularly those in oceanic regions.

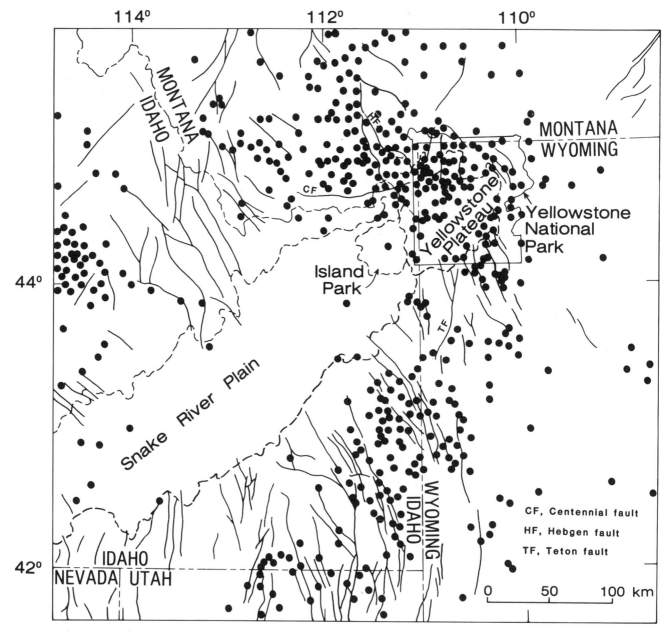

FIGURE 6.1 Yellowstone Plateau, Island Park, and the eastern Snake River Plain, showing locations of late Cenozoic faults (heavy lines) and epicenters of earthquakes (dots) (modified from R. B. Smith, 1978).

EVOLUTION OF THE YELLOWSTONE MAGMATIC SYSTEM

The rhyolitic volcanism of the Yellowstone region has evolved through three nearly identical cycles, each beginning and ending with long periods of intermittent lava eruptions but climaxing with catastrophic eruptions of explosively ejected material that formed voluminous ash flows. The hot, mobile ash flows spread for many tens of kilometers and welded to form dense rhyolite that, over the three cycles, covered thousands of square kilometers (Figure 6.2). Partial emptying of the sub-surface magma chambers by these eruptions caused their roofs to collapse to form gigantic calderas, tens of kilometers across (see Figure 6.4). During the explosive eruptions, fragments of volcanic glass and crystals were injected high into the atmosphere and carried for thousands of kilometers; deposits from these eruptions are found as far away as Saskatchewan, Mississippi, and California (Izett and Wilcox, 1981). Subsequent eruptions of rhyolitic lavas partly filled the calderas. Throughout this rhyolitic evolution, basalts of relatively minor but significant volume have continued to erupt intermittently around the margins of the active rhyolitic system (Figures 6.2 and 6.3).

Geologic mapping and stratigraphy (Christiansen and Blank, 1972), calibrated with K-Ar dating by J. D. Obradovich (U.S. Geological Survey, written communication), indicate that the first and most voluminous volcanic cycle began by about 2.2 m.y. B.P. with small-volume eruptions of both basalts and rhyolites, followed by the first climactic ash-flow eruption about 2.0 m.y. B.P. This eruption formed a single cooling unit, the Huckleberry Ridge Tuff, with a volume greater than 2500 km³. Internal features of this great ash-flow sheet, including the complete absence of erosional channeling or accumulation of even minor sediments, as well as its unitary cooling history, indicate that it must have been emplaced within a period of no more than a few hours or days. The eruption of such a large volume of igneous material in so brief a time clearly demonstrates that a large continuous chamber filled by magma existed in the upper crust at the time of eruption. The roof collapsed as the chamber partly emptied, but much of the resulting caldera, extending across both Island Park and the Yellowstone Plateau, is now obscured by younger volcanic rocks (Figures 6.2 and 6.4). Postcollapse rhyolitic volcanism at least partly

filled the first-cycle caldera, and some of the resulting lavas remain exposed at the western edge of the volcanic field (Figure 6.2).

The second volcanic cycle was the smallest of the three. In some respects the second cycle appears to represent a renewal of major activity within the first-cycle magmatic system. Volcanism of the second cycle was entirely nested within the first-cycle caldera in the northern part of Island Park. The climactic ash flows, forming the Mesa Falls Tuff, erupted 1.3 m.y. B.P. and had a volume of more than 280 km³. As did the Huckleberry Ridge Tuff, this single cooling unit was erupted in such a short time that the removal of support from the magma-chamber roof caused collapse, which formed a caldera 25 km across. Postcollapse rhyolitic lavas were erupted in a northwest-trending zone that extends across the second-cycle caldera and overlaps southeastward into the larger first-cycle caldera in which the second was nested (Figure 6.2).

After the first two volcanic cycles had run their courses, the area of rhyolitic activity shifted entirely away from Island Park to the present Yellowstone Plateau. As long as rhyolitic magma

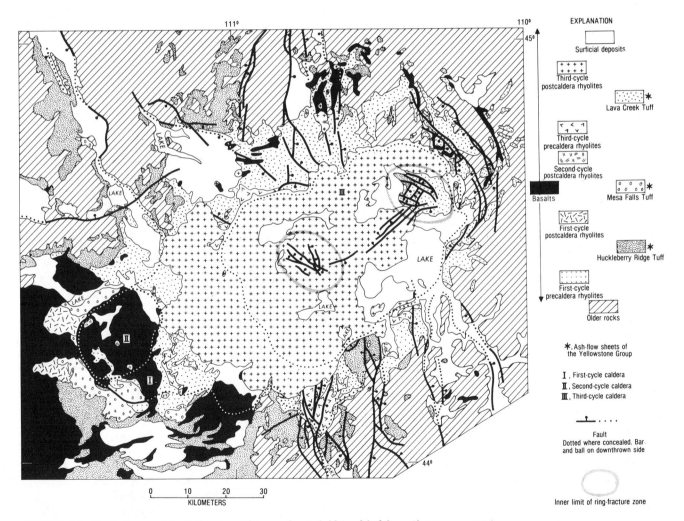

FIGURE 6.2 Geologic map of the Yellowstone Plateau volcanic field (modified from Christiansen, 1979).

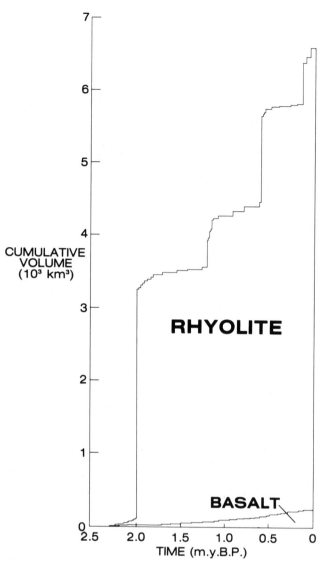

FIGURE 6.3 Estimated cumulative magmatic volumes of rhyolite and basalt erupted in the Yellowstone Plateau volcanic field. Because there are many sources of error in both the volume and the age data (by factors of 2 or more) this diagram should be considered only as a schematic indication of the gross relationships between the climactic caldera-forming eruptions and other eruptive events and between the activity of the various cycles, not as an indication of the relationships between most specific events.

fractures that allowed basalts to erupt through those fractures onto the caldera floors (Figure 6.2).

The record of the third volcanic cycle, which is far more complete than those of the first two, provides a model for interpreting the older cycles. The third cycle began about 1.2 m.y. B.P. For about 600,000 yr, rhyolitic lavas erupted intermittently on the Yellowstone Plateau from a slowly forming set of arcuate fractures that eventually outlined the area that would later collapse as the third-cycle Yellowstone caldera. Thus, a large magma chamber apparently was forming at shallow crustal levels, doming and stretching its roof and intermittently sagging it to form the arcuate fractures. By 630,000 yr B.P., rhyolitic lavas had erupted through every major sector of the ring-fracture system (see Figure 6.5). Also by that time, chemical evolution of the voluminous silicic magma body had advanced to a high degree and reached nearly the maximum state of differentiation to be recorded in the subsequent ash-flow eruptions (see Figure 6.6). With the generation of these ring fractures and the advanced chemical evolution of the magmatic system, conditions were set for a climactic eruption that could utilize the ring fractures as volcanic vents.

Eruptions during the preclimactic stages were generally nonexplosive, presumably because magma, rising relatively slowly through successively shallower zones, tended to equilibrate its volatile content to successively lower pressures. Episodic magmatic leaks to the surface produced rhyolitic eruptions but quickly exhausted the exsolved volatile constituents and allowed most of the magma to extrude as rather viscous lavas. By contrast, there must have been some triggering mechanism to begin a massive degassing of the main magma chamber, represented by the climactic eruption of the Lava Creek Tuff.

The triggers for such climactic pyroclastic eruptions are not well understood. It is clear, however, that the highly evolved rhyolitic magma of the third-cycle Yellowstone magma chamber developed a vertical compositional zonation just before the climactic eruption (Figure 6.6). As discussed by Hildreth (1981), such a zoned magma body should have a marked upward concentration of volatile constituents. In addition, deeply circulating hydrothermal waters may enter such a rising magma body and help bring it to volatile saturation. Possibly, in such a system, downward propagation of the ring fractures from the stretching surface above the rising magma body eventually penetrates the main magma chamber, where a fault-displacement event causes a transient pressure drop in the magma and exsolution of a gas phase from the dissolved volatile constituents. Such a sequence of events might initiate a chain reaction of saturation, pressure reduction, volatile exsolution and degassing, gas-lift of the frothing magma through the opening ring fractures, and further pressure reduction. Whatever the trigger may be, such a degassing process, once begun, probably continues at a rate constrained only by the dimensions of the fractures that connect the magma to the surface until pressure equilibrium is restored in the magma chamber or until downward tapping of the zoned magma body reaches the limits of the physical properties that allow rapid magma flowage (R. L. Smith, 1979). During this brief explosive degassing a significant proportion of the magma in the chamber expands by frothing then chills and shatters by explosive pressure relief. The

lay beneath the Island Park area, no basalts erupted through the major rhyolitic source area, although they did erupt episodically around its margins; a similar situation prevails around the rhyolitic source area of the still-active Yellowstone Plateau. Presumably, the denser basaltic magmas continued to rise from the mantle but were precluded from reaching the surface within the rhyolitic source area by less dense rhyolitic magma in their paths. The rhyolitic magma chambers beneath Island Park have completely solidified during the past million years, however, and in the past 200,000 yr they have been broken by tectonic

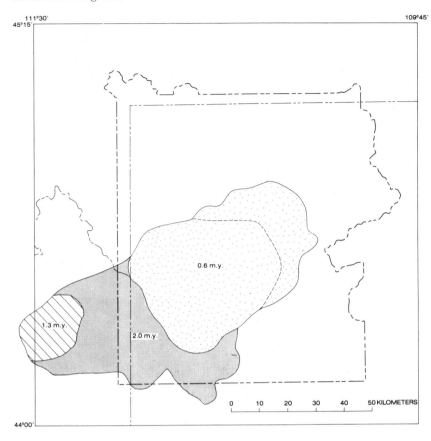

FIGURE 6.4 Locations of calderas of the Yellowstone Plateau volcanic field: first-cycle caldera (2.0 m.y.), second-cycle caldera (1.3 m.y.), and third-cycle caldera (0.6 m.y.). Outline of Yellowstone National Park shown for reference (see Figure 6.1).

quenched magma is expelled to the surface at high velocities as a pumiceous froth of glass and crystal fragments, along with fragments of rock torn from the walls of the venting fissures.

The climactic eruption of the third volcanic cycle emplaced more than 1000 km^3 of the Lava Creek Tuff 630,000 yr B.P. As noted above for the earlier two climactic eruptions, successive ash flows of the Lava Creek erupted so quickly (certainly within hours or days, not decades or centuries) that there was not even slight erosion, aqueous reworking, or accumulation of interbedded extraneous materials. Detailed stratigraphic study shows that the Lava Creek Tuff actually consists of two ash-flow sheets that erupted more or less consecutively but so quickly that they welded and crystallized as a single cooling unit. Furthermore, the caldera that formed by collapse of the Lava Creek magma-chamber roof encompassed two partly intersecting ring-fracture zones (see Figures 6.5 and 6.7); distributions and internal features of the two subsheets of the Lava Creek Tuff indicate that they were erupted respectively from these two ring-fracture zones. Thus, the ring-fracture system, which formed during the insurgence of magma early in the third volcanic cycle and through which the Lava Creek Tuff probably was erupted, appears to reflect the formation of two broad high-level culminations on the top of a very large underlying body of magma.

The rapid partial emptying of the Lava Creek magma chamber caused its roof segments to collapse along the two ring-fracture zones. The collapse probably occurred both during the eruption and by continued slumping of the walls as the ash flows cooled. A compound caldera basin as much as 45 km across and 75 km long was formed (Figures 6.2, 6.4, and 6.7).

The unbroken central block of each caldera segment soon responded by uplift to the resurgence of magma in the partially drained chamber, and two structural domes were formed (Figure 6.7); such resurgence is common in the largest rhyolitic calderas, as documented by R. L. Smith and Bailey (1968). Uplift of the floor of the Yellowstone caldera, as in all resurgent calderas yet studied, occurred soon after collapse; ash-flow eruption, caldera formation, resurgent doming, and eruption of the first postresurgence lavas all occurred within a time too short to resolve by K-Ar dating, probably no more than a few thousand years.

In the last 630,000 yr, both segments of the Yellowstone caldera have filled with sediments and with rhyolitic lavas that have continued to erupt intermittently; much of the caldera basin is covered by rhyolitic flows erupted during the past 150,000 yr (Figures 6.2 and 6.3). During the past 150,000 yr, lavas covered the western resurgent dome, localized uplift of that dome was renewed, and the western caldera segment was flooded by nearly 1000 km^3 of rhyolitic lavas that completely buried the west rim of the caldera. This youngest rhyolite was erupted in three short episodes, about 150,000, 110,000, and 70,000 yr B.P. The sources of the eruptions are two long zones of vent fissures (Figure 6.7) that extend across the caldera from fault zones outside (Figure 6.2). These faults have, with few

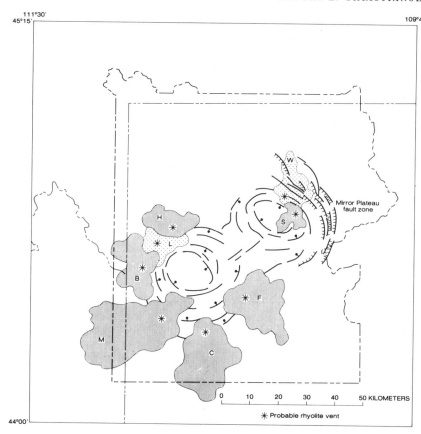

FIGURE 6.5 Reconstruction of precaldera lavas and ring structures of the third volcanic cycle; precaldera lavas are Wapiti Lake flow (W), Stonetop Mountain flow (S), Flat Mountain flow (F), Lewis Canyon flow (C), Moose Creek Butte flow (M), Big Bear Lake flow (B), Harlequin Lake flow (L), and Mount Haynes flow (H). Outline of Yellowstone National Park shown for reference (see Figure 6.1).

exceptions, vented rhyolitic magma only where they intersect the caldera ring-fracture system. This pattern suggests that magma at the top of the Yellowstone chamber has been crystallizing and solidifying during the last 150,000 yr; as the solidified crust becomes rigid enough to allow fracturing, it is broken by tectonic faults that cross the caldera on regional trends and allow deeper magma to rise and erupt to the surface.

Using the third and best understood cycle as a model, we can interpret the history of each of the three cycles of the volcanic field most readily as repeating a consistent pattern. In each cycle the voluminous pyroclastic eruptions, which occurred within hours or days, were preceded by a period of 200,000 to 600,000 yr of intermittent rhyolitic eruptions as a high-level crustal magma chamber evolved and slowly leaked upward beneath the site of eventual major pyroclastic eruption. The growth of such a chamber was marked by the development of surficial ring-fracture systems and the episodic eruption of rhyolitic lavas aggregating hundreds to thousands of cubic kilometers. A climactic eruptive episode was followed by eventual postcaldera cooling and consolidation of the magma body, reflected in a protracted period of intermittent but probably gradually declining eruptions of rhyolitic lavas (Figure 6.3) and partial filling of the caldera.

Perhaps 100 to 200 km³ of basalt has erupted marginally to the active rhyolitic source areas and directly through the formerly active calderas of the Island Park area. Furthermore, the Snake River Plain represents a prolonged basaltic evolution

after a 15-m.y. progression of predominantly rhyolitic systems, of which Yellowstone is but the present-day expression (Christiansen and McKee, 1978; R. B. Smith and Christiansen, 1980). Large rhyolitic systems more or less comparable to the Yellowstone Plateau volcanic field have evolved, solidified, and eventually cooled in areas situated successively farther northeastward over time along the axis of the eastern Snake River Plain. As the large subsurface magmatic bodies that sustained each of these rhyolitic fields eventually crystallized and cooled, the overlying surface subsided. Basalts have eventually erupted through each of these rhyolitic systems and have flooded the entire axial zone of the eastern Snake River Plain as it subsided. In the long-term evolution of this persistent regional tectonomagmatic association, basalt accounts for about as much volume as rhyolite.

ORIGIN OF YELLOWSTONE MAGMAS

Several classes of hypotheses have been advanced to explain such bimodal rhyolite-basalt associations as Yellowstone's, including (1) origin of the rhyolitic magmas by fractional crystallization of the basaltic magmas (e.g., Bowen, 1928), (2) fractional melting of a mantle source material (e.g., Yoder, 1973), (3) separation of primary intermediate liquids into both less and more silicic fractions that are immiscible (e.g., Hamilton,

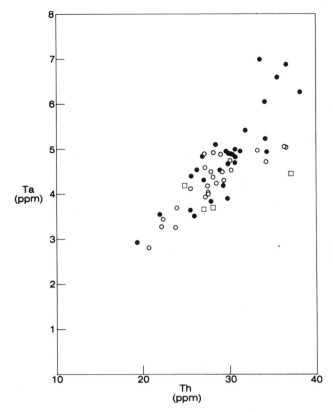

FIGURE 6.6 Ta versus Th contents in third-cycle volcanic rocks. Relationships are representative of those between numerous other elements and element ratios. As both Ta and Th are enriched roofward within various zoned silicic magmatic systems (Hildreth, 1981), the plot illustrates the range of compositional zonation in the magma chamber for Lava Creek Tuff (dots) and nearly comparable degrees of differentiation in compositions of preclimactic third-cycle lavas (squares) and intracaldera postclimactic lavas (circles). Data from W. Hildreth and R. L. Christiansen (U.S. Geological Survey, unpublished data, 1982).

1965), and (4) partial melting of different source materials to produce two primary liquid compositions (e.g., Holmes, 1931).

The absence of intermediate compositions in the volcanic field, the early appearance and volumetric predominance of rhyolite (as noted above), and the systematically different Pb- and Sr-isotopic compositions of the basalt and rhyolite (Doe *et al.*, 1982) virtually rule out the first hypothesis for Yellowstone. Numerous geologic relationships in Yellowstone favor the last hypothesis over the others; these relationships include the presence of both rhyolitic and basaltic magmas from the inception of the volcanic field throughout its development and the complete absence of magmas with compositions appropriate to give the observed compositions by immiscible-liquid separation, despite the common occurrence of such magmatic types elsewhere. All the known geologic and petrologic relationships at Yellowstone either directly support or are consistent with the idea that the basalt and rhyolite represent partial melts of different source materials.

Although the basaltic magmas clearly are partial melts of the

mantle, the source of the rhyolitic magmas is perhaps somewhat less certain. Pb- and Sr-isotopic ratios (Doe *et al.*, 1982) appear to indicate that, although the rhyolitic magmas did interact extensively with upper-crustal materials, they did not form by partial melting of once-sedimentary upper-crustal rocks. These isotopic ratios may be taken to reflect the ultimate parentage of the rhyolitic magmas by partial melting of intermediate-composition high-grade metamorphic rocks of the lower crust.

By any reasonable model, the generation of basaltic magmas by partial melting of the upper mantle is the fundamental process that drives the Yellowstone system. In this region of tectonic extension, at a locus of three intersecting zones of concentrated differential stress, continued extensional reduction of vertical stress and thermally driven regional uplift must reflect lateral mass transport and upward displacement of mantle material. As the lithosphere is stretched and uplifted, the reduction in pressure brings those components of the upper mantle that have the lowest melting temperatures within their melting range. Just as in the ocean basins, where the mantle is at a depth of only a few kilometers, the product of such partial melting is basaltic; the rise and addition of this basaltic magma to the extending crust transports heat upward. Lower-crustal materials are more silicic in composition, at lower pressures than the mantle below, and, on partial melting, yield more silicic magmas. Because these silicic melts are at lower temperatures and have lower densities and higher viscosities than basaltic melts, the buoyant silicic magmas tend to rise slowly and to accumulate as large shallow-crustal masses.

COMPARISON WITH OTHER VOLUMINOUS PYROCLASTIC SYSTEMS

Although the emphasis of this paper is on Yellowstone as an example of a magmatic system that has evolved to produce superexplosive eruptions, it is important to compare it, at least briefly, with a few other systems to delineate elements that are common in their various courses of development. Such systems, though not well represented in the written historical record, are geologically common in several regions of the Earth, such as Japan, New Zealand, the Central Andes, Mexico, and parts of the western United States.

The Jemez Mountains and their surrounding plateaus in northern New Mexico constitute a large volcanic field that has evolved both cyclically and secularly for the past 10 m.y. to erupt a complex sequence of volcanic products (R. L. Smith *et al.*, 1970). Among these are numerous basaltic eruptive centers, whose ages span the development of the field, and voluminous silicic volcanic rocks. Evolution of the field climaxed in two major pyroclastic eruptions that formed the Bandelier Tuff 1.4 m.y. B.P. and 1.1 m.y. B.P., associated with collapse to form the Toledo and Valles calderas, respectively. The precollapse lavas of the caldera area record the early development of a ring-fracture system. These lavas and the Bandelier Tuff record the evolution of a magma chamber with a marked vertical compositional zonation before each of the climactic ash-flow eruptions (R. L. Smith, 1979).

The Long Valley area of eastern California (Bailey *et al.*,

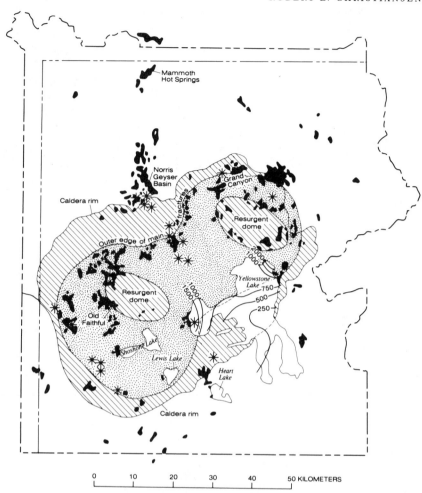

FIGURE 6.7 Third-cycle Yellowstone caldera (light stippling), showing locations of ring-fracture system (dark stippling), postcollapse intracaldera rhyolitic vents (stars), active and recently active hydrothermal features (black areas), and contours of heat flow in Yellowstone Lake (labeled in units of mW/m²). Outline of Yellowstone National Park shown for reference (see Figure 6.1).

1976) is also the focus of a long-lived magmatic system, whose early development was marked mainly by more than 2 m.y. of basaltic and related eruptions in a region focused on the eventual site of the Long Valley caldera. For a period of about a million years before the climactic eruption of the Bishop Tuff, the rhyolitic lavas of Glass Mountain were erupted from an arcuate zone that appears to reflect early formation of a ring-fracture system. The more than 170 km³ of the Bishop Tuff, erupted about 700,000 yr B.P., gives evidence for the existence at that time of a large magma body with a strong vertical compositional and thermal zonation (Hildreth, 1979), whose evolution climaxed with that voluminous pyroclastic eruption.

A very different tectonomagmatic setting is evident in the evolution of the San Juan Mountains volcanic field of southwestern Colorado, although its evolutionary pattern reveals important similarities to Yellowstone and the other two systems just noted. The earliest and simplest caldera complex of the San Juan Mountains field is the Platoro caldera (Lipman, 1975), within which andesitic to dacitic eruptions, over a period of about 5 m.y. beginning 35 m.y. B.P., were focused in the area of subsequent caldera-forming eruptions. The climactic events were a multicyclic group of ash-flow eruptions of rhyodacitic composition. The entire San Juan Mountains volcanic field

(Lipman et al., 1978) shows a repeated pattern of early focusing of predominantly andesitic stratocone volcanoes over a period of several million years, followed by the accumulation of a batholithic magmatic system and the rise of large compositionally zoned silicic magma bodies to shallow levels beneath these foci, which eventually produced major caldera-forming rhyodacitic to rhyolitic eruptions.

Mount Mazama in the Cascades of southern Oregon, as shown in the classic study of Williams (1942) and the comprehensive ongoing restudy by Bacon (1983), comprises a complex of overlapping predominantly andesitic stratocone volcanoes that grew over a period of more than 400,000 yr. Major intermediate-composition cone-building activity ceased about 50,000 yr B.P., and a brief climactic eruptive sequence led to the formation of Crater Lake caldera 6800 yr B.P. The immediate precursors of the climactic pyroclastic eruptions were a group of rather highly evolved rhyodacitic lava flows, associated with relatively minor pyroclastic materials, that were concentrated in an arc about the north flank of the stratocone complex during a period that may have been as short as 100 to 200 yr. The climactic eruption ejected a magmatic volume of about 50 km³ as pyroclastic fallout and flow deposits from a magma chamber with a strong vertical compositional zonation.

The pattern that thus emerges from a study of common elements in the volcanic evolution previous to many well-studied major pyroclastic eruptions indicates that a protracted crustal magmatic evolution must precede explosive volcanism on the largest scales. Sustained energy transfer from the mantle to the crust by the generation and rise of mafic magmas at a locus of long-lived magmatism in the upper mantle can lead to the formation of more silicic crustal magma chambers containing a large magmatic volume (Hildreth, 1981). As such a large rhyolitic magma body accumulates at a relatively high level in the crust, it is likely to undergo various differentiation processes that reflect, at least in part, the accumulation of volatile constituents near its roof (R. L. Smith, 1979; Hildreth, 1981). It is probably this protracted chemical as well as physical evolution that ultimately prepares such a magmatic system for a great pyroclastic eruption.

CURRENT ACTIVITY OF THE YELLOWSTONE SYSTEM

In light of the idea that a coherent pattern of volcanism and tectonism might reveal the evolution of a magmatic system that could lead to an episode of superexplosive volcanism, it is illuminating to consider the recent and current state of the Yellowstone Plateau volcanic field. Although no eruptive activity has occurred in this region for about the past 70,000 yr, the vigorously active hydrothermal system of Yellowstone reflects convective heat flow from a high-temperature crustal heat source. Extensive chemical and isotopic studies of Yellowstone's hydrothermal features suggest the possible presence of one or a few essentially uniform and continuous reservoirs of hot water from which numerous distinct, shallower systems are derived. The main groups of hot springs and geysers occur in basins and represent the surficial discharge of hot water that has circulated deeply along systems of fractures, including the ring fractures of the Yellowstone caldera and tectonic fractures that intersect the ring fractures more or less radially (Figure 6.7).

Total convective heat flow from the main hydrothermal areas of the Yellowstone caldera has been estimated by Fournier *et al.* (1976) by measuring the total discharge of streams flowing from the area and equating the Cl^- concentrations in these streams with the enthalpies of fluids in the deep hydrothermal system. The thermal-power output of all the hydrothermal features in the Yellowstone caldera is estimated to be about 5.3 $\times 10^9$ watts. Because the area of the Yellowstone caldera is about 2500 km², its average convective thermal-energy flux is about 1800 mW/m², more than 30 times the mean global heat flow.

The thermal energy transferred to the surface by the convective hydrothermal system for at least the past 20,000 yr (and probably much longer) requires more than just cooling of the volcanic rocks of the Yellowstone Plateau. Clearly, magma has existed until at least recently in the Yellowstone chamber; it can reasonably be inferred to still be present (Eaton *et al.*, 1975; Lehman *et al.*, 1982; see also Chapter 7 of this volume). The high thermal flux in the caldera represents the cooling of

a body of rhyolitic magma at a depth of a few kilometers, at least partial crystallization of that body, and circulation of meteoric waters to great depths in the caldera fractures to drive hydrothermal convection.

A regional zone of seismicity extends eastward north of the eastern Snake River Plain to the north rim of the Yellowstone caldera (Figure 6.1). Near the caldera rim, earthquakes are generally as deep as about 10 km, and the largest shocks have magnitudes greater than 6; the seismic zone bends southeastward and continues into the caldera, where earthquake activity becomes more episodic and swarmlike, magnitudes are generally smaller, and maximum focal depths are generally less than about 6 km (see Chapter 7 of this volume). South of the caldera, earthquakes are less continuously distributed but continue regionally in a zone along the margin of the Middle Rockies-Great Basin region (Figure 6.1). Fault trends and seismic focal mechanisms near the Yellowstone caldera indicate changes in the directions of crustal stresses associated with the lithospheric extension concentrated near the Yellowstone Plateau volcanic field (R. B. Smith, 1978; see also Chapter 7 of this volume).

Geodetic measurements reveal uplift along the axis between the 2 resurgent domes of the Yellowstone caldera area by more than 700 mm within the last 50 yr relative to areas outside the caldera (Pelton and Smith, 1979; see also Chapter 7 of this volume). Such high uplift rates, which are among the largest known for continental areas and nearly comparable to some measured on active volcanoes, may reflect upward movement of magma in the Yellowstone system.

Various additional geophysical data help define further the current state of the Yellowstone Plateau magmatic system. Iyer *et al.* (1981) described an area of anomalous teleseismic P-wave delays in the Yellowstone caldera and modeled the data to suggest a 15 percent reduction in P-wave velocities within the crust and a 5 percent reduction in the upper mantle to depths of 250 km. An upper-crustal body, which produces relatively high seismic-wave attenuations and is characterized by a P-wave velocity of about 5.7 km/sec, underlies the Yellowstone caldera in the place of a regional layer having a P-wave velocity of 6.0 km/sec, and areas of even lower seismic velocities and higher attenuation occur locally within or adjacent to the caldera (Lehman *et al.*, 1982; R. B. Smith *et al.*, 1982; see also Chapter 7 of this volume). The Yellowstone gravity field shows a large negative anomaly of 60 mgal over the caldera and its immediate surroundings (Blank and Gettings, 1974). The pattern of this anomaly is such that it cannot be due solely to low-density fill in the caldera; a significant part of the anomaly could be related to a solidified, or even a partially molten, body of magma at shallow levels in the crust. Aeromagnetic data have been modeled to suggest depths to the Curie temperature of about 10 km beneath the Yellowstone Plateau in general and as shallow as 6 km beneath some areas of the caldera (Bhattacharyya and Leu, 1975). Magnetotelluric surveys suggest that Yellowstone is underlain at relatively shallow depths by materials of high electrical conductivity (Stanley *et al.*, 1977).

Taken together, numerous geophysical data delineate an anomalous crustal and upper-mantle structure beneath Yellowstone (see Chapter 7). Low seismic velocities, high attenuations, low densities, high electrical conductivity, a shallow depth to the Curie temperature, and a large area of surface uplift all indicate an upper-crustal body at high temperature. These anomalous properties clearly reflect Yellowstone's magmatic plumbing system; although none of these parameters uniquely proves that liquid magma still exists in a shallow chamber, each parameter is consistant with such an inference, based on geologic data.

The voluminous rhyolitic lavas that have filled the Yellowstone caldera during the past 150,000 yr—locally to overflowing—are highly differentiated in composition, and have chemical and isotopic characteristics suggesting that an underlying magma body is nearly as far evolved as the magmas that produced the three major ash-flow sheets of the volcanic field (Figure 6.6). The geologic, petrologic, and geophysical picture of the Yellowstone Plateau volcanic field strongly suggests that the field retains the capacity to produce future eruptive activity. The scale of past volcanic activity, the lengths of previous volcanic repose periods, hydrothermal and geophysical evidence for a vigorous heat source at shallow crustal levels, a geologic record of renewed localized uplift and voluminous rhyolitic volcanism in the Yellowstone caldera during the past 150,000 yr, rapid present uplift, and swarm seismicity are all consistent with that suggestion.

It is unclear whether the current state of the Yellowstone system should properly be thought of as a declining stage of the third volcanic cycle or as an early stage of a fourth cycle, and the data available do not resolve that issue. It may be that relatively small eruptions of rhyolitic or basaltic lavas could occur around the margins of the Yellowstone Plateau or that rhyolite could erupt within the Yellowstone caldera in a protracted overall decline of activity. It is equally possible, however, that the volcanic system is heading toward another major ash-flow eruption. It remains unclear whether the Yellowstone Plateau volcanic field has entered a state that could be triggered into a fourth major pyroclastic eruption. Despite the uncertainties, the present state of the system may be comparable to the renewal of activity that occurred within the continuous magmatic system between the first and second volcanic cycles.

ACKNOWLEDGMENTS

This paper reports work done over many years with several colleagues, and I am indebted to all of them. H. R. Blank, Jr., collaborated with me in much of the geologic mapping upon which the study rests; Wes Hildreth is my colleague in studies of the petrology and geochemistry of Yellowstone Plateau volcanism; and R. B. Smith has shared abundant geophysical data and interpretations. W. Hildreth, J. Donnelley-Nolan and F. R. Boyd gave helpful reviews of the manuscript.

REFERENCES

Bacon, C. R. (1983). Eruptive history of Mount Mazama, U.S.A., *J. Volcanol. Geotherm. Res. 6*.
Bailey, R. A., G. B. Dalrymple, and M. A. Lanphere (1976). Volcanism, structure, and geochronology of Long Valley caldera, Mono County, California, *J. Geophys. Res. 81*, 725-744.
Bhattacharyya, B. K., and L. K. Leu (1975). Analysis of magnetic anomalies over Yellowstone National Park: Mapping of Curie point isothermal surface for geothermal reconnaissance, *J. Geophys. Res. 80*, 4461-4465.
Blank, H. R., Jr., and M. E. Gettings (1974). Complete Bouguer gravity map of the Yellowstone-Island Park region, *U.S. Geol. Surv. Open-File Map 74-22*.
Bowen, N. L. (1928). *The Evolution of the Igneous Rocks*, Princeton U. Press, Princeton, N.J., 334 pp.
Boyd, F. R. (1961). Welded tuffs and flows in the rhyolite plateau of Yellowstone National Park, Wyoming, *Geol. Soc. Am. Bull. 72*, 387-426.
Braile, L. W., R. B. Smith, J. Ansorge, M. R. Baker, M. A. Sparlin, C. Prodehl, M. M. Schilly, J. H. Healy, S. Mueller, and K. H. Olsen (1982). The Yellowstone-Snake River Plain seismic profiling experiment: Crustal structure of the eastern Snake River Plain, *J. Geophys. Res. 87*, 2597-2610.
Christiansen, R. L. (1979). Cooling units and composite sheets in relation to caldera structure, in *Ash-flow Tuffs*, C. E. Chapin and W. E. Elston, eds., Geol. Soc. Am. Spec. Pap. 180, pp. 29-42.
Christiansen, R. L., and H. R. Blank, Jr. (1972). *Volcanic Stratigraphy of the Quaternary Rhyolite Plateau in Yellowstone National Park*, U.S. Geol. Surv. Prof. Pap. 729-B, 18 pp.
Christiansen, R. L., and E. H. McKee (1978). Late Cenozoic volcanic and tectonic evolution of the Great Basin and Columbia Intermontane regions, in *Cenozoic Tectonics and Regional Geophysics of the Western Cordillera*, R. B. Smith and G. P. Eaton, eds., Geol. Soc. Am. Mem. 152, pp. 283-311.
Doe, B. R., W. P. Leeman, R. L. Christiansen, and C. E. Hedge (1982). Lead and strontium isotopes and related trace elements as genetic tracers in the Upper Cenozoic rhyolite-basalt association of the Yellowstone Plateau volcanic field, U.S.A., *J. Geophys. Res. 87*, 4785-4806.
Eaton, G. P., R. L. Christiansen, H. M. Iyer, A. M. Pitt, H. R. Blank, Jr., I. Zietz, D. R. Mabey, and M. E. Gettings (1975). Magma beneath Yellowstone National Park, *Science 188*, 787-796.
Fournier, R. O., D. E. White, and A. H. Truesdell (1976). Convective heat flow in Yellowstone National Park, in *Second United Nations Symposium on Development and Use of Geothermal Resources*, U.S. Government Printing Office, Washington, D.C., pp. 731-739.
Hamilton, W. (1965). *Geology and Petrogenesis of the Island Park Caldera of Rhyolite and Basalt, Eastern Idaho*, U.S. Geol. Surv. Prof. Pap. 504-C, 37 pp.
Hildreth, W. (1979). The Bishop Tuff: Evidence for the origin of compositional zonation in silicic magma chambers, in *Ash-flow Tuffs*, C. E. Chapin and W. E. Elston, eds., Geol. Soc. Am. Spec. Pap. 180, pp. 43-75.
Hildreth, W. (1981). Gradients in silicic magma chambers: Implications for lithospheric magmatism, *J. Geophys. Res. 86*, 10153-10192.
Holmes, A. (1931). The problem of the association of acid and basic rocks in central complex, *Geol. Mag. 68*, 241-255.
Iyer, H. M., J. R. Evans, G. Zandt, R. M. Stewart, J. M. Coakley, and J. N. Roloff (1981). A deep low-velocity body under the Yellowstone caldera, Wyoming: Delineation using teleseismic *P*-wave residuals and tectonic interpretation, *Geol. Soc. Am. Bull. 92*, Part II: 1471-1646.
Izett, G. A., and R. E. Wilcox (1981). Map showing localities and inferred distributions of the Huckleberry Ridge, Mesa Falls, and Lava Creek ash beds (Pearlette family ash beds) of Pliocene and Pleistocene age in the western United States and southern Canada, *U.S. Geol. Surv. Misc. Invest. Map I-1325*.

Kirkham, V. R. D. (1931). Snake River downwarp, *J. Geol. 39*, 456-483.

Lehman, J. A., R. B. Smith, M. M. Schilly, and L. W. Braile (1982). Upper crustal structure of Yellowstone from seismic and gravity observations, *J. Geophys. Res. 87*, 2713-2730.

Lipman, P. W. (1975). *Evolution of the Platoro Caldera Complex and Related Volcanic Rocks Southeastern San Juan Mountains, Colorado*, U.S. Geol. Surv. Prof. Pap. 852, 128 pp.

Lipman, P. W., B. R. Doe, C. E. Hedge, and T. W. Steven (1978). Petrologic evolution of the San Juan volcanic field, southwestern Colorado: Pb and Sr isotope evidence, *Geol. Soc. Am. Bull. 89*, 59-82.

Myers, W. B., and W. Hamilton (1964). Deformation accompanying the Hebgen Lake, Montana earthquake of August 17, 1959, in *U.S. Geol. Surv. Prof. Pap. 435*, pp. 55-98.

Pelton, J. R., and R. B. Smith (1979). Recent crustal uplift in Yellowstone National Park, *Science 206*, 1179-1182.

Smith, R. B. (1978). Seismicity, crustal structure, and intraplate tectonics of the interior of the western Cordillera, in *Cenozoic Tectonics and Regional Geophysics of the Western Cordillera*, R. B. Smith and G. P. Eaton, eds., Geol. Soc. Am. Mem. 152, pp. 111-144.

Smith, R. B., and R. L. Christiansen (1980). Yellowstone Park as a window on the Earth's interior, *Sci. Am. 242, No. 2*, 84-95.

Smith, R. B., and M. L. Sbar (1974). Contemporary tectonics and seismicity of the western United States, with emphasis on the Intermountain Seismic Belt, *Geol. Soc. Am. Bull. 85*, 1205-1218.

Smith, R. B., M. M. Schilly, L. W. Braile, H. A. Lehman, J. Ansorge, M. R. Baker, C. Prodehl, J. Healy, S. Mueller, and R. W. Greenfelder (1982). The 1978 Yellowstone-Eastern Snake River Plain seismic profiling experiment: Crustal structure of the Yellowstone region and experiment design, *J. Geophys. Res. 87*, 2583-2596.

Smith, R. L. (1979). Ash-flow magmatism, in *Ash-flow Tuffs*, C. E. Chapin and W. E. Elston, eds., Geol. Soc. Am. Spec. Pap. 180, pp. 5-27.

Smith, R. L., and R. A. Bailey (1968). Resurgent cauldrons, in *Studies in Volcanology*, R. R. Coats *et al.*, eds., Geol. Soc. Am. Mem. 116, pp. 613-662.

Smith, R. L., and H. R. Shaw (1975). Igneous-related geothermal systems, in *Assessment of Geothermal Resources in the United States— 1975*, D. E. White and D. L. Williams, ed., U.S. Geol. Surv. Circ. 726, pp. 58-83.

Smith, R. L., R. A. Bailey, and C. S. Ross (1970). Geologic map of the Jemez Mountains, New Mexico, *U.S. Geol. Surv. Misc. Invest. Map I-571*.

Stanley, W. D., J. E. Boehl, F. X. Bostick, and H. W. Smith (1977). Geothermal significance of magnetotelluric sounding in the eastern Snake River Plain-Yellowstone region, *J. Geophys. Res. 82*, 2501-2514.

Williams, H. (1942). Geology of Crater Lake National Park, Oregon, with a reconnaissance of the Cascade Range south to Mount Shasta, *Carnegie Inst. Washington Publ. 540*, 162 pp.

Yoder, H. S., Jr. (1973). Contemporaneous basaltic and rhyolitic magmas, *Am. Mineral. 58*, 153-171.

Crustal Structure and Evolution of an Explosive Silicic Volcanic System at Yellowstone National Park

7

ROBERT B. SMITH
University of Utah

LAWRENCE W. BRAILE
Purdue University

ABSTRACT

Large volumes of Quaternary silicic volcanics (~6700 km³), associated explosive caldera-forming eruptions, and high heat flow (in excess of 1800 mW/m²) infer the presence of silicic magmas within the crust and upper mantle beneath the Yellowstone Plateau. Seismic refraction-reflection data, analyses of earthquake hypocenters, and seismic attenuation have revealed a laterally inhomogeneous upper crust with low P-wave velocities but a more seismically homogeneous lower crust. The upper crust beneath the Yellowstone caldera is characterized by P-wave velocities of 5.7 km/sec and 4.0 km/sec—values that are anomalously low compared with that of the surrounding thermally undisturbed crystalline basement of 6.0 km/sec. The 5.7 km/sec body generally underlies the Yellowstone caldera (35 km × 65 km) and concides with a regional −60 mgal gravity low, suggesting concomitant low density and low velocity. The 5.7 km/sec low-velocity body is interpreted to represent a hot but relatively solid body approximately 8 to 10 km thick that was probably the reservoir for the silicic magmas. A ~4.0 km/sec low-velocity body located beneath the northeast boundary of the caldera coincides with a local −20 mgal gravity low and has a tenfold increase in seismic attenuation— properties that can be interpreted to result from a steam-water-dominated system to a body of 10 to 50 percent silicic partial melt. The P-wave velocity of the upper 100 to 250 km of the mantle beneath the Yellowstone region, analyzed from teleseismic arrivals, is reduced by ~5 percent, suggesting the presence of a basaltic partial melt that is probably the source of heat that drives the Yellowstone hydrothermal system. In comparison, the lower crust of the Yellowstone region appears seismically homogeneous to the horizontally propagating refracted rays and similar to that of the thermally undisturbed lower crust of the surrounding Rocky Mountains. This suggests that the seismic properties of the lower crust were relatively unaffected by the ascension of the parental basaltic magmas that are hypothesized to have intruded and partially melted the crust producing the voluminous rhyolite and ash-flow tuffs of the Yellowstone Quaternary volcanic system. Maximum focal depths of earthquakes in Yellowstone systematically shallow from ~20 km outside the caldera to ~5 km beneath the caldera, suggesting the influence of high-temperature, abnormal pore pressure, and compositional changes that restrict brittle failure to the upper crust. Orthometrically corrected reobservations of level lines across the Yellowstone caldera show an area of crustal uplift, up to ~15 mm/yr, that generally coincides with the outline of the 5.7 km/sec low-velocity layer. These data are consistent with a model in which an upper-crustal low-velocity/low-density layer, 75 km × 25 km, appears to be plastically deforming. Taken together with the geologic data this crustal model is interpreted to reflect the structure and properties of a thermally deforming Archean crust and the initial stages of the bimodal rhyolitic/basaltic volcanism of the Yellowstone-Snake River Plain volcano-tectonic system. While the interpretations are not unique, the youthfulness and volume of Quaternary volcanism, the high heat flow, the high rates of contemporary uplift, and the upper-crustal low-velocity layers infer the presence of hot crustal material and possible partial melts that underlie the Yellowstone Plateau. These properties cannot yet be evaluated to indicate temporal variations in volcanism, but the geologic record and the new geophysical models suggest future volcanic activity in the Yellowstone Plateau.

96

INTRODUCTION

The heat from the numerous geysers, hot springs, and fumaroles of Yellowstone National Park represents only a small fraction of the total thermal energy of this major silicic volcanic system. Voluminous Quaternary volcanism with rhyolite flows as young as 50,000 yr, 3 explosive caldera-forming eruptions, and very high heat flow (in excess of 1800 mW/m²) infer that large magma bodies have existed within the upper crust beneath the Yellowstone Plateau, which now provides the heat and energy that drives the hydrothermal and tectonic systems of Yellowstone (see Chapter 6 in this volume for the Quaternary volcanic history of Yellowstone). However, the longevity and large volumes of the Quaternary silicic volcanics suggest early and continuous involvement of basaltic magmas from lithospheric sources. An evaluation of earthquakes, crustal structure, and contemporary deformation of the Yellowstone Plateau when interpreted within the framework of Quaternary history can provide insight into the dynamics of a thermally evolving continental crust. Interpretations of these data can be used to assess the locations and physical states of possible magma bodies—information required to identify precursors that could precede eruptions.

The Quaternary history of the Yellowstone Plateau is characterized by rather continuous eruptions of rhyolites and ash-flow tuffs that accumulated over 6500 km³ but was marked by 3 explosive eruptions and subsequent caldera-forming collapses of the crust 2.0, 1.3, and 0.6 million years (m.y.) ago (Christiansen, in press; see also Chapter 6 in this volume). While the young volcanism and high heat flow reflect a major thermal episode, they are but the primary phase of a temporal-spatial propagating silicic system—the Yellowstone-Snake River Plain (Y-SRP) volcano-tectonic province that is now hydrothermally active at Yellowstone. It is generally agreed that the Y-SRP province, which is characterized by a bimodal suite of rhyolites and basalts, is related to the systematic progression of silicic volcanic centers that migrated northeasterly at about 4 cm/yr beginning about 15 m.y. ago from southwestern Idaho to the present location of the Quaternary volcanism at the Yellowstone Plateau (Christiansen and Blank, 1972; Armstrong *et al.*, 1975; Christiansen and McKee, 1978). Several hypotheses have been proposed to explain the spatial-temporal progression of the silicic volcanism along the axis of the Y-SRP, including (1) the passage of the North American Plate across a convective mantle plume or a hot spot related to heat-generating radioactive material in the upper mantle, (2) the propagation of a lithospheric fracture, (3) the spatial progression of thermal instabilities produced by friction at the base of the lithosphere, and (4) temporal progressive volcanism along a transform fault.

An understanding of the dynamics of the Yellowstone volcanic system is crucial to an evaluation of the thermal modification of a continental Archean crust, particularly to an understanding of the kinematics of lithospheric deformation accompanying crustal differentiation from intrusion and partial melting. Models derived from such properties as size, depth, and physical state of magma-generating bodies may be inferred from seismic velocities, seismic-wave attenuation, and density. These properties, when interpreted within the geologic frame

work, can be useful in predicting locations and future volcanic eruptions.

The presence of shallow crustal magmas in other continental settings has been inferred on the basis of seismic *P*-wave delays, crustal structure, and the attenuation of seismic waves; for example, beneath the Rio Grand Rift, New Mexico (Sanford *et al.*, 1977); beneath the Long Valley caldera, California (Hill, 1976); and beneath the Coso geothermal area, California (Reasenberg *et al.*, 1980). There was a lack of distinct seismic evidence for a shallow magma chamber beneath the Mount St. Helens volcano prior to the 1980 eruption (S. Malone, University of Washington, personal communication, 1982), which is suggestive of a different mechanism for upper-crustal magma emplacement.

A relative perspective of Quaternary volcanism at Yellowstone is shown in Figure 7.1, where volumes of several well-known eruptions (Williams and McBirney, 1979; Decker and Decker, 1981) are shown for comparison. The 1980 Mount St. Helens eruption, which dominates our current concept of explosive eruptions, is on the order of a ten thousandth of the volume of the large explosive Yellowstone eruptions. And while the distribution of airborne ash from Mount St. Helens produced millions of dollars of damage, airborne ash flows from the Quaternary Yellowstone eruptions were measured as far away as Saskatchewan and Mississippi. A comparison with other large explosive eruptions, such as Katmai in 1920, Krakatoa in 1882, Tambora in 1815, and Mazama (about 7000 yr B.P.) dramatizes the fact that these catastrophic eruptions were hundreds of times smaller than the Quaternary Yellowstone explosive eruptions and emphasizes the importance of understanding this major silicic volcanic system.

The Quaternary tectonic framework of the Yellowstone Plateau is portrayed in Figure 7.2, which shows the distribution of volcanic flows, collapse calderas (including the most recent 600,000-yr-old caldera), the Sour Creek (northeast) and the Mallard Lake (southwest) resurgent domes, and Cenozoic faulting related both to tumescence during thermal expansion and collapse and contraction of the crust following the explosive eruptions. The youthfulness and large volumes of Quaternary volcanics at Yellowstone imply that intermediate- to upper-crustal heat sources may still be present.

Geophysical experiments designed to characterize the crustal properties of the Yellowstone system were conducted during the last decade. Earthquake monitoring on both a regional and local scale has been conducted by the University of Utah and the U.S. Geological Survey (USGS) (Eaton *et al.*, 1975; Trimble and Smith, 1975; Smith *et al.*, 1977; Pitt, 1981). Aeromagnetic surveys and gravity compilations have been used to infer the distribution of shallow silicic melts and caldera infill (Blank and Gettings, 1974; Eaton *et al.*, 1975; Smith *et al.*, 1974, 1977). *P*-wave delays of up to 1.8 sec from teleseismic arrivals were interpreted by Iyer *et al.* (1981) to represent up to a 20 percent velocity reduction in the crust and a 5 percent reduction in the upper mantle to depths of 250 km. These delays were interpreted to reflect zones of possible melt, deep in the lithosphere beneath the Yellowstone caldera. While conductive heat-flow measurements are difficult to obtain in Yellowstone, because of the widespread shallow-water circulation, signs of

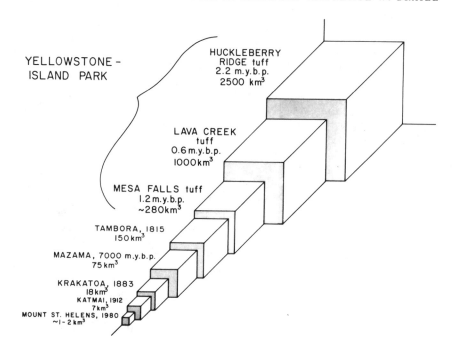

FIGURE 7.1   Relative volumes of some well-known volcanic eruptions (from Williams and McBirney, 1979; Decker and Decker, 1981).

conductive heat flow from marine-type probes in Yellowstone Lake (Morgan *et al.*, 1977) and the Cl geochemical budget (Fournier *et al.*, 1976) show that the average heat flux through the Yellowstone caldera is approximately 1800 mW/m²—about 60 times the world average. These data and inferences made from the geologic record suggest that crustal sources of heat are driving the Yellowstone hydrothermal systems and that they provide much of the energy for the contemporary tectonic deformation.

In 1978 a major seismic refraction/reflection experiment was conducted in the eastern Snake River Plain-Yellowstone region as a means of evaluating the lithospheric structure of this continental tectono-volcanic province (Braile *et al.*, 1982; Smith *et al.*, 1982). In this project up to 225 vertical-component seismographs were located on profiles and in two-dimensional arrays throughout the region. Within Yellowstone, five explosions external to the caldera were recorded across the Yellowstone Plateau. Interpretations of *P*-wave refraction and wide-angle reflections from these profiles yielded average crustal velocity structures (Schilly *et al.*, 1982). Three-dimensional velocity distributions of the upper crust were determined from delay-time analyses (Lehman *et al.*, 1982), from simultaneous inversion of travel times from local earthquake and refraction-data measurements for the upper-crustal structure (Benz and Smith, 1981), and from University of Utah unpublished refraction data of the southwest Yellowstone Plateau recorded in 1980. Information on the attenuation of *P*-waves was determined from inversion of spectral amplitude ratios obtained from the refraction data (Clawson and Smith, 1981). Together these studies provided a seismic image of the crustal velocity structure, with a highest resolution of a structure to a few kilometers—near the scale of surface tectonic features. Longer-range refraction/wide-angle reflection profiles (Smith *et al.*, 1982) and *S*-wave

delay and surface-wave dispersion investigations (Daniel and Boore, 1982) were used to evaluate the averaged *P*- and *S*-wave structure of the crust and upper mantle.

Interpretations of the new seismic refraction-reflection data served as the basis for this discussion and together with other geophysical and geologic data provide an evaluation of the locations of possible bodies of hot rock, zones of partial melt, and magma in the crust beneath the Yellowstone Plateau. Seismic models are synthesized with new hypocenter data, with evaluations of contemporary crustal deformation from level-line reobservations (Pelton and Smith, 1979, 1982), and the Quaternary geologic record (Christiansen, Chapter 6 of this volume). Inferences of physical state, including composition and temperature, are then used to evaluate the sources of the Quaternary volcanism and make speculations about future volcanism at Yellowstone.

GEOPHYSICAL AND GEODETIC MEASUREMENTS

Earthquake Data

Yellowstone National Park and the adjacent Hebgen Lake region to the west have been the most seismically active areas in the Rocky Mountains in historic time, with 1 earthquake of magnitude 7.1 (the 1959 Hebgen Lake earthquake) and 7 magnitude-6+ events (see Figure 7.3). Earthquakes have occurred primarily in the western Yellowstone Park area and along the adjacent Hebgen Lake fault zone. The seismicity in the Yellowstone region is part of the larger Intermountain Seismic Belt, which marks the location where the north-south-trending earthquake zone to the south turns toward the northwest. A

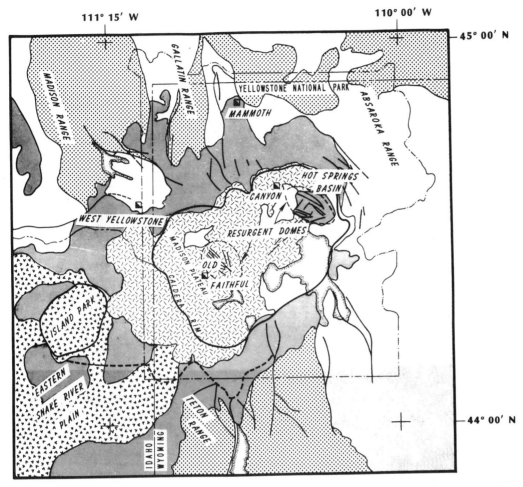

FIGURE 7.2 Generalized Late Cenozoic tectonic map of Yellowstone, showing distribution of main volcanic flows, collapse calderas, resurgent domes, and Cenozoic faulting (modified from R. L. Christiansen, U.S. Geological Survey, personal communication, 1978). Heavy line corresponds to boundary of the 0.6-m.y.-old caldera.

secondary east-west zone of seismicity, including that of the Hebgen Lake fault, intersects this regional trend at Yellowstone. This intersection of seismicity zones suggests a location of intraplate stress reorientation.

Crustal earthquakes are thought to result from stress drops across shear fractures in a domain of brittle failure. Thus, their occurrence and distribution can be used to infer the relative state of stress. Maximum focal depths may mark a distinction between brittle and nonbrittle mechanical behavior. To evaluate the contemporary seismicity of the Yellowstone region, a new compilation of hypocenter data was developed from data from the University of Utah and the USGS (Smith *et al.*, 1977; Smith and Christiansen, 1980; Pitt, 1981). The new data were

selected on the basis of an average root-mean-square location error equivalent to less than ±0.2 sec. Most epicenters were located with 6 or more stations and assumed flat-layered velocity models. More than 3000 epicenters were compiled from over 50 seismograph stations that were located throughout the Yellowstone region for the period 1971 to 1979. The earthquake map (Figure 7.3) demonstrates several important characteristics of the Yellowstone seismicity: (1) The seismicity south of the east-west-trending Hebgen Lake fault zone extends eastward toward the Yellowstone caldera, then turns southeastward along zones that continue 2 to 4 km into the caldera. (2) Diffuse seismicity occurs throughout the caldera and is characterized by numerous swarms but decreases in frequency of occurrence.

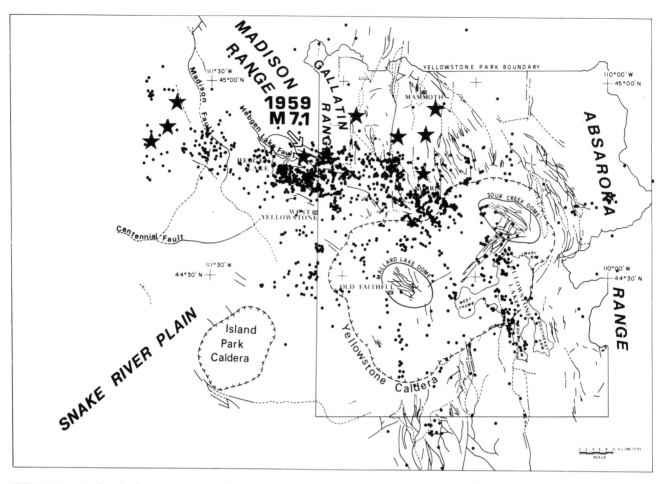

FIGURE 7.3   Earthquake distribution of the Yellowstone-Hebgen Lake region (1971-1979). Epicenters are those accurately determined from data of the University of Utah and the U.S. Geological Survey. Faults are taken from a compilation by R. L. Christiansen (U.S. Geological Survey, personal communication, 1978). Calderas are shown corresponding to their respective ages: I, 2.0 m.y.; II, 1.3 m.y.; and III, 0.6 m.y. Large stars correspond to epicenters of magnitude 6 to 7.1 historical earthquakes.

Scattered events were distributed throughout the southwest caldera and along north-south and east-west zones that extend radial to the Sour Creek resurgent dome. (3) Epicenters near Yellowstone Lake occurred on north-south trends, and the regional seismicity extends southward beyond the caldera boundary along a zone probably related to the Teton fault system.

The state of stress cannot be determined directly from the earthquake observations; however, directions of the principal components of the stress field can be inferred from the directions of *P*- and *T*-axes of fault-plane solutions (Smith, 1978; Smith and Lindh, 1978; Weaver *et al.*, 1979). These data suggest that the regional stress distribution of the Yellowstone system is dominated by north-south extension in the Hebgen Lake-Norris Geyser Basin area, but the direction of maximum extension changes to northeasterly in the Yellowstone caldera and eventually rotates to east-west extension in the southern caldera.

While the earthquake data collected so far are not sufficient to evaluate the stress around possible magma bodies, the directions of the minimum compression (*T*-axes) and maximum

compression (*P*-axes) are consistent with a concept in which the Yellowstone Plateau is being uplifted, producing horizontal minimum compressive stress and primarily vertical components of maximum compressive stress associated with the gravitational loading. The stress field is also consistent with that inferred for the circular ring fractures that mark the boundary of the Yellowstone caldera as normal faults, downdropped toward the center of the caldera.

Gravity Field and Teleseismic P-Wave Delays

Excess heat and advective magma transport from deep lithospheric sources can generate zones of low density and low-seismic velocities, parameters that produce measurable gravity anomalies and delays in the travel times of seismic waves. The Earth's gravitational field in the Yellowstone region is shown in Figure 7.4, which is a complete Bouguer gravity map contoured from data compiled by Blank and Gettings (1974). A striking feature of the gravity data map is the negative (−60 mgal) regional anomaly that closely coincides with the mapped

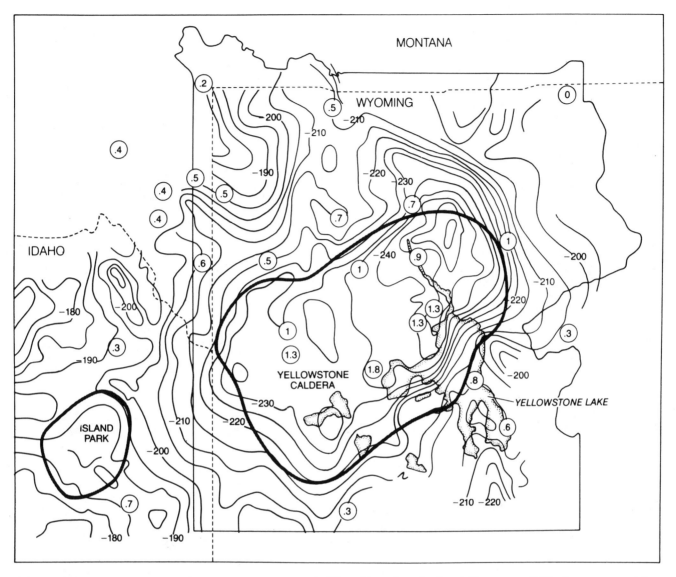

FIGURE 7.4 Bouguer gravity anomaly map at a contour interval of 5 mgal (map from Smith and Christiansen, 1980); data recontoured from Blank and Gettings, 1974). Open circles correspond to locations of seismograph stations that recorded teleseismic *P*-wave delays. Average delays are shown within the circles in seconds (from Iyer *et al.*, 1981). *P*-wave delays were normalized to the seismograph station at the northeast side of Yellowstone National Park.

caldera. This large negative anomaly corresponds to a mass deficiency and may reflect shallow low-density material in the crust or larger low-density bodies that extend into the upper mantle. Modeling of the gravity data in conjunction with a velocity model from the seismic refraction-reflection data will be discussed later.

Of particular importance is the delineation of an additional gravity low of − 20 mgal at the northeast side of the Yellowstone caldera, which corresponds in location in part with the Sour Creek resurgent dome but extends ~20 km north beyond the caldera. As will be shown later, this anomaly coincides with an upper-crustal low-velocity body.

Superimposed on the gravity map are the locations of the regional seismograph stations that were used to determine teleseismic *P*-wave delays by Iyer *et al.* (1981). The station locations are shown by large circles that include the *P*-wave delays (in seconds) relative to a uniform-velocity Earth model. These data clearly show that the Yellowstone Plateau is underlain by material of relatively low seismic velocities necessary to produce *P*-wave seismic delays of up to 1.8 sec. Iyer *et al.* (1981) interpret these delays to represent velocity decreases averaging 10 to 20 percent throughout the upper crust and 5 to 10 percent in the lower crust and upper mantle to depths as great as 250 km.

In a joint inversion of the Yellowstone gravity data and the teleseismic *P*-wave delays, density and velocity decreases were modeled by Evoy (1977) and demonstrated that the regional gravity low and teleseismic *P*-wave delays could be produced by a concomitant low-density, low-velocity body that corresponds in lateral shape to the outline of the caldera and extends to depths of 100 km. These combined interpretations suggest that the Yellowstone Plateau is underlain by anomalous low-density, low-velocity material that may extend into the asthenosphere and reflect deep magmatic sources. Because of the nonuniqueness of the potential field interpretations, the gravity anomalies could also be produced by shallower density anomalies—a topic to be discussed further.

Contemporary Crustal Deformation

If the Yellowstone volcano-tectonic system is tectonically active, measurements of surface deformation could reveal the source of the causative body. To examine this property, first- and second-order level-line benchmarks were reobserved after a 55-yr span to calculate relative deformation rates (Pelton and Smith, 1979, 1982). Figure 7.5 shows a map of orthometric height variations determined from the releveling data portrayed in units of vertical velocity (millimeters/year). These data

were determined from level-line reobservations (Pelton and Smith, 1979, 1982) that were converted to displacement rates by fixing a benchmark at the east side of Yellowstone Park as a relative zero-base outside the effects of the thermal and seismic deformation. The uplift data were corrected for systematic and random errors and were evaluated for errors resulting from correlation with topography and poor rod calibrations (Pelton and Smith, 1982). The results (Figure 7.5) demonstrate anomalously high uplift rates over the Yellowstone Plateau with uplift contours that generally parallel the outline of the Yellowstone caldera. Uplift rates averaged greater than about 5 mm/yr within the outer edge of the ring-fracture zone and are greatest, up to 15 mm/yr, in the northeast caldera area adjacent to the Sour Creek resurgent dome. An increase from ~7 mm/yr to ~15 mm/yr northeast across the caldera toward the Sour Creek dome suggests an asymmetry or tilt and may reflect mass transport in the northeast caldera area. The anomalous crustal uplift rates measured in Yellowstone are similar to rates measured in other active volcanic areas of the world, such as Hawaii and Iceland.

Over the 55-yr monitoring period a symmetrical uplift of up to 700 mm occurred, coincident with the Yellowstone caldera,

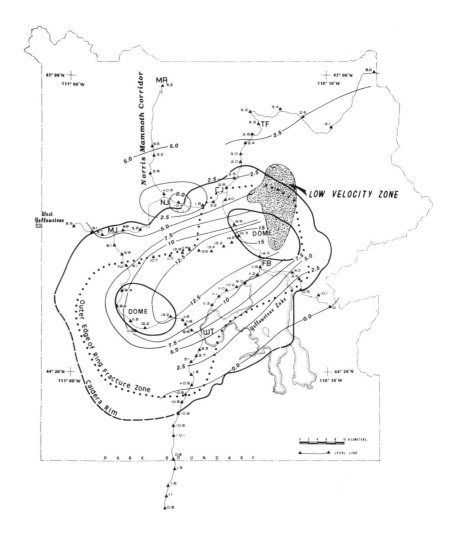

FIGURE 7.5   Contour map of relative uplift rates over the Yellowstone Plateau. Contours are in millimeters/year and were computed from the data of Smith and Pelton (1979, 1982). Benchmark locations shown by triangles; MA, Mammoth; NJ, Norris Junction; CJ, Canyon Junction; FB, Fishing Bridge; WT, West Thumb; TF, Tower Falls.

which raises the question of the longevity of this rapid elevation change. If these high uplift rates continue, within a few million years the inner caldera would be elevated on the order of kilometers, obviously an unrealistic value. We consider the current uplift to be ephemeral, as demonstrated by the occurrence of lake terraces along the northwest shores of Yellowstone Lake, which show little evidence of large-scale warping (Pelton and Smith, 1982) and suggest that the deformation in the region is fairly recent. The size and relatively large vertical strain rates that characterize the Yellowstone uplift likely correspond to a causative body in the upper crust that would have a lower viscosity than the cooler, brittle crustal material. Thus, the upper crust beneath the caldera appears to behave plastically or visoelastically to long-term strain.

UPPER-CRUSTAL SEISMIC VELOCITY STRUCTURE

Perhaps the most definitive characteristic of the upper crust beneath the Yellowstone Plateau is the low *P*-wave seismic velocity corresponding to the crystalline layer. Information on this layer has been interpreted from delay-time analyses (Lehman *et al.*, 1982) of the critically refracted branch, Pg, that were recorded in the 1978 Y-SRP seismic refraction project. This method utilized travel times from a two-dimensional distribution of ray paths that crossed the Yellowstone caldera at various azimuths between the seismic stations and the sources.

The upper-crustal crystalline layer shows an elongated northeast-trending velocity distribution (see Figure 7.6). The 6.0 km/sec velocity layer external to the caldera is interpreted to represent the thermally undisturbed crystalline basement that corresponds in composition to rocks like the Precambrian granitic gneiss of the nearby Beartooth Mountains. The boundary of the large 5.7 km/sec body (marked by a 6 percent decrease compared with the thermally undisturbed basement) closely coincides with the outline of the southern Yellowstone caldera but extends to the north approximately 15 km beyond the boundary. Two additional low-velocity zones of 30 percent decrease to ~4.0 km/sec were identified in the southwestern and the northeastern caldera areas.

The crystalline velocity layering is a diagnostic parameter of the upper crust beneath the Yellowstone caldera. The outline of the 5.7 km/sec layer closely corresponds to the large, regional

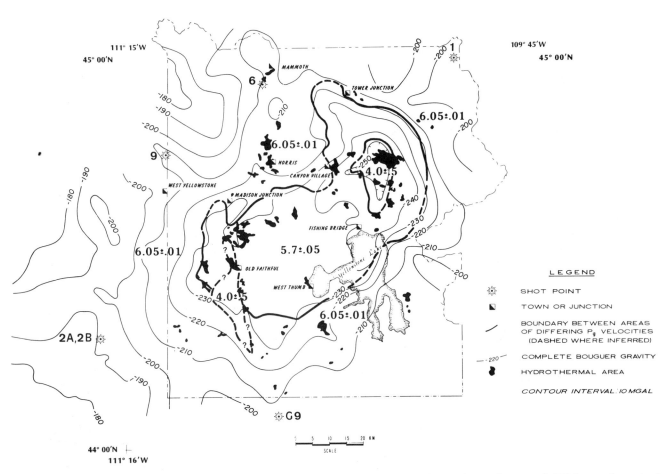

FIGURE 7.6   Areal extent of crystalline upper-crustal layer (2- to 10-km depth). *P*-wave velocities from Lehman *et al.* (1982) were determined from delay-time analyses. Velocities are in kilometers/second.

gravity anomaly of −60 mgal (Figure 7.4) and thus suggests a concomitant density decrease corresponding to the low-velocity body.

The ~4.0 km/sec low-velocity layer at the northeast side of the caldera is by itself unusual, but an examination of the Bouguer gravity map (Figure 7.4) shows that this anomalous body coincides with an additional (−20 mgal) local gravity low, suggesting an additional concomitant low-density, low-velocity volume, whereas the low-velocity body to the southwest shows no corresponding gravity anomaly.

To compare the causative bodies that produced the density and velocity anomalies, three-dimensional gravity modeling was done (Lehman *et al.*, 1982) with the constraints that the density layers corresponded to the general thickness and outline of the velocity layers determined from the seismic refraction data. The seismic data showed that the 5.7 km/sec body did not appear to extend to depths beyond about 10 km.

For the northeast ~4.0 km/sec low-velocity body, the assumption of a normal velocity-density decrease was not sufficient to model both parameters, and an additional constraint was required, i.e., that the causative body, in order to have a concurrent 30 percent velocity reduction and a 10 percent density reduction, required a fluid constituent or an unlikely compositional variation. Theoretical models of seismic-wave propagation in fluid-filled material (O'Connell and Budiansky, 1977; Mavko, 1980) and experimental measurements of *P*-wave velocities in melts (Murase and Suzuki, 1966; Murase and McBirney, 1973) suggest a decrease in the velocity due to partial melting of the order reported here. The gravity and seismic modeling of the anomalous northeast low-velocity body provide an additional constraint—the density/velocity ratio requires a major percentage of fluid that may range from 10 to 50 percent of water or silicic partial melt. While this interpretation is not unique, it would be unlikely that low velocities from a highly porous gas- or steam-filled system would produce the relatively small density change. A large compositional variation to low-density, low-velocity material, such as high-porosity sediments, could partly account for the velocity and density anomalies, but the size and depth extent of the anomalous body make that interpretation also unlikely. As a working hypothesis we consider that the northeast caldera low-velocity, low-density body may correspond to an upper-crustal fluid-filled (steam or partial melt) silicic body.

The southwestern low-velocity body, on the other hand, does not have an accompanying gravity anomaly and was interpreted by Lehman *et al.* (1982) to be more representative of a fractured, fluid-saturated zone producing attenuation, velocity reduction, or both. It may represent a northward extension of a Cenozoic fault system beneath the Yellowstone Plateau.

To examine the anomalous physical parameters of the Yellowstone system, Figure 7.7 shows a northwest-southeast upper-crustal cross section through the Yellowstone-Hebgen Lake region that portrays (1) average *P*-wave delays from teleseismic arrivals (Iyer *et al.*, 1981), (2) geodetically measured crustal deformation (Pelton and Smith, 1979), (3) depths of the most accurately determined hypocenters, and (4) averaged *P*-wave velocities from the combined results of the Y-SRP seismic experiment following the interpretations of Lehman *et al.* (1982),

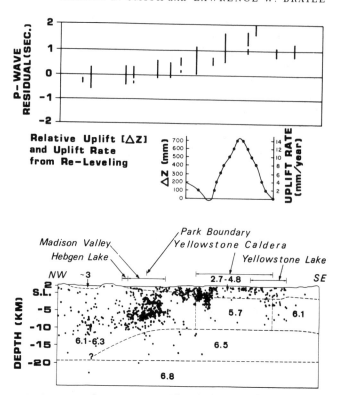

FIGURE 7.7 Generalized northwest-southeast cross section of crustal parameters across the Yellowstone-Hebgen Lake region. Teleseismic *P*-wave delays from Iyer *et al.* (1981); relative uplift and uplift rates from Pelton and Smith (1979); focal depths from new earthquake compilation (Benz and Smith, 1981; Pitt, 1981); seismic velocities from interpretations of Schilly *et al.* (1982), Smith *et al.* (1982), and Lehman *et al.* (in press). Focal depths are shown by black dots.

Schilly *et al.* (1982), and Smith *et al.* (1982). These data show a correspondence for the area of marked increase in *P*-wave teleseismic delays to greater than 1.0 sec with the area of greatest uplift; maxima of both parameters appear centered along the axis of the Yellowstone caldera.

The maximum depths of earthquakes (Figure 7.7) shallow from ~20 km in the Madison Valley west of Yellowstone to 5 km or less beneath the Yellowstone caldera. The lateral extent of deepest foci shallow markedly toward the caldera and appear to decrease in depth to 1 to 2 km near the top of the 5.7 km/sec low-velocity layer. Note that relatively few earthquakes occurred within the 5.7 km/sec layer, with the exception of a cluster at the northwest edge of the caldera that corresponds to aftershocks of the magnitude-6.0 30 June 1975 earthquake (Figure 7.3).

Modeling of plastic-elastic deformation based on strain-rate calculations of quartz-olivine systems led Smith (1981) to conclude that maximum depths of earthquakes in continental crustal environments such as Yellowstone may be controlled by temperature and composition, where cessation of brittle failure is modeled at a critical temperature of 350 ± 50°C at which point the strain release is in the form of plastic deformation. If this model is applicable to the upper crust of Yellowstone, the maximum depth of earthquakes may be used to infer the depth of this boundary isotherm. The shallow focal depths thus appear to be constrained to a shallow layer characterized by brittle failure to depths marked by an isotherm of approximately 350°C (Smith, 1981). This boundary also appears to mark the lateral transition from the higher-velocity, thermally undisturbed 6.0 km/sec crystalline layer to the lower-velocity, 5.7 km/sec layer. The data on this profile do not directly map magma bodies, but they show that the upper 10 km of the crust beneath the Yellowstone caldera are marked by a low *P*-wave velocity and apparently a hot crust that restricts brittle fracture.

In Figure 7.8 a southwest-northeast upper-crustal cross section parallels the axis of the Yellowstone caldera. In this profile rhyolite flows and caldera fill are shown from the determinations of Lehman *et al.* (1982) and Schilly *et al.* (1982). The ~4.0 km/sec body is interpreted to be of granitic composition on the basis of low velocity and low density and correlation with the inferred source for the rhyolite flows. To the northeast the Precambrian granitic basement and possible buried Paleozoics

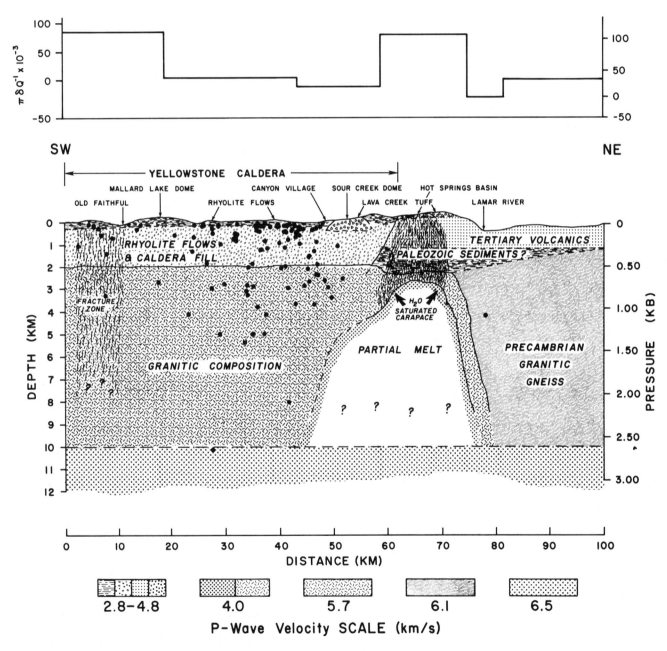

FIGURE 7.8 Generalized southwest-northeast upper-crustal cross section parallel to the axis of the Yellowstone caldera. Focal depths (shown by black dots) were projected from outside the low-velocity body. Velocities correspond to sources in Figure 7.7. Relative *P*-wave seismic attenuations are averaged from data of Clawson and Smith (1981).

sediment are overlain by Tertiary volcanics. Also plotted is the relative *Pg* attenuation that has been determined by the spectral ratio technique (Clawson and Smith, 1981) for the laterally varying velocity structure. This averaged apparent attenuation profile is shown at the top of Figure 7.8 in a block diagram. Note that the apparent-*Q* is high over the northeast low-velocity body and the southwest caldera.

LOCATION AND STATE OF POSSIBLE MAGMA SOURCES

The upper- and intermediate-depth crustal structure of the Yellowstone Plateau is shown in Figure 7.9. Locations of anomalous bodies are based on the inference of shallow granitic volumes from the relationship to surface geology, from the seismic velocity structure, from the high heat flow, and by distribution of hydrothermal features of the Yellowstone Plateau. Also included is an interpretation of the state and composition, based on gravity-velocity modeling. The pervasive 5.7 km/sec layer is interpreted to represent a hot but plastically deforming body of granitic composition, most likely a residuum of the Quaternary Yellowstone rhyolite flows and ash-flow tuffs. The ~4.0 km/sec body near the southwest caldera boundary appears to be related to a water-saturated fracture system,

perhaps a buried extension of the Teton fault zone. However, the ~4.0 km/sec low-velocity body of the northeast caldera boundary is interpreted to correspond to a volume of material that is consistent with a silicic body that could range in composition from a highly porous steam-water system to a 10 to 50 percent partial melt—an ambiguous model that is not resolvable with the existing data but that is considered as a working hypothetical model. It is important to note that the 5.7 km/sec and the 4.0 km/sec layers appear to be bounded at a depth of about 10 km by a 6.5 km/sec layer that has been shown to extend along the length of the Snake River Plain as a pervasive intermediate crustal layer, perhaps the remnants of the undersaturated mafic upper-crustal material.

The sources of magma and crustal evolution accompanying the development of the Y-SRP volcanic-tectonic system have been interpreted by Smith (1981), Braile *et al.* (1982), and Smith *et al.* (1982), based on the seismic velocity structure. The Quaternary volcanic history and petrologic models are discussed by Christiansen (Chapter 6 of this volume) and Leeman (in press). These data suggest that the *P*-wave velocity structure of the crust along the Y-SRP volcanic system is characterized by systematic variations that are consistent with a model in which mantle-derived basalts ascend buoyantly to the base of a cold Archean crust heating, partially melting the upper mantle and producing a *P*-wave, low-velocity body. In our

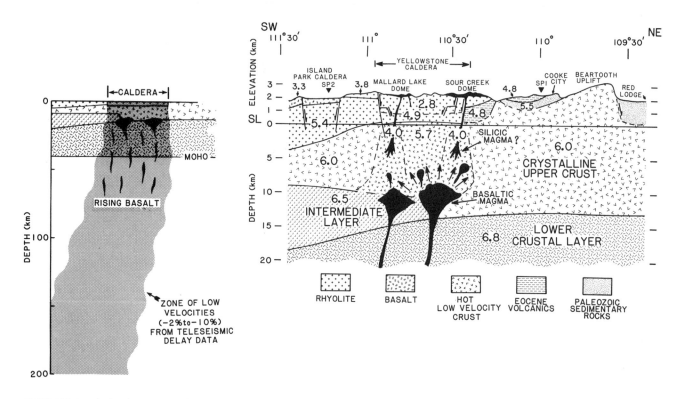

FIGURE 7.9   Idealized NE-SW geologic-seismic velocity model for the crustal structure of the Yellowstone-Island Park-Snake River Plain region. *P*-wave velocities are in kilometers/second (from Braile *et al.*, 1982; Smith *et al.*, 1982). Geologic interpretations are based on constraints of the Quaternary history and the petrologic models of Christiansen (U.S. Geological Survey, personal communication, 1978, and in press) and Leeman (in press). Low velocities in lower crust and upper mantle inferred from teleseismic delay studies (Iyer *et al.*, 1981). Model for basaltic melt rising from the upper mantle into the crust is from Hildreth (1981).

seismic model for the crust the basaltic magma ascends into the lower crust but apparently without significant perturbation of the seismic velocity structure. This conclusion is based on the similarity of the lower-crustal (6.8 km/sec) layer beneath the Y-SRP system and that of the surrounding undisturbed Rocky Mountain crust. It suggests that the mantle-derived basalt ascended into the lower crust without mechanically deforming or modifying the seismic velocity structure.

On ascent into the upper crust, partial melts transport heat advectively upward and in turn could produce additional silicic components that further ascend toward the surface. The final stage produces tumescence and thermal expansion prior to explosive eruptions. A residual mafic body may be marked by the ~6.5 km/sec layer that appears well developed beneath the Snake River Plain at a depth of 10 to 20 km, whereas beneath Yellowstone it is less clearly defined. The shallow 5.7 km/sec layer is interpreted as a body of granitic composition accumulated from numerous crustal intrusions. As the parental magma source is depleted in its low-temperature components by eruptions the crust is cooled and becomes refractory, thereby reducing the surface elevation and completing the main silicic volcanic cycle. Finally, basaltic parent magmas follow the silicic components and are extruded, giving the bimodal rhyolite/basalt complement; however, we have no evidence for shallow basaltic bodies from the geophysical data gathered at Yellowstone.

Petrologic models by Christiansen (Chapter 6 of this volume) and Leeman (in press) suggest that the Yellowstone rhyolitic magmas were derived from one or more of the following sources: (1) fractional melting, (2) separation of intermediate composition liquids into silicic fractions, and (3) partial fusion of the lower crust producing a felsic melt. Lead and strontium isotope ratios (Doe *et al.*, 1982) indicate that the rhyolitic magmas at Yellowstone did not interact extensively with upper-crustal materials, i.e., they did not form by partial melting of once-sedimentary upper-crustal rocks but rather by partial melting of high-grade metamorphic rocks of the lower crust.

An important implication of the nonunique relationship between composition and velocity is in the lower crust of Yellowstone, where the process (as modeled petrologically) did not significantly alter the velocity structure as measured by the critically refracted (horizontally propagating) head waves. A quandry thus exists; the lower crust shows normal 6.8-km/sec *P*-wave velocity (no percentage reduction) from the seismic refraction data but a 5 to 10 percent decrease in the lower crust from generalized inversion of teleseismic *P*-wave delays of vertically incident rays (Iyer *et al.*, 1981). Possible explanations of the differences in the seismic and petrologic implications are that (1) the residual lower-crustal layer may have a composition different than the original thermally undisturbed crust but with the same *P*-wave velocity; (2) mafic magmas from the mantle may have intruded the lower crust along vertically oriented plexes, which are not large enough to be resolved by the horizontally propagating waves; (3) the lower crust may be segregated into horizontally layered, sill-like low-velocity bodies in a high-velocity matrix that produces an apparent seismic anisotropy, i.e., refracted waves propagate at 6.8 km/sec, whereas the vertically incident teleseismic rays pass through a residuum

of variable-composition low- and high-velocity material with an average vertical velocity 5 to 20 percent slower than normal; (4) true vertical-horizontal-oriented seismic anisotropy produced by oriented silicate layering; and (5) frequency-dependent *P*-wave velocity—frequencies of teleseismic waves had energy in the 0.1- to 1-Hz band compared with the frequency band of 3 to 12 Hz for the refraction/wide-angle reflections.

The last explanation has important implications to velocity determination of hot material, as discussed by Daniel and Boore (1982), who determined averaged crustal shear velocities at Yellowstone from Raleigh-wave dispersion and teleseismic *P*-wave delay. Daniel and Boore (1982) fixed the *P*-wave velocity at 6.8 km/sec, and utilizing *S*-wave velocity delays of 2 to 9 sec, determined Poisson's ratios of 0.32 for the lower crust and 0.42 for the upper crust—values that can normally be attributed to the presence of fluids (water or magma) or to subsolidus relaxation (Bonner *et al.*, 1981). Daniel and Boore (1982) point out that frequency-dependent velocity and attenuation can be produced by (1) melt squirts and phase changes, (2) frequency-dependent softening, (3) subsolidus grain boundary relaxation, and (4) thermally activated lattice defects. The recent laboratory work by Bonner *et al.* (1981) suggests frequency-dependent moduli of partly melted quartz monzonite in the seismic frequency band at simulated crustal pressure and temperature and may account for mechanisms 1 and 2. However, mechanisms 3 and 4 may permit the existence of a shear-wave velocity reduction in hot but solid rock—a hypothesis not thoroughly investigated but important to seismic velocity determination in thermal environments.

SPECULATION ON POSSIBLE FUTURE EXPLOSIVE ERUPTIVE ACTIVITY AT YELLOWSTONE

The primary question with respect to explosive volcanism at Yellowstone is the potential for future eruptions. The seismic velocity structure (Figures 7.7, 7.8, and 7.9), taken together with the Quaternary geologic record, suggests that melting and crystallization of the upper crust has generated a large (5.7 km/sec) body that was the source for the Quaternary rhyolites and ash flows. On the basis of the crustal deformation, this body appears to be relatively solid but deforming plastically or viscoelastically.

The pocket of low-velocity, ~4.0 km/sec material beneath the northeast caldera boundary suggests that a fluid-filled body of steam-water, partial melt, or both has existed recently. Whether it is in a stage prior to an eruption cannot be determined from the available data. However, the trend of increasing crustal uplift (at rates greater than 15 mm/yr and in excess of normal nonthermal processes) toward the northeast points toward this low-velocity body. The coincidence of the low-density, low-velocity body near an area of increasing uplift suggests that the northeast caldera could be a candidate for eruption. However, on the basis of the relatively small volume of the low-velocity layer, an eruption from a source of the size of the 4.0 km/sec body would most likely not be as large as the earlier catastrophic eruptions. On the other hand, the state of stress in the crust may reflect "breathing," in which the influx and solidification of upper-crustal melts vary with time, producing inflation and deflation without eruption. We do not have data that can differentiate these models, but we should be aware of the potential hazard they suggest.

The detailed crustal structure from seismic measurements and the implication of the volcanic evolution of the Yellowstone silicic system in relation to its well-known Quaternary history suggest that the volcanic evolution has not ceased. There is not yet enough information to make specific predictions, but the past volcanism, contemporary uplift, length of time between volcanic eruptions, swarms of earthquakes, recent volcanism in the caldera, high rates of uplift, and presence of shallow heat sources in the upper crust all point to a future of continued volcanism in the Yellowstone Plateau. The very high heat flow, the teleseismic *P*-wave delays, the anomalous upper-crustal low-velocity layers, and the apparent shallowing of an isotherm separating brittle fracture from stable sliding are properties that are consistent with the probable existence of shallow hot bodies, at least one of which may contain a large volume of fluid (steam-water, partial melt, or both).

The evidence does not indicate what form the future volcanism might take; it could consist of small eruptions of rhyolite and basalts at the margins of the Yellowstone Plateau, medium-size eruptions of rhyolitic magma within the caldera, or a major ash flow, as suggested by Smith and Christiansen (1980). We believe that any renewal of volcanism within the Yellowstone Plateau would probably be preceded by such phenomena as localized earthquake activity, variation in seismic velocity, increased gas emission, and increased crustal deformation—properties that can be assessed in a continuous manner using modern instrumentation.

ACKNOWLEDGMENTS

This paper synthesized geophysical and geologic information obtained by us and our colleagues. We appreciate their cooperation and discussions. In particular, R. L. Christiansen has kindly shared his geologic perspectives, provided preprints, and broadened our geophysical interpretations. H. Benz, M. Brokaw, and S. Clawson assisted in data compilations and preparation of artwork. A. M. Pitt shared some U.S. Geological Survey earthquake data with us. W. P. Leeman provided a preprint that was helpful for our discussions. Personnel of the National Park Service, Yellowstone National Park (J. Townsley, W. Hamilton, A. Mebane, J. Phillips, and M. Meagher), were very helpful in coordinating and planning our field work. We thank R. L. Christiansen, W. P. Nash, G. A. Thompson, and H. M. Iyer for their critical reviews and thoughtful comments on this manuscript. Support for the research discussed here came from the National Science Foundation (grant EAR 77-23707 to the University of Utah and grant EAR 77-23357 to Purdue University) and from the U.S. Geological Survey, Geothermal Research Program (grant 14-08-0001-G-771 to the University of Utah and grants 14-08-0001-G-532 and 14-08-0001-G-674 to Purdue University).

REFERENCES

Armstrong, R. L., W. P. Leeman, and H. E. Malde (1975). K-Ar dating, Quaternary and Neogene volcanic rocks of the Snake River Plain, Idaho, *Am. J. Sci.* 275, 225-251.

Benz, H. M., and R. B. Smith (1981). Simultaneous inversion for lateral velocity variations and hypocenters in the Yellowstone region using earthquake and controlled source data, *EOS* 62, 470.

Blank, H. R., Jr., and M. E. Gettings (1974). Complete Bouguer gravity map: Yellowstone-Island Park Region, *U.S. Geol. Surv. Open-File Map 74-22*.

Bonner, B. P., L. Thigpen, and G. W. Hedstrom (1981). An elastic creep of partly melted quartz monzonite with implications for the frequency dependence of *Q* (abstr.), *Earthquake Notes 52*, 33.

Braile, L. W., R. B. Smith, J. Ansorge, M. R. Baker, M. A. Sparlin, C. Prodehl, M. M. Schilly, J. H. Healy, S. T. Muller, and K. H. Olsen (1982). The Yellowstone-Snake River Plain seismic profiling experiment: Crustal structure of the eastern Snake River Plain, *J. Geophys. Res.* 87, 2597-2609.

Christiansen, R. L. (1978). *Quaternary and Pliocene Volcanism of the Yellowstone Rhyolite Plateau Region of Wyoming, Idaho, and Montana*, U.S. Geol. Surv. Prof. Pap. (preprint).

Christiansen, R. L., and H. R. Blank, Jr. (1972). *Volcanic Stratigraphy of the Quaternary Rhyolite Plateau in Yellowstone National Park*, U.S. Geol. Surv. Prof. Pap. 729-B, 18 pp.

Christiansen, R. L., and E. H. McKee (1978). Late Cenozoic volcanic and tectonic evolution of the Great Basin and Columbia Intermountain regions, in *Cenozoic Tectonics and Regional Geophysics of the Western Cordillera*, R. B. Smith and G. P. Eaton, eds., Geol. Soc. Am. Mem. 152, 283-311.

Clawson, S. R., and R. B. Smith (1981). The lateral *Q*-structure of the upper crust in Yellowstone National Park from seismic refraction data, *EOS* 62, 970.

Daniel, R. G., and D. M. Boore (1982). Anomalous shear wave delays and surface wave velocities at Yellowstone caldera, Wyoming, *J. Geophys. Res.* 87, 2731-2744.

Decker, R., and B. Decker (1981). *Volcanoes*, W. H. Freeman, San Francisco, Calif., 255 pp.

Doe, B. R., W. P. Leeman, R. L. Christiansen, and C. E. Hedge (1982). Lead and strontium isoptopes and related trace elements as genetic tracers in the Upper Cenozoic rhyolite-basalt association of the Yellowstone Plateau volcanic field, U.S.A., *J. Geophys. Res.* 87, 4785-4806.

Eaton, G. P., R. L. Christiansen, H. M. Iyer, A. N. Pitt, H. R. Blank, Jr., I. Zeitz, D. R. Mabey, and M. E. Gettings (1975). Magma beneath Yellowstone National Park, *Science 188*, 787-796.

Evoy, J. A. (1977). Precision gravity reobservations and simultaneous inversion of gravity and seismic data for subsurface structure of Yellowstone, M.S. thesis, U. of Utah, 212 pp.

Fournier, R. O., D. E. White, and A. H. Truesdell (1976). Convective heat flow in Yellowstone National Park, in *Second United Nations Symposium on Development and Use of Geothermal Resources*, U.S. Government Printing Office, Washington, D.C., pp. 731-739.

Hildreth, W. (1981). Gradients in silicic magma chambers: Implications for lithospheric magmatism, *J. Geophys. Res.* 86, 10153-10192.

Hill, D. P. (1976). Structure of Long Valley caldera, California, from a seismic refraction experiment, *J. Geophys. Res.* 81, 745-753.

Iyer, H. M., J. R. Evans, G. Zandt, R. M. Stewart, J. M. Cookley, and J. N. Roloff (1981). A deep low velocity body under the Yellowstone caldera, Wyoming; Delineation using teleseismic *P*-wave residuals and tectonic interpretations: Summary, *Geol. Soc. Am. Bull.* 92, 792-798, 1471-1646.

Leeman, W. P. (in press). Development of the Snake River Plain-Yellowstone Plateau province: An overview and petrologic model, *Idaho Bur. Mines and Geol. Bull.*

Lehman, J. A., R. B. Smith, M. M. Schilly, and L. W. Braile (1982). Upper crustal structure of the Yellowstone caldera from delay time analyses and gravity correlations, *J. Geophys. Res. 87*, 2713-2730.

Mavko, G. M. (1980). Velocity and attenuation in partially molten rocks, *J. Geophys. Res. 85*, 5173-5189.

Morgan, P., D. D. Blackwell, R. E. Spafford, and R. B. Smith (1977). Heat flow measurements in Yellowstone Lake and the thermal structure of the Yellowstone caldera, *J. Geophys. Res. 82*, 3719-3732.

Murase, T., and A. R. McBirney (1973). Properties of some common igneous rocks and their melts at high temperatures, *Geol. Soc. Am. Bull. 84*, 3563-3592.

Murase, T., and T. Suzuki (1966). Ultrasonic velocities of longitudinal waves in molten rocks, *J. Fac. Sci., Hokkaido U. Ser. 11 VII 2*, 272-285.

O'Connell, R. J., and B. Budiansky (1977). Viscoelastic properties of fluid-saturated cracked solids, *J. Geophys. Res. 82*, 5719-5735.

Pelton, J. R., and R. B. Smith (1979). Recent crustal uplift in Yellowstone National Park, *Science 206*, 1179-1182.

Pelton, J. R., and R. B. Smith (1982). Contemporary vertical surface displacements in Yellowstone National Park, *J. Geophys. Res. 87*, 2745-2751.

Pitt, A. M. (1981). Catalogs of earthquakes in Yellowstone National Park—Helsgen Lake region from November to December 1975, *U.S. Geol. Surv. Open-File Rep. 80-2006.*

Reasenberg, P., W. Ellsworth, and A. Walter (1980). Teleseismic evidence for a low-velocity body under the Coso geothermal area, *J. Geophys. Res. 85*, 2471-2483.

Sanford, A. L., R. P. Mott, Jr., P. J. Shuleski, E. J. Rinehart, F. J. Caravella, R. M. Ward, and T. C. Wallace (1977). Geophysical evidence for a magma body in the crust in the vicinity of Socorro, New Mexico, in *The Earth's Crust*, J. G. Heacock, ed., Am. Geophys. Union Geophys. Monogr. 20, pp. 385-403.

Schilly, M. M., R. B. Smith, L. W. Braile, and J. Ansorge (1982). The 1978 Yellowstone-eastern Snake River Plain seismic profiling experiment: Data and detailed crustal structure of the Yellowstone region, *J. Geophys. Res. 87*, 2692-2704.

Smith, R. B. (1978). Seismicity, crustal structure, and intraplate tectonics of the interior of the western Cordillera, in *Cenozoic Tectonics and Regional Geophysics of the Western Cordillera*, R. B. Smith and G. P. Eaton, eds., Geol. Soc. Am. Mem. 152, pp. 111-144.

Smith, R. B. (1981). Listric faults and earthquakes, What evidence? *EOS 62*, 961.

Smith, R. B., and R. L. Christiansen (1980). Yellowstone Park as a window on the Earth's interior, *Sci. Am. 242*, 104-117.

Smith, R. B., and A. Lindh (1978). A compilation of fault plane solutions of the western United States, in *Cenozoic Tectonics and Regional Geophysics of the Western Cordillera*, R. B. Smith and G. P. Eaton, eds., Geol. Soc. Am. Mem. 152, pp. 107-110.

Smith, R. B., R. T. Shuey, R. Freidline, R. Otis, and L. Alley (1974). Yellowstone hot spot: New magnetic and seismic evidence, *Geology 2*, 451-455.

Smith, R. B., R. T. Shuey, J. R. Pelton, and J. P. Bailey (1977). Yellowstone hot spot: Contemporary tectonics and crustal properties from earthquakes and aeromagnetic data, *J. Geophys. Res. 82*, 3665-3676.

Smith, R. B., M. M. Schilly, L. W. Braile, H. A. Lehman, J. Ansorge, M. R. Baker, C. Prodehl, J. H. Healy, S. Muller, and R. W. Greenfelder (1982). The 1978 Yellowstone-eastern Snake River Plain seismic profiling experiment: Crustal structure of the Yellowstone region and experiment design, *J. Geophys. Res. 87*, 2583-2596.

Trimble, A. E., and R. B. Smith (1975). Seismicity and contemporary tectonics of the Hebgen Lake-Yellowstone Park region, *J. Geophys. Res. 80*, 733-741.

Weaver, C. S., A. M. Pitt, and D. P. Hill (1979). Crustal spreading direction of the Snake River Plain-Yellowstone system (abstract), *EOS 60*, 946.

Williams, H. P., and A. R. McBirney (1979). *Volcanology*, Freeman, Cooper and Co., San Francisco, Calif., 397 pp.

Explosive Eruptions in Space and Time: Durations, Intervals, and a Comparison of the World's Active Volcanic Belts

8

TOM SIMKIN *and* LEE SIEBERT
Smithsonian Institution

ABSTRACT

The explosive volcanism that dominates colliding plate boundaries supplies only 10 to 13 percent of the magma reaching the surface of the Earth each year (Nakamura, 1974) but accounts for roughly 84 percent of the world's known historical eruptions. A computerized data bank, compiled over the last 12 yr at the Smithsonian Institution, allows summaries to be made of Holocene volcanism. However, the record of most volcanoes is poor before the last 100 yr and some eruptions pass unreported even today.

Calculations of the time interval since the previous eruption is possible for 4835 of the 5564 compiled eruptions. The median interval is 5.0 yr, but much longer intervals commonly precede unusually violent eruptions. For the 25 most violent eruptions in the file (with known preceding interval), the median interval is 865 yr. Of the historical eruptions in this group, 50 percent resulted in fatalities.

The interval between an eruption's start and its most violent paroxysm may be measured in months or years but is usually short. Of the 205 larger eruptions for which there are data, 92 had the paroxysmal event within the first day of the eruption, allowing little time for emergency preparations after the eruption's opening phase.

To compare the recent vigor of different volcanic belts, we calculated the number of years in which each volcano was active in the last 100 yr, summed these for each belt, and divided by the belt length. Another index of recent vigor is the number of recognized Holocene volcanoes in each belt divided by its length. A third index is the number of large explosive eruptions (VEI $\geq$ 3) of the last 100 yr, again normalized by belt length. These three measures correlate reasonably well, serving to contrast vigorous belts such as Kamchatka, Central America, and New Zealand with relatively quiet belts such as the Cascades, South Sandwich Islands, Greece, and southern Chile. Relatively quiet volcanism generally accompanies the subduction of young (<10 million yr) crust, gently dipping seismic zones, low aseismic slip rates, and plate motions highly oblique to the volcanic belt.

INTRODUCTION

Explosive volcanism at colliding plate boundaries, even though it supplies only 10 to 13 percent of the magma that reaches the Earth's surface each year (Nakamura, 1974), has long been the principal focus of volcanological interest. The reason for this is simply that the other types of volcanism, at spreading plate boundaries and intraplate hot spots, commonly pass unnoticed on the seafloor or remote islands. Because of its visibility, the subduction-zone volcanism of continental margins and island arcs makes up 63 percent of the world's known Holocene volcanoes and 84 percent of all historical eruptions. Subduction-zone volcanism, being both largely explosive and largely subaerial, has had dramatic impact on human populations (it is responsible for 88 percent of known fatal eruptions, for example), and it is therefore the only major type for which there is a substantial historical record. This paper examines that historical record, augmented by geologic and other evidence for Holocene volcanism, in an effort to provide a perspective of global volcanism during the last few thousand years. A quantitative assessment of contemporary volcanism should assist understanding of the Earth's dynamic processes today, and this, in turn, should help both interpretations of the geologic past and preparations for the volcanological future.

VOLCANOLOGICAL RECORDS

In 1917 the German volcanologist Sapper, building on the efforts of previous compilers, produced a monumental catalog of the world's volcanoes. He used these records to define the major volcanic belts and to compare their eruptive histories. Later, in a long-term project of the International Association of Volcanology, a series of catalogs devoted to individual volcanic regions was produced between 1950 and 1975, and the final two volumes (Alaska and Iceland) should be completed in the near future. In vigorous volcanic regions, however, such compilations are swiftly outdated, and in 1960 the Volcanological Society of Japan began recording the new eruptions of the world by publishing a yearly *Bulletin of Volcanic Eruptions*. More immediate notification of eruptions, and other short-lived natural events, began in 1968 when the Smithsonian Institution started the Center for Short-Lived Phenomena. Reinstituted as the Scientific Event Alert Network (SEAN) by the Smithsonian in 1975, this group now issues a monthly *SEAN Bulletin* and irregular mid-month reports in *EOS* (Transactions of the American Geophysical Union).

As a natural outgrowth of our involvement with both reporting and research on contemporary volcanism, the Smithsonian began compiling a volcanological data bank in 1971. Data from the above-mentioned sources were entered in a computer format that could be easily manipulated and updated, both with current eruption reports and with additional information from the scientific literature. Much effort has gone into reconciling the frequently subjective and qualitative reports of ancient eruptions with the unblinkingly quantitative nature of a computer data bank. The format has been revised repeatedly to accommodate new data elements, and the data have been published in book form (Simkin *et al.*, 1981). We hope that we will receive additions and corrections not only from the regional geologists, who know individual volcanoes better than we can ever know them, but also from historians, anthropologists, and others interested in the events of the last few thousand years. In the first 18 months after publication of this book, we added 62 volcanoes and 363 eruptions to the data file, and we look forward to continued improvement of the volcanological record.

VOLCANISM IN TIME: TRENDS, DURATIONS, AND INTERVALS

Changes in volcanism with time—particularly changes that might display some pattern that could assist predictions—have intrigued scientists for centuries. One of our first exercises with the data bank, therefore, was to plot the 5564 eruptions chronologically. As a measure of volcanism we plotted the number of volcanoes active per year rather than the number of new eruptions per year. This procedure recognizes the eruptions that continue for many years and deemphasizes the common reporting (by untrained or infrequent observers) of multiple "eruptions" that would have been recognized as phases of a single ongoing eruption by better-trained (or more frequent) observers. We have also excluded from the plot all eruptions with a dating uncertainty greater than ±1 yr.

Figure 8.1 shows what is, at first glance, an astonishing (and frightening) recent increase in global volcanism. However, dramatic increases in the reporting of volcanic (and other) events have also marked the 600-yr period shown, so we have also plotted both the growth of the Earth's human population and the increase in the number of known volcanoes on the same figure. Only 27 volcanoes produced historically documented eruptions by A.D. 1000. That number climbed to 59 by A.D. 1500, 246 by A.D. 1800, and 530 today (429 of which are subduction-zone volcanoes). We believe that the increases shown are in reporting rather than in volcanism and that historical developments, such as the invention of the printing press, the early Spanish-Portuguese maritime explorations, and the technological advances of the industrial revolution, have had recognizable effects on the volcanological record. Furthermore, careful study of the record for the last 100 yr shows apparent drops in global volcanism during periods in which editors and scientists were preoccupied with other things, such as world wars, and apparent increases during periods of unusually strong interest in volcanism, such as the years immediately following the catastrophic eruption disasters of Krakatau and Mont Pelée (Simkin *et al.*, 1981). We should expect to find years of increased volcanism even with a random distribution, and episodic plate movement probably accounts for regional concentrations of activity, such as the three major 1902 eruptions on the margins of the Caribbean plate and the 1974 eruptions in the western Bismarck arc (Cooke *et al.*, 1976). However, worldwide pulsations in volcanism are not clearly seen in the historical record. In his compilation of historical volcanism in the Aleutian arc, Coats (1951) noted that major pulses of reported eruptions corresponded precisely with large natural history expeditions along this active, but little explored, volcanic arc.

Any study of historical volcanism must remember the fact that eruptions are reported by people: human distribution in space and time has always been highly irregular, the ability to report volcanism has varied enormously from person to person, and the likelihood that reports survive in the scientific literature is influenced by another host of factors quite capable of frustrating the most vigorous efforts of the best-trained observers.

Once an eruption has begun, there is immediate and intense interest in knowing how long it will last and what changes might be expected. Dates of cessation are recorded for only 2139 of the 5564 eruptions in the file, largely because the end of an eruption is less spectacular than its beginning. Figure 8.2 shows the distribution of these durations for both arc (subduction zone) and non-arc settings. The median duration for 1779 arc eruptions is 65 days but is only 31 days for 360 non-arc eruptions. These durations are misleadingly short because of differing uses of the word *eruption* by different reporters. Careful study of a single eruption commonly shows pulsations of activity over months or years that would be reported as separate events

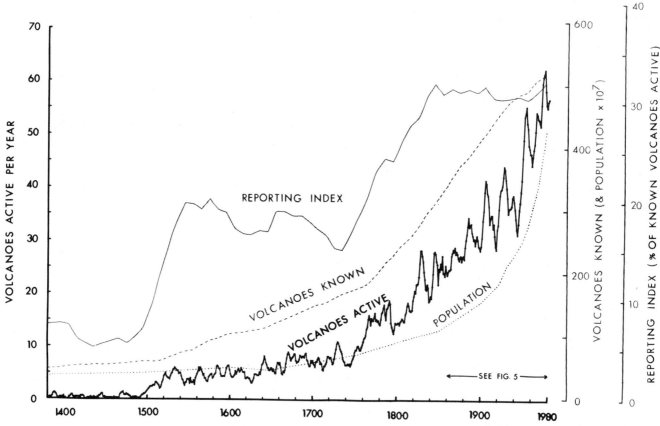

FIGURE 8.1 Last 600 yr of volcanism reporting (Simkin *et al.*, 1981). "Volcanoes active" is the 5-yr running average of the total number of volcanoes active per year. "Volcanoes known" is the total number known to have had historical eruptions. "Population" is the world's estimated human population.

by distant, untrained, or different observers. Most older eruption reports came from such observers. Even recognizing this bias, however, the most striking feature of Figure 8.2 is the great range of durations shown—some eruptions have ended in minutes, while others, like Yasour or Stromboli, have continued intermittently for several centuries or more. The historical record is not very helpful in predicting when an eruption will end.

An important element of eruption duration, and one with great significance to human populations, is the fact that an eruption's paroxysmal phase may take place months or years after its beginning. We attempted to explore this fact using the more explosive historical eruptions in the data file. For most eruptions, we list a Volcanic Explosivity Index (VEI), developed by Newhall and Self (1982), that estimates the explosive magnitude, or "bigness," of the eruption on a scale from 0 to 7. We selected all eruptions with a dated paroxysmal event of VEI $\geq$ 3 ($>10^7$ m^3 of tephra, cloud column height >3 km, and described as "severe" or "violent") for which we have data on the time period between that paroxysm and the start of eruptive activity. The data for 205 eruptions are plotted as a histogram in Figure 8.3. Clearly, the largest number of these paroxysmal eruptions (92, or 45 percent, including 8 that killed more than

100 people each) came within the eruption's first day. Where data are available, half of these first-day paroxysms came within the eruption's first hour.

History's largest documented explosive eruption, the devastating 1815 event at Tambora, Indonesia, culminated nearly 3 yr after its start and killed 92,000 people. Several prominant eruptions, such as Krakatau (1883) and Mount St. Helens (1980), began months before the paroxysmal event. However, other major eruptions (e.g., Tarawera in 1886, Bandai-san in 1888, Hekla in 1947, Sheveluch in 1964, Usu in 1977) reached their climax within only an hour of their start. *It is foolish to assume either that the worst is over after an initial explosive phase or that days, weeks, or months are always available in which to take necessary protective measures before a paroxysmal phase.*

The time interval between past eruptions is of interest to anyone considering the time interval before the next eruption of any particular volcano. This former interval is easily calculated for the 4875 eruptions in the data bank for which we have a dated previous eruption. We calculate the interval from the start of one eruption to the start of the next, rather than the "repose" interval (from the end of one eruption to the start of the next), for two reasons: we know the end of only one third of the eruptions in the data file, and for most of these the

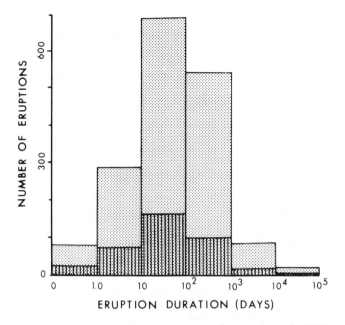

FIGURE 8.2 Eruption duration. Duration data are shown for 1779 island-arc eruptions (stippled pattern) and 360 non-arc eruptions (dark pattern). The median duration for arc eruptions is 65 days and only 31 days for non-arc eruptions (but see text discussion).

duration is only a small proportion of the much larger interval before the next eruption.

Figure 8.4 is a histogram of these 4875 eruptions from the data bank—again a logarithmic time axis is used, and data are shown for both arc (subduction zone) and non-arc eruptions. The median interval for both groups of eruptions is, coincidently, 5.0 yr, but this figure is very misleading. It suffers from

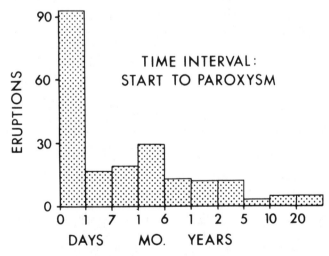

FIGURE 8.3 Time from eruption start to paroxysm. The time interval between the beginning of an eruption and its ensuing paroxysmal phase is shown for 205 eruptions with a VEI of 3 or greater (including 10 eruptions with a VEI ≥ 5). The time axis is only quasi-logarithmic in order to show convenient time intervals.

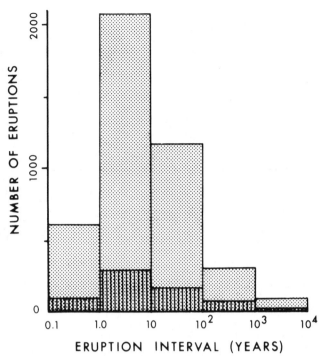

FIGURE 8.4 Interval between eruptions. The distribution of intervals between eruptions of the same volcano is shown for 4246 arc eruptions and (superimposed dark pattern) for 629 non-arc eruptions. The median interval for both groups is 5.0 yr.

the problem of different definitions of the word *eruption*, the short historical record in many regions precludes the recording of many long intervals, and the very nature of the plot emphasizes those volcanoes with many closely spaced eruptions. The growing number of eruptions dated by radiocarbon, or other nonhistorical techniques (many from a few carefully studied volcanoes), is rapidly showing that intervals measured in thousands of years are not uncommon between significant eruptions. The long historical record of a volcano like Ischia, which erupted about A.D. 300 and only once since, in 1301, is easily forgotten amongst its vigorous Italian neighbors Stromboli and Etna. The longer intervals between eruptions are significant because they commonly precede unusually violent eruptions. For the 25 most explosive eruptions in the data file (with known preceding interval), the median interval is 865 yr. Of the historical eruptions in this group, 50 percent have caused fatalities.

To look more carefully at the relationship between explosivity and preceding interval, we plotted the interval data by VEI group in Figure 8.5. This shows in a striking way the shifting proportion of eruptions in each interval group with progressively more explosive eruptions. The human significance of these more explosive eruptions is underlined by the fatality statistics. Using the much longer geologic record of several well-studied volcanoes, Smith (1979) and Smith and Leudke (Chapter 4 of this volume) have shown that this trend continues, with gigantic ash-flow eruptions (vastly larger than any known historical eruption) following quiet intervals measured in many thousands of years. This relationship between explosivity and

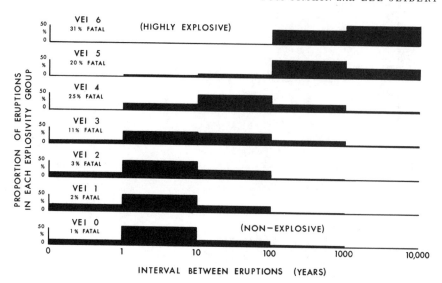

FIGURE 8.5 Explosivity and time intervals between euptions. For each VEI group the percentage of eruptions in each logarithmic time interval (from start of previous eruption) is shown. The number of eruptions in VEI groups 0 to 6 are, respectively, 354, 338, 2882, 617, 102, 19, and 8. For each group the percentage of historical eruptions that have been fatal is also shown.

preceding interval is enormously important in any consideration of volcanic hazards. It adds sobering perspective to a data file covering only 10,000 yr and a historical record that is, for most volcanic regions, quite poor beyond the last 100 yr. Of the 21 most explosive historical eruptions in the data file (VEI ≥ 5, or at least as large as the 1980 Mount St. Helens eruption), 17 were the *first* historical eruption for that volcano. Volcanoes with no historical activity are often termed *inactive*, and many are not even recognized to be volcanoes by people who live on their flanks. But the chronological record of explosive volcanism shows that a volcano's long repose is more likely to be a cause for concern than reassurance. Clearly, more attention must be paid to "inactive" and unrecognized volcanoes that could erupt, with devastating consequences.

VOLCANISM IN SPACE: A COMPARISON OF THE WORLD'S VOLCANIC BELTS

The distribution of volcanoes in narrow belts has long been recognized, and the relationship of these belts to plate boundaries and hot spots is now well known. Figure 8.6 is a world map showing the locations of 1343 Holocene volcanoes from the Smithsonian data file. We divided the principal volcanoes into 46 belts, in some cases separating apparently unbroken belts into segments based on changes in volcanism or plate-tectonic setting. The Bismarck arc, for example, has been broken into east and west segments, following Johnson's (1976) petrologic-tectonic distinction, and the continental margin volcanoes of Alaska are separated from those of the Aleutian Islands, but we tried to maintain large groups of reasonably comparable character. These 46 groups include more than 80 percent of the volcanoes and more than 94 percent of the eruptions in the data bank. Their belt lengths, measured between the terminal volcanoes in each group (see Table 8.1), total 33,366 km, nearly equal to the Earth's circumference. Assuming an average belt width of 100 km, however, yields a total volcanic belt area that is only 0.7 percent of the Earth's

surface. Volcanism is concentrated in small segments of the Earth's crust, and there are large variations in both style and frequency of volcanism between those small segments.

Knowing that some volcanic belts are more vigorous than others, we compiled in Table 8.1 data that permit quantitative comparison. We use the word *vigor* in the sense of forcefulness, physical energy, and active strength, rather than such words as *magnitude* or *intensity*, long used in seismology and recently introduced to volcanology by Walker (1980) for individual eruptions. *Vigor*, in contrast, refers to the activity of a coherent volcano group over a substantial length of time. We developed three measures of volcanic vigor utilizing the data in Table 8.1.

The first of these measures is simply the number of Holocene volcanoes in each belt divided by belt length. This measures the concentration of Holocene volcanoes within a belt but makes no allowance for differences in eruptive vigor between different volcanoes in the belt. It suffers from regional differences in the amount of geologic field work needed to recognize Holocene volcanoes that have had no historical activity and from varying definitions of the word *volcano* (see discussion in Simkin *et al.*, 1981), but it seems to be the simplest way to give consideration to a 10,000-yr time period rather than the short historical record.

The second measure utilizes frequency of known eruptions, despite the cautions emphasized in the preceding section of this paper. We again used the number of years a volcano was active rather than the number of eruptions (see earlier discussion of Figure 8.1) for each volcano and summed these for each belt before dividing by belt length. Table 8.1 shows these data for post-1880, post-1800, and post-1500 time periods. Half of the eruptions in the data bank began in the last 100 yr, and inspection suggests that this period has the most complete record for most regions. In 1880 the global human population was one third what it is today, the first telephone exchange had just opened, the first newspaper photograph was printed, and neither the automobile nor commercial radio had been invented. The last 100 yr are an apallingly short time interval in volcanology, but it is the longest period for which the historical record permits reasonably standardized comparisons of different parts of the world. We used the post-1880 data, except for 9 belts in which a longer time period (underlined in Table 8.1) resulted in a higher figure when normalized on a per-100-yr basis. The data in column 9 of Table 8.1 (and plotted in Figure 8.7) can be thought of as the maximum number of eruption years per 100 yr per 100 km of belt length.

The third measure incorporates explosivity data in an attempt to assess the vigor of the larger eruptions that should have been more consistently reported (and that have had more impact on humans). We added the number of eruptions with maximum VEI of 3 or more during the last 100 yr and again normalized by belt length. Eruptions with VEI $\geq$ 3 (generally indicating $\geq 10^7$ m^3 of ejected tephra) make up only 16 percent of the data file, and the generally long time interval between these larger eruptions (Figure 8.5) decreases the likelihood that they will be present in meaningful numbers during the last 100 yr, particularly in short belts with few volcanoes. For longer belts, however, this measure filters out the smaller, less explosive events and provides a useful complement to our other measures of volcanic vigor. The explosivity data in column 11 of Table 8.1 are displayed as quasi-logarithmic circle sizes in Figure 8.7. Rittmann (1962), using the data of Sapper (1917), developed another measure of explosivity that consists of the proportion of post-1500 eruptive products that are explosive. We have not extended this approach because reliable estimates of eruptive product volumes are available for so few of the eruptions in our file.

The correlation between these three measures of volcanic vigor, as shown in Figure 8.7, is surprisingly good, particularly when attention is focused only on volcanoes of the typically more explosive arc, or subduction zone, setting. Many of the belts combining high eruption frequency with low concentration of Holocene volcanoes (e.g., Ryukyus, Izu-Marianas, Tonga) include substantial stretches of seawater, and more marine exploration might reveal additional Holocene centers that would bring these belts closer to the general trend.

Before further discussion of the arc belts, a few points should be made about the oceanic hot spots and spreading-center belts. These show a wide range of volcanic vigor, but (with the exception of Iceland and Galápagos) explosivity is characteristically low. This low explosivity is also reflected in the generally high proportion of eruptions with lava flows (Table 8.1, column 14). Iceland and Galápagos are also distinctive for their high concentration of Holocene volcanoes. The explosivity value for Iceland would have plotted off of Figure 8.7 if we had divided the 63 Holocene volcanoes in the data file by belt length, but recent work in this rifting region has repeatedly shown subsurface connections between individual "volcanoes" along rift; therefore, we used the 29 "volcano systems" recognized by Jakobsson (1980) in calculating the volcano concentration for Iceland.

Returning to the data for arc, or subduction-zone, volcanoes, we made preliminary comparisons with other geophysical parameters for various belts in the hope of finding simple explanations for the large variation in volcanic vigor shown here for the active belts of the world. Subduction-zone geometry is correlated with volcanic vigor at one extreme, as evidenced by

Barazangi and Isacks's recognition (1976) that the prominent volcanic gaps in western South America (Figure 8.6) overlie anomalously shallow-dipping subduction zones. Nur and Ben-Avraham (1981) further suggested that these shallow dips result from the overriding of thick aseismic ridges at the two gaps, and Sacks (Chapter 3 of this volume) interprets these zones as horizontal movements of subducted crust under the continental lithosphere. In comparing the data with the dips compiled by Uyeda and Kanamori (1979), we find low dips generally associated with relatively quiet volcanic belts, but the overall correlation is poor. As the 65° dip of the Izu-Volcano Islands belt turns to 90° under the Marianas we see a very slight decrease in what is already relatively quiet volcanism.

Thickness of the overriding crustal plate, using the data compiled by Coulon and Thorpe (1981), also appears to correlate with volcanism. Their plate thicknesses of less than 20 km (Izu-Marianas, Tonga, Kermadec, and South Sandwich Islands) are all marked by relatively quiet volcanism, and the most vigorous belts of New Zealand, Kamchatka, and Central America are all on thick (25 to 35 km) crust. The unusually thick crust of the central and northern Andes, however, does not show comparably extreme volcanism.

In Toksöz's (1975) classification of subduction zones, type C belts (slow subducting belts, such as Greece, West Indies, and South Sandwich Islands) are among the least vigorous by our volcanic measures, and types D and E (bent, weak, or broken slab geometry, such as the Solomons, Izu-Marianas, and Ryukyus) both show uniformly low concentrations of Holocene volcanoes along belt length, despite some wide variation in eruptive frequency. Toksöz's types A and B (more normal geometry, such as the northwest Pacific, Central America, and Sunda-Sumatra), however, show broad variation in volcanic vigor.

Like several other parameters, age of the subducted crust appears to correlate with volcanic vigor at only one end of the spectrum. Unusually young crust is associated with relatively weak volcanism, as shown particularly by changes in two regions. The southern portion of the central Chile volcanic belt (here labeled Chile-S) is markedly less vigorous than its northern portion (Chile-C) and the principal difference seems to be the much younger crust being consumed under the southern portion (Figure 8.6). The change is even more dramatic from Central America to the Mexican belt, under which much younger crust is being destroyed. The Cascades of western North America form another example of relatively quiet volcanism associated with young down-going crust, but no age/volcanism correlation is recognized for crust older than 20 m.y.

Carr (1977) called attention to the relationship between volcanism and great earthquakes, and Acharya (1981) tied this to plate convergence rates, noting that volcanism is strong where aseismic slip rates are high and weak where great earthquakes indicate strong coupling between converging plates. Our data file tends to support Acharya's observation and even to extend it. For example, Acharya noted that large earthquakes occur on the western and eastern ends of the main Indonesian arc, with high aseismic slip rates in the middle: the data show vigorous volcanism in Java decreasing to east (Sunda) and west (Sumatra). More calculations of aseismic slip rates are needed.

No fully satisfying correlations have emerged from our pre-

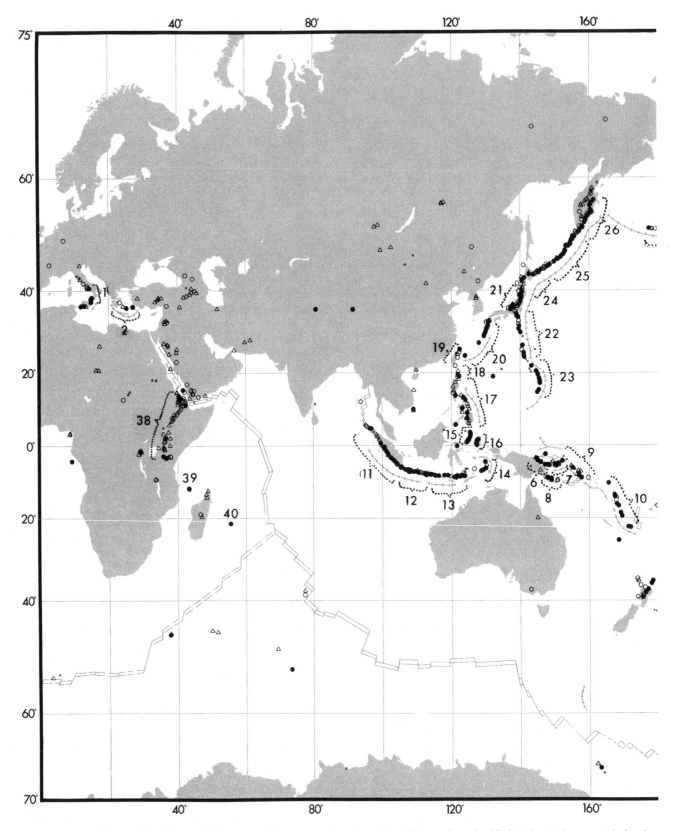

FIGURE 8.6 Map of the world's volcanoes. Volcanoes with known eruptions since A.D. 1880 are shown by filled circles. Volcanoes with dated eruptions, but none known since 1880, are shown by open circles, and those with undated but probable Holocene eruptions are shown by triangles.

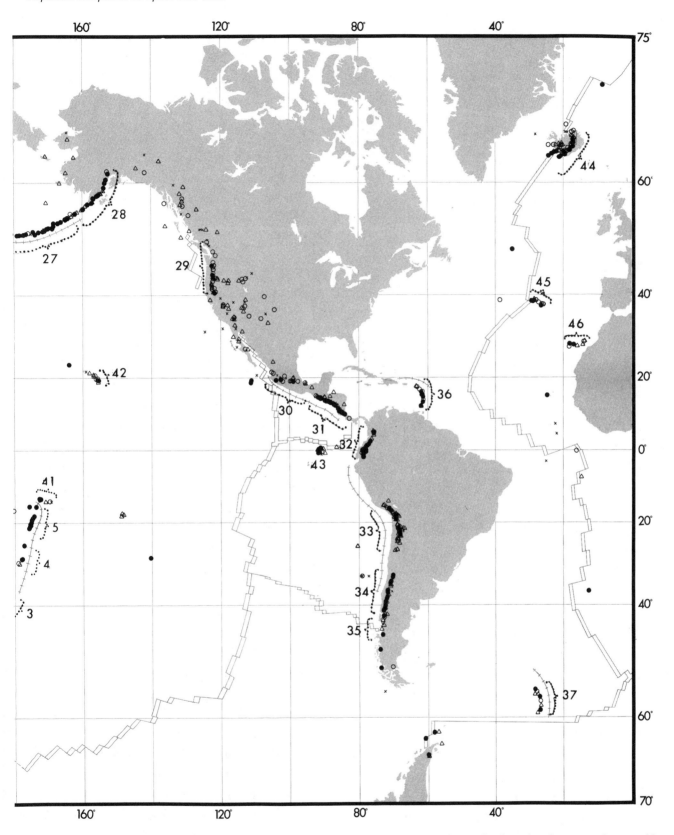

Volcanoes with uncertain or only solfataric activity are shown by a small x. Volcanic belts are marked and numbered corresponding to Table 8.1. Spreading centers (from Halbouty *et al.*, 1981, and Heirtzler, 1968) are indicated by light double lines and oceanic trenches by light crossed lines.

TABLE 8.1 Comparative Volcanic Belt Data

	1	2	3	Holocene Volcanoes		Eruptions, Years Active as of Last				VEI >3 (>1880)		Eruptions		
				4	5	6	7	8	9	10	11	12	13	14
	Volcano Numbers	Belt Name	Kilometers	Total	per 100 km	100	180	480	per 100 km	Total	per 100 km	Total	Fatal	Percent with Lava
1.	0101-01 –0101-06	Italy-South	352	7	2.0	241	441	978	69.6	21	6.0	388	13	52
2.	0102-01 –0102-06	Greece	390	6	1.5	8	15	22	2.1	0		18	2	50
3.	0401-02 –0401-27	New Zealand	295	27	9.1	136	156	160	46.1	10	3.4	187	4	7
4.	0402-01 –0402-05	Kermadecs	530	5	0.9	5	9	9	0.9	0		9	0	0
5.	0403-01 –0403-101	Tonga	686	10	1.5	29	41	46	4.2	2	0.3	45	0	20
6.	0501-00 –0502-01	Bismarck-West	477	11	2.3	123	142	155	25.8	15	2.5	117	4	19
7.	0502-02 –0502-13	Bismarck-East	231	12	5.2	19	19	20	8.2	2	0.9	20	0	30
8.	0503-00 –0503-07	New Guinea	956	10	1.0	10	11	11	1.0	2	0.2	6	2	33
9.	0504-01 –0505-07	Solomons–New Ireland	1037	13	1.3	53	70	71	5.1	8	0.8	36	3	52
10.	0507-00 –0507-10	New Hebrides	708	13	1.8	180	267	396	25.4	12	1.7	75	3	21
11.	0601-01 –0602-01	Sumatra	1741	30	1.7	150	195	199	8.6	3	0.2	155	2	4
12.	0603-01 –0603-35	Java	732	36	3.9	279	442	494	30.0	10	1.1	436	45	13
13.	0604-01 –0604-27	Sunda	965	30	3.1	112	152	158	11.6	6	0.6	144	8	18
14.	0605-04 –0605-09	Banda	373	6	1.6	9	17	51	2.9	0		40	5	18
15.	0606-02 –0607-05	Sangihe	359	17	4.7	102	130	142	28.4	4	1.1	126	11	12
16.	0608-01 –0608-07	Halmahera	162	9	5.6	84	117	147	51.9	1	0.6	86	10	19
17.	0701-01 –0703-081	Philippines	1148	30	2.6	56	112	139	5.4	10	0.9	116	21	23
18.	0703-082 –0704-06	Philippines-North	486	9	1.9	9	20	23	2.3	0		18	1	6
19.	0801-02 –0801-05	Taiwan	501	7	1.4	1	5	5	0.6	0		4	0	0
20.	0802-01 –0802-13	Ryukyu–Kyushu	1212	14	1.2	177	198	277	14.6	14	1.2	281	13	3
21.	0803-01 –0803-29	Honshu	732	36	4.9	143	175	290	19.5	11	1.5	288	19	3
22.	0804-01 –0804-133	Izu–Volcano Island	1477	19	1.3	116	134	207	7.9	4	0.3	151	6	26
23.	0804-134–0804-21	Marianas	710	11	1.5	35	43	46	4.9	2	0.3	39	0	36

	1	2	3	4	5	6	7	8	9	10	11	12	13	14
24.	0805-01 -0805-09	Hikkaido	556	15	2.7	72	97	109	12.9	6	1.1	101	10	2
25.	0900-01 -0900-39	Kuriles	1078	50	4.6	68	114	137	6.3	14	1.3	118	1	15
26.	1000-01 -1000-27	Kamchatka	716	47	6.6	203	234	267	28.4	26	3.6	225	1	18
27.	1101-01 -1101-39	Aleutians	1457	40	2.8	132	238	306	9.1	13	0.9	203	1	18
28.	1102-01 -1103-06	Alaska Peninsula	944	33	3.5	100	126	157	10.6	27	2.9	120	0	8
29.	1200-17 -1203-06D*	Cascades	1152	38	3.3	8	45	53	2.2	2	0.2	141	1	36
30.	1401-03 -1401-112	Mexico	1043	22	2.1	78	98	156	7.5	4	0.4	64	0	23
31.	1401-13 -1407-01*	Central America	1254	79	6.3	337	556	804	26.9	30	2.4	352	10	25
32.	1501-01 -1502-09	Colombia–Ecuador	875	23	2.6	186	344	509	21.8	24	2.7	236	9	26
33.	1504-00 -1505-14	Peru–Chile-North	1363	63	4.6	43	72	86	3.1	1	0.1	69	1	4
34.	1507-01 -1508-02	Chile-Central	975	26	2.8	146	217	235	16.0	8	0.9	171	4	22
35.	1508-021 -1508-056	Chile-South	469	11	2.4	7	11	12	1.5	1	0.2	11	1	18
36.	1600-01 -1600-17	West Indies	632	17	2.7	25	38	51	4.0	5	1.4	59	3	5
37.	1900-07 -1900-14	South Sandwich	395	9	2.3	9	16	16	2.3	0		15	1	7
38.	0201-01 -0202-16	Africa-East	2188	82	3.7	59	82	85	2.7	2	0.1	59	2	42
39.	0303-01 -0303-011	Comores	72	2	3.6	9	22	22	12.6	0		22	2	77
40.	0303-02	Reunion	—	1	—	62	100	147	—	0		122	0	93
41.	0404-04 -0404-04	Samoa	309	4	1.3	8	9	10	2.6	0		4	0	75
42.	1302-01 -1302-06	Hawaii	174	6	3.4	86	165	168	52.8	0		114	4	87
43.	1503-01 -1503-12	Galapagos	235	15	6.4	43	55	57	18.3	4	1.7	53	2	92
44.	1701-01 -1703-14	Iceland	473	29	6.1	37	60	113	7.8	8	1.7	149	8	38
45.	1802-01 -1802-10	Azores	312	10	3.2	7	11	27	2.2	0		33	5	36
46.	1803-01 -1803-07	Canaries	432	7	1.6	3	4	31	1.4	0		20	1	75

NOTE.: Eruption data are shown for 37 island-arc belts and 9 non-arc belts (bottom). Column 1 marks the beginning and ending volcano number (Simkin et al., 1981) of each volcanic belt and column 3 the belt length in kilometers. Belt length is the great circle distance between the most distant volcanoes in each group, but in the highly arcuate Banda belt we have measured distance along the arc. The asterisks in column 1 warn that 3 volcanoes in eastern Oregon are excluded from the Cascades belt and 2 Honduras volcanoes are excluded from Central America. Columns 4 and 5 display the total number of Holocene volcanoes and the number of volcanoes normalized per 100 km of belt length, respectively. Columns 6 to 8 contain the total number of years active since 1880, 1800, and 1500, respectively, and column 9 contains the maximum number of eruption years per 100 km of belt length (when a time period longer than the last 100 yr gives a higher figure the italicized figure is used and normalized to a per 100 yr basis). Columns 10 and 11 display the total number of eruptions with a VEI ≥ 3 since 1880 and the number normalized per 100 km, respectively. Column 12 shows the total number of eruptions (as opposed to years active) in our file for each volcanic belt, column 13 shows the number of fatal eruptions in each belt, and the percentage of eruptions producing lava flows is shown in column 14.

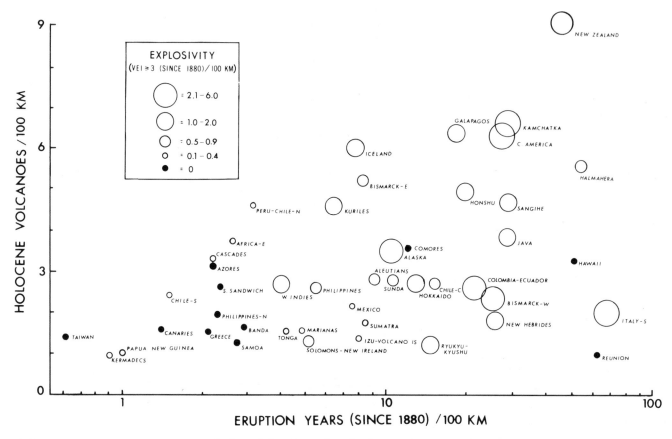

FIGURE 8.7 Comparative eruptive vigor of volcanic belts. The number of Holocene volcanoes, number of years in which volcanoes have been active since 1880, and number of large explosive eruptions (VEI ≥ 3) since 1880, all normalized per 100 km of belt length, are plotted for 45 volcanic belts. The data for Reunion's single volcano are not normalized for belt length. For 9 belts, relatively less active during the last 100 yr than during earlier portions of the historical record, the post-1800 or post-1500 data (normalized per 100 yr) are given in Table 8.1.

liminary work, but some generalizations are intriguing, and we offer these quantitative measures of volcanic vigor as a contribution to those more actively engaged in the fascinating new field described by Uyeda (1981) as "comparative subductology."

ACKNOWLEDGMENTS

We benefited from discussions with Jason Morgan, R. L. Smith, Dick Fiske, Bob Decker, and Chris Newhall, among others, and the manuscript was improved by the reviews of Stephen Self, M. J. Carr, Lindsay McClelland, and F. R. Boyd. For help in preparation of the manuscript we especially thank Kathy Auer Mihm, Mary McGuigan, and the Smithsonian Institution's Automatic Data Processing Office. The volcano data bank project received valuable support from the Volcano Hazards Program of the U.S. Geological Survey.

REFERENCES

Acharya, H. (1981). Volcanism and aseismic slip in subduction zones, *J. Geophys. Res. 86*, 335-337.

Barazangi, M., and B. Isacks (1976). Spatial distribution of earthquakes and subduction of the Nazca plate beneath South America, *Geology 4*, 686-692.

Carr, M. J. (1977). Volcanic activity and great earthquakes at convergent plate margins, *Science 197*, 655-657.

Coats, R. R. (1951). Volcanic activity in the Aleutian arc, *U.S. Geol. Surv. Bull. 947-D*, 35-47.

Cooke, R. J. S., C. O. McKee, V. F. Dent, and D. A. Wallace (1976). Striking sequence of volcanic eruptions in the Bismarck volcanic arc, Papua, New Guinea, in 1972-75, in *Volcanism in Australasia*, R. W. Johnson, ed., Elsevier, Amsterdam, pp. 149-172.

Coulon, C., and R. S. Thorpe (1981). Role of continental crust in petrogenesis of orogenic volcanic associations, *Tectonophysics 77*, 79-93.

Halbouty, M. T., *et al.* (1981). *Plate Tectonic Map of the Circum-Pacific Region (Southeast, Northeast, Southwest, Northwest and Antarctic Quadrants)*, Circum-Pacific Coun. Energy Min. Resour., Am. Assoc. Petrol. Geol., Tulsa, Okla.

Heirtzler, J. R. (1968). Sea-floor spreading, *Sci. Am. 219*, No. 6, 60-70.

Jakobsson, S. P. (1980). Petrology of Recent basalts of the eastern volcanic zone, Iceland, *Acta Nat. Isl. 26*, 103 pp.

Johnson, R. W. (1976). Late Cainozoic volcanism and plate tectonics at the southern margin of the Bismarck Sea, Papua New Guinea, in *Volcanism in Australasia*, R. W. Johnson, ed., Elsevier, Amsterdam, pp. 101-116.

Nakamura, K. (1974). Preliminary estimate of global volcanic production rate, in *Utilization of Volcanic Energy*, J. Colp and A. S. Furimoto, eds., U. of Hawaii and Sandia Corp., Hilo, Hawaii, pp. 273-286.

Newhall, C. G., and S. Self (1982). The volcanic explosivity index (VEI): An estimate of explosive magnitude for historical volcanism, *J. Geophys. Res. 87* (Oceans and Atmospheres), 1231-1238.

Nur, A., and Z. Ben-Avraham (1981). Volcanic gaps and the consumption of aseismic ridges in South America, in *Nazca Plate: Crustal Formation and Andean Convergence*, L. D. Kulm, J. Dymond, E. J. Dasch, D. M. Hussong, and R. Rodderick, eds., Geol. Soc. Am. Mem. 154, pp. 729-740.

Rittmann, A. (1962). *Volcanoes and Their Activity* (translated from 2d German edition by E. A. Vincent), Wiley, New York, 305 pp.

Ruff, L., and H. Kanamori (1980). Seismicity and the subduction process, *Phys. Earth Planet. Inter. 23*, 240-252.

Sapper, K. (1917). *Katalog der Geschichtlichen Vulkanausbrüche*, Karl J. Trübner, Strasburg, West Germany, 358 pp.

Simkin, T. (1976). Historic volcanism and eruption forecasting in Latin America, in *Geophysics in the Americas*, J. G. Tanner, and M. R. Dence, eds., Energy, Mines Resour. Canada, Publ. Earth Phys. Branch 46, pp. 79-84.

Simkin, T., L. Siebert, L. McClelland, D. Bridge, C. Newhall, and J. H. Latter (1981). *Volcanoes of the World: A Regional Directory, Gazatteer, and Chronology of Volcanism During the Last 10,000 Years*, Hutchinson Ross, Stroudsburg, Pa., 240 pp.

Smith, R. L. (1979). Ash-flow magmatism, in *Ash-flow Tuffs*, C. E. Chapin and W. E. Elston, eds., Geol. Soc. Am. Spec. Pap. 180, pp. 5-27.

Toksöz, M. N. (1975). The subduction of the lithosphere, *Sci. Am. 233, No. 5*, 89-98.

Uyeda, S. (1981). Subduction zones and back arc basins—a review, *Geol. Rundsch. 70*, 552-569.

Uyeda, S., and H. Kanamori (1979). Back-arc opening and the mode of subduction, *J. Geophys. Res 84*, 1049-1061.

Walker, G. P. L. (1980). The Taupo Pumice: Product of the most powerful known (ultraplinian) eruption? *J. Volcanol. Geotherm. Res. 8*, 69-94.

Explosive Eruptions of Kilauea Volcano, Hawaii

9

ROBERT W. DECKER *and* ROBERT L. CHRISTIANSEN
U.S. Geological Survey

ABSTRACT

Although most Kilauea eruptions produce effusive basaltic lavas, about 1 percent of the prehistoric and historical eruptions have been explosive. Multiple steam explosions from Halemaumau Crater in 1924 followed subsidence of an active lava lake. A major hydromagmatic explosive eruption in 1790 deposited most of the Keanakakoi Formation—a blanket of pumice, vitric ash, and lithic tephra that is locally more than 10 m thick around Kilauea's summit area. The Keanakakoi was deposited in multiple air-fall and pyroclastic-surge phases, probably accompanied by caldera subsidence. The Uwekahuna Ash, exposed near the base of the present caldera cliffs and on the southeast flank of Mauna Loa, was formed by a major sequence of explosive eruptions about 1500 yr before present (B.P.). Beneath the Uwekahuna Ash in a few localities are two to three similar pyroclastic deposits. The Pahala Ash, extensive on the south flank of Kilauea and on adjacent Mauna Loa, reflects many explosive eruptions from about 25,000 to 10,000 yr B.P. Although it is not clear whether parts of the much weathered and reworked Pahala are of lava-fountain or hydromagmatic origin, much of it appears to be hydromagmatic. Pyroclastic deposits are present in the Hilina Formation on the south flank of Kilauea near the coast; about six of these deposits are estimated to be 40,000 to 50,000 yr old, and others are both younger and older. None of the deposits older than 2000 yr is well dated, but if we assume generally uniform growth rates for Kilauea's shield during the past 100,000 yr, an average, but not periodic, recurrence of major explosive eruptions is about every 2000 yr; minor explosive eruptions may be more frequent. The occurrence of relatively rare but dangerous explosive eruptions probably relates to sudden disruptions of equilibrium between subsurface water and shallow magma bodies, triggered by major lowering of the magma column.

INTRODUCTION

The eruptions of Hawaiian volcanoes have a reputation for being nonexplosive and relatively benign. They produce spectacular lava fountains, sometimes more than 500 m high, but by far the bulk of the erupted material forms effusive basaltic flows of relatively low viscosity. Although these flows commonly destroy property, they move slowly enough that they seldom threaten the lives of people. Since 1823, the time of the first well-recorded Hawaiian eruptions, only one person is known to have been killed by a volcanic eruption in Hawaii.

The major eruptions with high lava fountains do produce pyroclastic materials, including basaltic pumice, small drops of basaltic glass (known as Pele's tears), and threads of volcanic glass (known as Pele's hair). Such pyroclastic deposits are purely magmatic in origin and commonly occur near and downwind from major vent areas. Among the most recent major deposits of this type is that associated with the Kilauea Iki eruption of 1959, which formed a pumice cone and downwind trail of pumice that together totaled about 2.5×10^6 m^3 in volume. The lava lake formed in that same eruption totaled about 39×10^6 m^3 in volume, and so the proportion of pyroclastic deposits to flows is about 6.4 percent by volume.

The overall proportion of pyroclastic deposits to flows for

122

Hawaiian volcanic products visible above sea level has been estimated at less than 1 percent (Macdonald, 1972, p. 355). This proportion of pyroclastic deposits to flows has been used as a measure of explosiveness for volcanoes throughout the world. For some island-arc volcanic provinces it exceeds 90 percent; the 1 percent in Hawaii is nearly the minimum. Why, then, is there any concern at all about explosive eruptions on Hawaii?

THE 1924 ERUPTION

In 1924 an unusual explosive eruption occurred from Halemaumau crater of Kilauea Volcano (see Figure 9.1; Finch and Jaggar, 1924; Jaggar and Finch, 1924). Before 1924, Halemaumau, an oval pit about 530 m across, had contained an active lava lake for many decades; extensive overflows of this lava lake covered large parts of the floor of Kilauea caldera in 1919 and 1921. The east rift zone of Kilauea had been inactive for nearly 40 yr, until a series of small eruptions broke out along it in 1922 and 1923. At the beginning of 1924 the lava-lake surface in Halemaumau was 50 m below the rim. During February 1924 the lava lake subsided and disappeared, and the floor of Halemaumau sank to a level 115 m below the rim. During March and April 1924 an earthquake swarm along the middle and lower east rift zone indicated that magma was being injected into the rift zone; the subsiding magma column at the summit was apparently related to this downrift intrusion. The earthquakes in the lower east rift, 50 km east of Halemaumau, increased in late April, and major ground cracks as much as 1 m wide, appeared in the Kapoho area. Subsidence of 1 to 3 m along a 1-km-wide graben parallel to the lower east rift indicated that major extension of the rift was taking place, possibly

associated with a submarine eruption but certainly associated with magmatic injection into the rift zone below sea level.

Sinking of the floor in Halemaumau was renewed in late April, and by May 6, 1924 the floor was 200 m below the rim. Subsidence continued, and on May 10 steam explosions began from Halemaumau. Up to 13 explosions per day, lasting from a few minutes to 7 h, continued for 18 days (see Figures 9.2 and 9.3). The largest explosions, which occurred on May 18, expelled ash clouds more than 2 km high; one of these explosions ejected blocks that fatally wounded a man who was 500 m southeast of Halemaumau, and other blocks, weighing as much as 8 tons, landed nearly 1 km from the crater. The dust-to block-size ejecta from all the explosions consisted largely of old lithic material from the walls and floor of Halemaumau and a few cored lava bombs. The blocks ranged in temperature from ambient to about 700°C (barely incandescent at night). Some large rock masses exposed in the walls of the collapsing crater were also observed to have a dull red glow. The gases propelling the explosions were not of the sulfurous variety normally emitted from the active lava lake. Most observers downwind of the explosion clouds considered these gases to be mainly odorless steam. Electrical storms and small showers of accretionary lapilli also were produced by the eruption clouds. Although the 18-day series of explosions deposited a field of blocks and ash about 1 km wide around Halemaumau, the total ejecta volume was estimated to be less than 1 percent of the volume of subsidence formed by the withdrawing magma column. In June 1924, after the subsidence and explosions, Halemaumau crater was 960 m wide and 400 m deep to its rubble-filled floor (see Figure 9.4).

In contrast to the 1959 Kilauea Iki pyroclastic deposits, which were entirely magmatic in origin, the pyroclastic material from Halemaumau in May 1924 was derived from older rocks, bro-

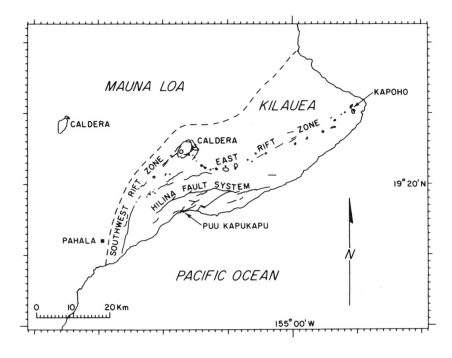

FIGURE 9.1   Index map of southeastern part of the island of Hawaii, showing all of Kilauea Volcano and part of Mauna Loa Volcano.

FIGURE 9.2 Explosive eruption from Halemaumau crater in Kilauea caldera about 8:20 A.M. (local time), May 24, 1924. Photograph by Tai Sing Loo.

ken up and expelled by steam explosions. The steam is thought to have come from subsurface water moving in to replace the subsiding magma column as it was lowered by injection far down the east rift zone; the water flashed to steam by contact with the hot but largely solid rocks that had surrounded the magma. Even though the crater was only partly clogged by collapsing and fall-back rubble, some intermittent steam generation was apparently rapid enough to cause strong explosions. In addition to steam generated by the rapid interaction of subsurface water and hot rocks, local hydrothermal systems near the high-level magma column may have flashed to steam as their hydrostatic pressures were suddenly reduced.

A 1262-m-deep well drilled for scientific information 1 km south of Halemaumau in 1973 (Zablocki *et al.*, 1974, 1976; Keller *et al.*, 1979) revealed a standing-water level in that area 491 m below the surface (611 m above sea level). This observation appears to confirm the hypothesis of Finch (1943) and Stearns (1946, p. 44) that the explosions of 1924 were produced not from a basal water table near sea level but from water confined at higher levels by dikes around the vent area. Assuming that similar conditions existed near Halemaumau in 1924 to those at the well in 1973, these explosions probably originated near a depth of 500 m below the rim of Halemaumau.

Though unusual and spectacular, the explosive eruptions of 1924 produced so little ejecta that they will probably not be evident in the eventual stratigraphic record of Kilauea, in sharp contrast to the extensive hydromagmatic deposits produced from the major explosive eruption in 1790.

THE 1790 ERUPTION

William Ellis, after an extensive tour of the island of Hawaii in 1823, wrote the first systematic account of Hawaiian erup-

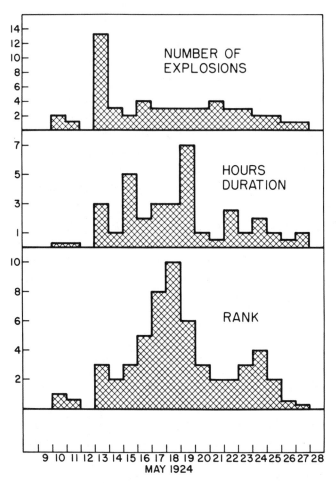

FIGURE 9.3 Characteristics of multiple explosions during 1924 explosive eruption of Kilauea. Top graph shows the number of separate explosions per day. Middle graph shows the total duration of explosions during each day. Bottom graph shows the relative ranking of various explosions on a scale of 1 to 10 (10 being the greatest) by geologists who witnessed the eruption (Finch and Jaggar, 1924).

tions, based on his own observations and on a lively oral tradition among the Hawaiian people (Ellis, 1827). Still vivid in memories at that time was an explosive eruption that had occurred at Kilauea in 1790, during the wars that first brought the Hawaiian Islands together under a single ruler. About a third of the warriors who were marching across Kilauea to oppose the dominant chief, Kamehameha, were killed in that eruption. Preliminary studies (Swanson and Christiansen, 1973; Christiansen, 1979) have reconstructed the events of the 1790 eruption and its immediate precursors and successors from their stratigraphic record—preserved in the Keanakakoi Formation (Wentworth, 1938)—and from early records of the traditional Hawaiian accounts.

Bedded pyroclastic deposits that make up the Keanakakoi Formation, locally more than 10 m thick, mantle the area around Kilauea caldera and can be traced more or less continuously for more than 20 km from the summit (see Figures 9.5 and

9.6). All who have studied these deposits agree that at least their upper part dates from the eruption of 1790. Although early researchers assumed that all of the Keanakakoi had formed in that eruption, H. Powers (1948) showed in a detailed study that these deposits include numerous unconformities. He concluded that only the uppermost layer of lithic ash and blocks (below a significantly younger layer of fallout pumice) dates from the 1790 explosive eruption.

Current study of the Keanakakoi shows that it contains both fallout and surge deposits. A generalized stratigraphy from bottom to top is as follows, with each major unit separated from its neighbors by surfaces of broad truncation (see Figure 9.7): (1) basal wind-redeposited Pele's hair and lapilli of basaltic pumice; (2) predominately well-sorted vitric ash, generally with more or less continuous planar mantle bedding; (3) somewhat less well-sorted lithic-vitric ash, commonly with wavy or lenticular bedding, accretionary lapilli in the finest layers, cross-stratification, and bedding sags beneath lithic blocks; (4) a thin local lava flow, erupted from a fissure circumferential to the caldera near its southwest margin and associated minor fallout deposits of Pele's hair and pumice; (5) lithic ash and blocks, commonly poorly sorted and with abundant accretionary lapilli, cross-stratification, and many discontinuous layers; (6) an uppermost deposit of fallout and wind-resorted pumice and Pele's hair. Units 2, 3, and 5 are the principal units of the Keanakakoi; the other units have smaller volumes. Units 1 and 6, which are wholly to partially reworked, clearly preceded and succeeded, respectively, the explosive eruptions that deposited the bulk of the formation.

Distinctive nonplanar bedding characteristics of many layers in the Keanakakoi Formation demonstrate that these layers were deposited by pyroclastic surges. It is now clear that most of the local unconformities recognized by H. Powers (1948) in the Keanakakoi have broad, open to U-shaped profiles (Figure 9.5) that reflect scouring during the emplacement of such surges. Only the surfaces between the principal units represent continuous widespread disconformities. These surfaces, however, also appear to represent mainly scour by strong pyroclastic surges, probably after pauses in the eruption sequence that may have allowed the formation of local gullies to guide the surge flows; any such gullies must have been completely modified by those surges. Careful search has revealed no clear evidence of stream erosion, channel gravel, or soil formation within the Keanakakoi section between the base of unit 2 and the top of unit 5, although groundwater oxidation at the tops of poorly permeable layers mimics soil colors locally. No remnants of vegetation have been found within the Keanakakoi other than carbonized material near its base that appears to represent vegetation killed in the eruption and buried in the initial deposit. Recent ^{14}C dating of this material (Kelley *et al.*, 1979) has consistently given nominal ages that range from less than 200 to about 350 radiocarbon years.

The features described above indicate that the bulk of the Keanakakoi Formation was deposited without significant time breaks and are consistent with its emplacement during the eruption of 1790. The traditional Hawaiian accounts of this eruption are interpreted to indicate that its most powerfully explosive phases occurred over a period of about 3 days and

FIGURE 9.4 Schematic cross sections showing conditions at Halemaumau crater before, during, and after the 1924 explosive eruption. On January 1, zone of subsurface water 1 is cooler than the H_2O boiling temperature, and zone 2 is at the H_2O boiling temperature for that depth; dash-dot line denotes approximate boundary between these two zones. As magma column subsides below the water table, equilibrium between subsurface water and magma is destroyed, and water from zone 1, rushing into broken hot rocks surrounding subsiding magma column, flashes into steam. Water from zone 2 also flashes into steam as hydrostatic pressures are suddenly reduced. Explosions are multiple because of progressive lowering of the magma column; the explosions stop when exhaustion of thermal energy above subsiding magma column leaves a collapsed crater on June 1. Drawings by Maurice Sako, U.S. Geological Survey.

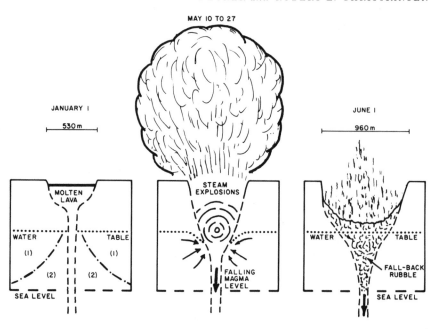

that the deaths of the Hawaiian warriors occurred during a rapidly emplaced pyroclastic surge (Swanson and Christiansen, 1973). The Keanakakoi Formation appears to contain the stratigraphic record of a nonrandom sequence, the principal events of which occurred within a fairly short period whose exact duration remains uncertain—possibly a few days to a few weeks. The sequence of units is interpreted as follows: (1) The magma column before the explosive eruption stood high and at times produced lava fountains in the caldera high enough to generate the extremely inflated variety of pumice called *reticulite* (or thread-lace scoria). This state probably was analogous to but more gas-rich than that of the high-standing magma column that generated the long-lived lava lakes of Halemaumau before the explosive eruption of 1924. During such a time the subsurface water would have developed a quasi-equilibrium boiling front in the heated rocks around the magma column. (2) Rapid lowering of the magma column, possibly related to the lava flows erupted about 1790 on Kilauea's lower east rift zone (Macdonald, 1941; Holcomb, 1980), allowed the subsurface water near Kilauea's summit to enter the vicinity of the emptied magmatic conduits; contact of that water with magma in the subsiding column chilled and fractured the magma, and explosive boiling of the water caused eruptions of the type called *hydromagmatic* and generated fallout deposits and some pyroclastic surges close to the vent. (3) As the magma level continued to drop with continuing injection into the rift zone, more water entered the conduit system and its surrounding envelope of hot rocks and the eruptions became more intense. Progressively more lithic material was erupted along with the vitric ash, and pyroclastic-surge deposition became prevalent.

FIGURE 9.5 Pyroclastic deposits of the Keanakakoi Formation in sand wash about 1 km southwest of Kilauea caldera. Average height of gully wall is about 5 m. Major disconformities, apparently caused by scouring from pyroclastic surges, are clearly visible.

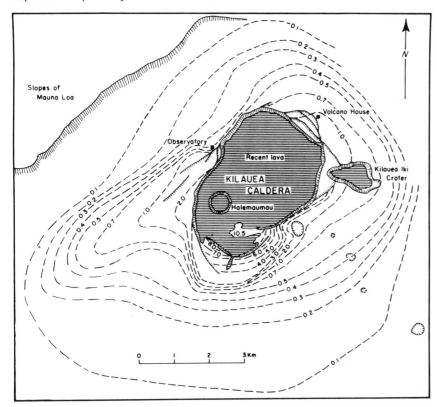

FIGURE 9.6 Isopach map of the approximate thickness (in meters) of the Keanakakoi Formation. Because of internal disconformities, thicknesses shown are generally less than reconstructed stratigraphic thicknesses. Areas shown as recent lava covering the Keanakakoi are those up to 1930; additional areas have been covered by lavas since 1930. Adapted from Stearns and Clark (1930).

(4) With surface subsidence of the summit area accompanying lowering of the magma column, magma from an isolated, shallow storage chamber was able to open a circumferential fracture and erupt as a small lava flow. (5) As magma dropped below the level of the explosions, increasing amounts of water entered the system, and violent steam blasts ensued; only lithic ash and blocks were ejected in these blasts, commonly in pyroclastic surges. The major caldera subsidence that followed destroyed the eruptive conduit and vent system, to end the eruptions. (6) Ultimately, but some time before the visit of William Ellis in 1823, magma returned to surface levels and again erupted high lava fountains in the caldera.

THE UWEKAHUNA ASH

The Uwekahuna Ash is a hydromagmatic deposit exposed near the base of the present caldera cliffs at Kilauea and on or near the surface of some areas on the southeast flank of Mauna Loa. These deposits mantle an irregular buried topography, including a steep slope on the west wall of Kilauea caldera, that may be part of an earlier caldera rim.

A 5-m-thick section of the Uwekahuna Ash along the west wall of Kilauea caldera was described by S. Powers (1916, p. 230) as consisting of ". . . yellow ash with some rock fragments 1-2 inches in diameter, lava droplets, thread-lace scoria, and a few bombs 6 inches in length." This exposure was buried by a lava flow in 1919.

Work in progress by D. Dzurisin, J. P. Lockwood, and T. J. Casadevall (U.S. Geological Survey) indicates that the Uwekahuna Ash, which is comparable to or greater in volume and extent to the Keanakakoi Formation, was formed by similar explosive eruptions. They recognize a basal deposit of lithic blocks and lapilli stained pink to dull red by an oxidized vitric matrix. This basal layer is overlain by fine-bedded vitric ash and dust with accretionary lapilli layers, followed by lithic ash containing scattered lithic blocks and cored lava bombs. Above this lithic ash is another unit of vitric ash and dust containing dense lithic and black glass fragments, followed by a coarser unit of lithic blocks mixed with some pumice and glassy-lava drops and scoria. The uppermost unit is dominantly pumice and glass fragments containing some lithic ash and blocks. Lateral variations in the stratigraphy, and internal cross-stratification and unconformities, indicate a pyroclastic-surge origin for parts of the Uwekahuna Ash.

A zone of 2.4 m of pyroclastic material at a depth of 34 m in a drill hole south of Halemaumau may be part of the Uwekahuna Ash.

Radiocarbon dates on charcoal from overlying and underlying flows, and on charcoal from the lower part of the Uwekahuna Ash near the town of Volcano, indicate an approximate ^{14}C age of 1500 yr for this episode of explosive hydromagmatic eruptions (Kelley *et al.*, 1979; J. P. Lockwood, U.S. Geological Survey, personal communication, 1982).

At two exposures of the Uwekahuna Ash—one in a roadcut near the town of Volcano and the other in a cesspool on the

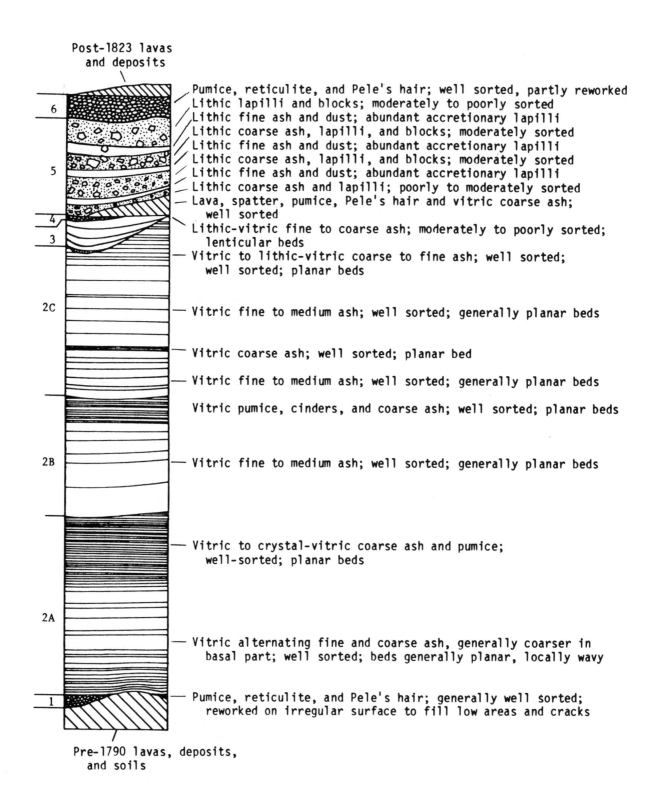

Post-1823 lavas
and deposits

Pumice, reticulite, and Pele's hair; well sorted, partly reworked
Lithic lapilli and blocks; moderately to poorly sorted
Lithic fine ash and dust; abundant accretionary lapilli
Lithic coarse ash, lapilli, and blocks; moderately sorted
Lithic fine ash and dust; abundant accretionary lapilli
Lithic coarse ash, lapilli, and blocks; moderately sorted
Lithic fine ash and dust; abundant accretionary lapilli
Lithic coarse ash and lapilli; poorly to moderately sorted
Lava, spatter, pumice, Pele's hair and vitric coarse ash;
   well sorted
Lithic-vitric fine to coarse ash; moderately to poorly sorted;
   lenticular beds
Vitric to lithic-vitric coarse to fine ash; well sorted;
   well sorted; planar beds

Vitric fine to medium ash; well sorted; generally planar beds

Vitric coarse ash; well sorted; planar bed

Vitric fine to medium ash; well sorted; generally planar beds

Vitric pumice, cinders, and coarse ash; well sorted; planar beds

Vitric fine to medium ash; well sorted; generally planar beds

Vitric to crystal-vitric coarse ash and pumice;
   well-sorted; planar beds

Vitric alternating fine and coarse ash, generally coarser in
   basal part; well sorted; beds generally planar, locally wavy

Pumice, reticulite, and Pele's hair; generally well sorted;
   reworked on irregular surface to fill low areas and cracks

Pre-1790 lavas, deposits,
and soils

FIGURE 9.7 Idealized stratigraphic section of the Keanakakoi Formation.

southeast flank of Mauna Loa, 7 km west of Halemaumau Crater—are similar pyroclastic deposits, with an aggregate thickness of about 1 m, which immediately underlie the Uwekahuna. Oxidized layers in these deposits suggest, but do not prove, that they represent two or three episodes of explosive hydromagmatic eruptions. Carbonized material from the uppermost of these units beneath the Uwekahuna Ash near the town of Volcano yields a ^{14}C age of about 2100 yr. In both these exposures the relatively unaltered pyroclastic deposits that underlie the Uwekahuna Ash clearly overlie the deeply weathered, more indurated pyroclastic material normally identified as the Pahala Ash.

THE PAHALA ASH

About 32 km southwest of Kilauea caldera is the town of Pahala. The rich agricultural soil in this area is developed on a volcanic-ash section, as much as 15 m thick (see Figure 9.8). The Pahala Ash in the area is apparently the partially reworked accumulation of several pyroclastic eruptions of Kilauea, deposited on old Mauna Loa lavas. Much of the Pahala ash has been modified by weathering and by wind and water reworking during its long period of accumulation, and its original character is difficult to decipher. Stearns and Clark (1930), Wentworth (1938), and Stearns and Macdonald (1946) concluded that most of the Pahala Ash is of lava-fountain origin but allowed the possibility of an explosive hydromagmatic origin. Stone (1926) and Fraser (1960) preferred hydromagmatic eruptions as the major source; we agree with their interpretation. Physical evidence for hydromagmatic origin includes accretionary lapilli and some lithic fragments as well as the predominant vitric ash, much of which consists of angular fragments like those of hydromagmatic deposits rather than of the Pele's hair and pumice glass shards typical of distal lava-fountain deposits.

The age of the Pahala Ash ranges from approximately 10,000 to 25,000 yr, on the basis of radiocarbon dates on charcoal from beneath lavas within or overlying all or part of the ash sequence (Kelley *et al.*, 1979; J. P. Lockwood and P. W. Lipman, U.S. Geological Survey, personal communication, 1981). Although a firm interpretation is not possible at this stage of investigation, it appears probable that the Pahala Ash south of Kilauea caldera includes the accumulated and reworked deposits of several major explosive hydromagmatic eruptions from Kilauea caldera. If this interpretation is correct, the eruptions that resulted in the Pahala Ash may have deposited material at distances considerably farther than did the eruptions responsible for the Keanakakoi Formation and Uwekahuna Ash.

PYROCLASTICS IN THE HILINA FORMATION

The previously named Hilina Volcanic Series, now renamed Hilina Formation, is an approximately 250-m-thick sequence of Kilauea lava flows and pyroclastic deposits that occurs beneath the Pahala Ash. These rocks form the oldest exposures of Kilauea's eruptive products, and their age is estimated by Easton and Garcia (1980) to range from about 25,000 to as much as 100,000 yr. The best exposures of the Hilina are on the fault escarpments of the Hilina Pali and on the faulted and wave-cut escarpment of Puu Kapukapu, 12 to 15 km south of Kilauea caldera.

Pyroclastic deposits are common in the Hilina Formation exposures. Easton (1978) mapped 8 horizons on the Hilina Pali, and 9 layers of ash beds are visible on the 300-m-high cliffs at Puu Kapukapu. The major pyroclastic unit capping the summit of Puu Kapukapu is the Pahala Ash, which is 12 m thick at that location.

Below the Pahala Ash, Easton and Garcia (1980) named several pyroclastic units in the Hilina Formation, which are separated by major sequences of lava flows. From youngest to oldest these pyroclastic units are as follows: (1) the Mo'o Member, a 50- to 250-cm-thick deposit of yellow-brown ash and palagonite (hydrated basaltic glass) containing some accretionary lapilli; (2) the Pohakaa Member, comprising as many as 6 pyroclastic layers of vitric ash, palagonite, and soil ranging in thickness from 1 to 4 m, separated by intercalated lava flows— some layers contain abundant lithic and crystal fragments and appear to be pyroclastic-surge deposits; (3) the Kahele Member, a 10- to 125-cm-thick deposit of crudely bedded red clay containing some palagonite and, in one exposure, a layer of glassy vesicular scoria and glass fragments; and (4) the Halape Member, a 10- to 50-cm-thick deposit of poorly bedded clay and palagonite. The age of the Pohakaa ash layers and intercalated flows is estimated at 40,000 to 50,000 yr.

The Hilina Formation in the areas of exposure consists of about 95 percent flows and 5 percent pyroclastic deposits (Easton and Garcia, 1980). This relatively high proportion of pyroclastics to flows, compared with Hawaiian subaerial volcanic products in general, may be due in part to the semidownwind location of exposures of the Hilina Formation in relation to

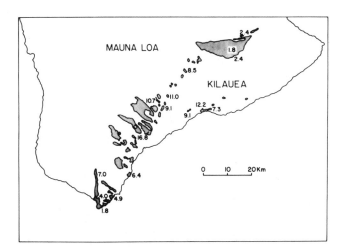

FIGURE 9.8 Exposures of the Pahala Ash in south half of Island of Hawaii. Thicknesses in meters. Deposits northeast of Kilauea's summit region may have originated mainly from Mauna Kea Volcano, situated in the northern part of the island; other exposures probably originated mainly from Kilauea Volcano. Adapted from Stearns and Macdonald (1946).

Kilauea's summit. Although there is no unequivocal evidence that the pyroclastic deposits within the Hilina Formation are of explosive hydromagmatic origin, the thickness of the layers and the occurrence of surge-bedded layers and locally abundant lithic material suggest that they are. Furthermore, there is no direct evidence that they are not hydromagmatic.

Although the exposed lavas and pyroclastic deposits of Kilauea are no older than about 100,000 yr, the general subsidence of the island of Hawaii implies that subaerial lavas and pyroclastic deposits probably continue well below sea level.

DISCUSSION

The general similarity of the older volcanic products of Kilauea to its eruptive products of the past 200 yr suggests that the 1790 Keanakakoi explosive eruption was not a unique event. It is difficult, however, to establish a recurrence interval with any precision. The total number of preserved and exposed pyroclastic deposits is about 15, but the Pahala Ash and some of the pyroclastic units within the Hilina Formation probably result from the accumulation of several episodes of major explosive eruptions. During accumulation of the Pahala Ash, few lava flows moved down the south slopes of Kilauea over a time span of about 15,000 yr. Perhaps flows were largely contained within a major summit caldera during this interval, or perhaps most of the eruptive activity was along the east rift zone at Kilauea. R. M. Easton (U.S. Geological Survey, personal communication, 1981) noted that the Pohakaa deposits, which resemble the Pahala Ash, were deposited over a period of 10,000 yr during which few flows reached the south flank of Kilauea.

Allowing for the multiple ash units, such as the Pahala Ash and the Pohakaa Member, the total number of major explosive hydromagmatic eruptions indicated by the exposed pyroclastic deposits is about 25 during the past 100,000 yr (see Table 9.1). Only the largest or most recent explosive eruptions are represented by these exposed pyroclastic deposits; materials from older explosive eruptions comparable in size to the Keanakakoi and Uwekahuna events are largely or completely buried by the copious lava flows from the summit and upper parts of the rift zones of Kilauea. In the deeper part of the drill hole south of Halemaumau (Zablocki *et al.*, 1974, 1976; Keller *et al.*, 1979), there are many intervals with larger than normal hole diameter; these intervals may have been formed by washing out of softer pyroclastic layers during drilling. The incomplete coring, however, and the contamination of drill cuttings with pyroclastic material from sections higher up in the well preclude any firm interpretation of the total number of pyroclastic units penetrated by the drill hole.

The number of buried and completely unexposed Kilauea pyroclastic deposits is probably at least equal to that of the exposed deposits (25). Therefore, a reasonable but not unequivocal estimate of the total number of explosive eruptions of Kilauea as large or larger than the 1790 Keanakakoi eruption is about 50 in the last 100,000 yr. On the basis of this estimate, major explosive eruptions have probably recurred from Kilauea's summit on an average of about once every 2000 yr during the past 100,000 yr or more. We have no reason to assume that this 2000-yr interval is even approximately periodic; very large explosive eruptions may have recurred more frequently during deposition of the Pahala Ash and the Pohakaa ash beds and less frequently before and since. Minor explosive

TABLE 9.1 Data on Exposed Pyroclastic Deposits of Apparent Explosive Origin from Kilauea's Summit Region[a]

Name	Number of Separate Deposits	Age (years)	Observed Maximum Thickness (meters)	Approximate Area in km²[b]	Approximate Volume in m³[b]
1924	1	58	trace	10	10^5
Keanakakoi Formation	1	192	11	300	10^8-10^9
Uwekahuna Ash	1 or 2	~ 1,500	5	300?	unknown
Unnamed	2 or 3	≳ 2,100	1	300?	unknown
Pahala Ash	5 to 10?	~10,000 to 25,000	16[c]	1000?	unknown
Mo'o Member[d]	1	~30,000	2.5	1000?	unknown
Pohakaa Member[d]	6	~40,000 to 50,000?	15[c]	1000?	unknown
Kahele Member[d]	1	~60,000?	1	1000?	unknown
Halape Member[d]	1	~70,000?	0.5	1000?	unknown

[a]Data from S. Powers (1916), Finch and Jaggar (1924), Stearns and Clark (1930), Easton (1978), and D. Dzurisin and J. P. Lockwood (U.S. Geological Survey, written communication, 1982).

[b]Area and volume estimates are for deposits on land.

[c]Cumulative thickness of pyroclastic layers.

[d]Nomenclature of Easton and Garcia (1980).

eruptions similar to the 1924 event, which were probably more frequent than the major explosive eruptions, are not evident in the stratigraphic record.

Although most major explosive eruptions at Kilauea appear to have originated in the summit area, some evidence for smaller explosive eruptions exists near sea level in the rift zones. Kapoho Cone and the Puulena craters along the lower east rift have associated hydromagmatic deposits, and part of the 1960 Kapoho eruption in this same area was characterized by nearly explosive ejection of black clouds of condensed steam laden with lithic ash. One adjacent segment of the same 1960 fracture that was erupting steam blasts was erupting incandescent lava fountains. This unusual eruption has been attributed either to contact of subsurface water with the intruding dike or to explosive boiling of an existing shallow hydrothermal reservoir through an extension of the eruption fracture.

The general cause of hydromagmatic eruptions at Kilauea appears to be closely connected with explosive boiling of subsurface water. The low gas content of Hawaiian magmas—about 0.5 percent by weight (Moore, 1970)—and their relatively low viscosity indicate that purely magmatic eruptions of pyroclastic material probably do not exceed in violence the spectacular but generally harmless ejection of incandescent lava fountains. Steam-blast eruptions from the summit area, such as the one in 1924, seem best explained by a sudden lowering of the magma column beneath Kilauea, which would cause intermittent explosions of steam from movement of subsurface water into the zone evacuated by the shallow magma. A larger version of this same process probably caused the 1790 Keanakakoi, the 1500 yr ago Uwekahuna, and perhaps earlier explosive eruptions. It is likely that sudden flashing of large hydrothermal systems beneath the summit of Kilauea also contributed to the major explosive eruptions. In all cases a sudden lowering of the magma column by a far-downrift intrusion or eruption probably triggered the instability needed to initiate the sudden conversion of thermal into mechanical energy.

Explosive release of superheated H_2O or CO_2 seems to be the fundamental mechanism of all explosive volcanic eruptions (see Chapter 12). Given the complexity of both magmatic and hydrologic systems, that fundamental mechanism can take many forms.

VOLCANIC HAZARDS IN HAWAII

The old proverb that "familiarity breeds contempt" is certainly true with regard to volcanic hazards in Hawaii. The historical record of eruptions covers only the last 160 yr, a period that has been one of many spectacular but relatively harmless effusive eruptions of lava fountains and flows. Only one small explosive eruption in 1924, more curious than dangerous, has occurred during this period. In contrast, the geologic record indicates that major explosive eruptions do occur, the latest one less than 200 yr ago. If the Hawaiian Volcano Observatory had been at its present location in 1790, it would have been destroyed. Our present concept of the onsets of these explosive

eruptions suggests that they can be identified in time to evacuate the danger area. So says the theory; in practice, we must remember to try to become more familiar with our subject without becoming contemptuous.

ACKNOWLEDGMENTS

This paper was reviewed by Donald W. Peterson, Robert I. Tilling, John P. Lockwood, R. Michael Easton, and F. R. Boyd; we thank them for their careful reading and helpful comments and suggestions.

REFERENCES

Christiansen, R. L. (1979). Explosive eruption of Kilauea Volcano in 1790 [abstr.], in *Hawaii Symposium on Intraplate Volcanism and Submarine Volcanism*, Hilo, Hawaii, July 16-22, 1979, p. 158.

Easton, R. M. (1978). The stratigraphy and petrology of the Hilini Formation, Master's thesis, Department of Geology, U. of Hawaii, Honolulu, 274 pp.

Easton, R. M., and M. O. Garcia (1980). Petrology of the Hilina Formation, Kilauea Volcano, Hawaii, *Bull. Volcanol. 43*, 657-674.

Ellis, W. (1827). *Journal of William Ellis: Narrative of a Tour of Hawaii, or Owyhee: With Remarks on the History, Traditions, Manners, Customs and Language of the Inhabitants of the Sandwich Islands* (reprinted in 1963), Advertising Publishing Co., Honolulu, 342 pp.

Finch, R. H. (1943). Lava surgings in Halemaumau and the explosive eruptions in 1924, *Volcano Lett., No. 479*, pp. 1-3.

Finch, R. H., and T. A. Jaggar (1924). Monthly Bulletin of the Hawaiian Volcano Observatory, U.S. Department of Agriculture, Weather Bureau, Washington, D.C., Vol. 12, No. 1-6, pp. 1-69.

Fraser, G. D. (1960). Pahala Ash—an unusual deposit from Kilauea Volcano, Hawaii, *U.S. Geol. Surv. Prof. Pap. 400-B*, pp. B354-B355.

Holcomb, R. T. (1980). Kilauea volcano, Hawaii: Chronology and morphology of the surficial lava flows, *U.S. Geol. Surv. Open-File Rep. 80-796*.

Jaggar, T. A., and R. H. Finch (1924). The explosive eruption of Kilauea in Hawaii, 1924, *Am. J. Sci., Ser. 5., Vol. 8*, 353-374.

Keller, G. V., L. T. Grose, J. C. Murray, and C. K. Skokan (1979). Results of an experimental drill hole at the summit of Kilauea Volcano, Hawaii, *J. Volcanol. Geotherm. Res. 5*, 345-385.

Kelley, M. L., E. C. Spiker, P. W. Lipman, J. P. Lockwood, R. T. Holcomb, and M. Rubin (1979). Radiocarbon dates XV: Mauna Loa and Kilauea Volcanoes, Hawaii, *Radiocarbon 21*, 306-320.

Macdonald, G. A. (1941). Lava flows in eastern Puna, *Volcano Lett., No. 474*, pp. 1-3.

Macdonald, G. A. (1972). *Volcanoes*, Prentice-Hall, Englewood Cliffs, N.J., 510 pp.

Moore, J. G. (1970). Water content of basalt erupted on the ocean floor, *Contrib. Mineral. Petrol. 28*, 272-279.

Powers, H. A. (1948). A chronology of the explosive eruptions of Kilauea, *Pac. Sci. 2*, 278-292.

Powers, S. (1916). Explosive ejectamenta of Kilauea, *Am. J. Sci., Ser. 4, Vol. 41*, 227-244.

Stearns, H. T. (1946). *Geology of the Hawaiian Islands*, Division of Hydrography, Territory of Hawaii, Bull. 8, 106 pp.

Stearns, H. T., and Clark, W. O. (1930). *Geology and Groundwater*

Resources of the Kau District, Hawaii, U.S. Geol. Surv. Water-Supply Pap. 616, 194 pp.

Stearns, H. T., and G. A. Macdonald (1946). Geology and groundwater resources of the Island of Hawaii, *Hawaii Division of Hydrography Bull.* 9, 363 pp.

Stone, J. B. (1926). *The Products and Structure of Kilauea*, B. P. Bishop Museum Bull. 33, Honolulu, 60 pp.

Swanson, D. A., and R. L. Christiansen (1973). Tragic base surge in 1790 at Kilauea Volcano, *Geology 1*, 83-86.

Wentworth, C. K. (1938). *Ash Formations on the Island of Hawaii*, Hawaiian Volcano Observ. Spec. Rep. 3, Hawaii Volcano Research Assoc., Honolulu, 183 pp.

Zablocki, C. J., R. I. Tilling, D. W. Peterson, R. L. Christiansen, G. V. Keller, and J. C. Murray (1974). A deep research drill hole at the summit of an active volcano, Kilauea, Hawaii, *Geophys. Res. Lett. 1*, 323-326.

Zablocki, C. J., R. I. Tilling, D. W. Peterson, R. L. Christiansen, and G. V. Keller (1976). A deep research drill hole at Kilauea Volcano, Hawaii, *U.S. Geol. Surv. Open-File Rep.* 76-538, 35 pp.

Chronology and Character of the May 18, 1980, Explosive Eruptions of Mount St. Helens

10

JAMES G. MOORE
U.S. Geological Survey

CARL J. RICE
The Aerospace Corporation

ABSTRACT

A study of satellite infrared sensor data and survivor photographs provides an extensive record of the events of the first few minutes of the May 18, 1980, explosive eruptions of Mount St. Helens. Timing of events and photographs is based on satellite observations combined with videotape and other surface information. The detachment of the giant rockslide-avalanche relieved confining pressure on the intruding cryptodome and triggered the first explosive episode at 08 hr 32.7 min PDT (15:32.7 UT). These explosions generated a pyroclastic surge that swept north at more than 90 m/sec, overriding the avalanche and overtaking its toe at about 15:33.8 UT. The hot fragmenting dacite within the avalanche continued emitting explosions as it was carried downslope. Block II of the avalanche, which contained the bulk of the dacite cryptodome, arrived 8 km north at the Toutle River and the southern margin of Spirit Lake at about 15:34.3 UT. A major northern explosion occurred near the Toutle River at this time, perhaps triggered when block II collided with the north wall of the valley. Alternately, the explosion could have resulted by mixing of the fragmented hot dacite with surface water in the valley of the North Fork of the Toutle River and perhaps with Spirit Lake. This northern explosion at 15:34.3 UT, nearly 2 min after the first emissions (1) was apparently the largest explosion of the eruption and caused much of the damage and tree blowdown; (2) generated a pressure pulse that triggered atmospheric condensation above 6-km elevation and partly obscured satellite observations of ground-level features; (3) produced an abrupt acceleration of the front of the preexisting pyroclastic surge to velocities greater than 150 m/sec; (4) caused an ash cloud to rise from the explosion locus at 70 m/sec, faster than any previous rise; and (5) contributed to the geographic distribution of damage and pyroclastic surge deposits. The concept of a series of explosions from the moving avalanche better explains features of the eruption than does the common interpretation of a single explosive episode.

INTRODUCTION

The precursor events culminating in the explosive eruptions of Mount St. Helens on May 18, 1980, began on March 20 with the onset of shallow earthquakes beneath the north flank of the mountain. This seismic activity rapidly intensified, and the first crater-forming explosion occurred on March 27. Similar small steam-blast (phreatic) explosions erupted nonjuvenile lithic ash intermittently until May 18. Beginning with the first explosion in March, an eastward-trending fracture system developed across the summit and bounded a region of the northern flank that was fractured and uplifted. Subsequent measurements indi-

cated that this area continued expanding, creeping steadily northward at a rate of about 2 m/day until May 18. This bulge on the northern flank was apparently caused by the intrusion of a shallow dacite cryptodome (see Figure 10.1), whose volume (based on photogrammetrically measured volumetric increase of the volcano) attained 0.11 km³ by May 18 (Moore and Albee, 1981).

At 15:32.2 UT (08:32.2 PDT) on May 18, a magnitude 5+ earthquake was immediately followed by landslide failure of the north flank and then the summit region, including the cryptodome system (Figure 10.1). By 15:32.7, explosions from the moving landslide and from the head of the slide near the

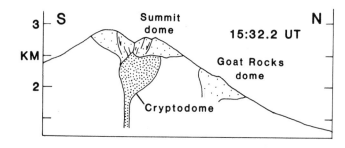

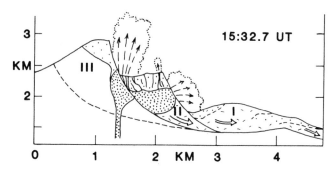

FIGURE 10.1 Sections through Mount St. Helens showing early development of rockslide-avalanche into three blocks (I, II, III) and explosions emitted primarily from block II. Fine stipple, preexisting domes; coarse stipple, 1980 cryptodome.

previous summit began to feed an ash-rich density flow (pyroclastic surge) that moved downslope primarily to the north. Ash clouds billowed up from the moving surge, which overtook the landslide about 6 km north of the summit and hence obscured further landslide movement and subsequent events within the developing amphitheater crater. The hot pyroclastic surge moved primarily west, north, and east for 15 to 25 km, blowing down trees over an area of 600 km² and leaving a thin layer (averaging about 30 cm thick) of surge deposits (see Figure 10.2).

Military satellite data, which became available subsequent to the initial investigations of the May 18 eruption, yield new insight into the explosive events. These data provide well-timed displays of the cloud margins and improve our knowledge of the absolute timing of many photographs taken from different vantage points.

PHOTOGRAPHIC MEASUREMENTS

Although automatic time-lapse cameras north of the crater were destroyed by the eruption, valuable photographic records were made by survivors beyond the limit of blast destruction. Among the most useful are those by Gary Rosenquist from 17 km northeast of the crater, those by John Christiansen and Vince Larson from Mount Adams (53 km east), those by Paul and Carol Hickson from 15 km east, and a videotape by Ed Hinkle from Silver Lake (48 km west-northwest). The timing of these photographs can be estimated by early onsite measurements;

by comparing cloud morphology between photographs; by comparing these times with the events shown on the videotape, which is internally timed but has an indefinite absolute beginning; and by correlating photographs and satellite images. The onsite measurements include the seismically recorded initial earthquake (15:32.2 UT) as well as the taped message of Gerry Martin, located 12 km to the north (Figure 10.7). Martin reported the earthquake and observed Coldwater II camp (10 km north) covered by the surge at 15:34.0-15:34.2; his radio transmission was cut at 15:34.3. The videotape shows the summit and south flank only; it indicates that the surge that swept the south flank emerged from the crater at 15:33.7, was well developed at 15:33.8, and reached the south base of the cone by 15:35.1. The videotape also shows a major collapse of the south crater rim at 15:33.4. This timing, provided by Stephen Malone (University of Washington, written communications, 1980, 1981), is preliminary and subject to revision. Timing of the first 0.5 min is based in part on the analysis of the Rosenquist photographs (Voight, 1981).

The photographic record indicates that, shortly after the earthquake at 15:32.2, landslide failure began and the north flank of the volcano broke into two giant blocks (designated blocks I and II in Figure 10.1), which moved north at velocities of 50 to 80 m/sec (Voight et al., 1981). About 30 sec after detachment, explosions were emitted from the top and toe of slide block II (Figure 10.1). These explosions intensified, coalesced, and fed a rising ash cloud and a northward-moving pyroclastic surge. The most vigorous explosions emanated from the steep toe of block II, which itself was moving as part of the landslide.

The timed photographs from Mount Adams (see Figures 10.3 and 10.4) show the movement of the pyroclastic surge and the development of ash clouds rising from it. The surge moved north in excess of 90 m/sec (see Figure 10.5), somewhat faster than the avalanche, which had a headstart of about 3 km, and overtook the avalanche toe at about 15.33.8, 8 km north (all distances are measured from the post-May 18 southern or back rim of the crater). At about 12 km north the surge accelerated to some 150 m/sec but then began slowing to less than 100 m/sec after reaching 20 km north.

The small pyroclastic surge that swept down the south flank was much slower, traveling at about 30 m/sec (Figure 10.5). We believe it was fed from explosions originating behind (south of) block III, which completed its detachment at about 15:33.4.

The Mount Adams photograph taken at 15:35.0 (Figure 10.3D) shows the abrupt appearance of a dark horizontal atmospheric cloud 14 km long (north-south) at an elevation of 6.5 km that was not present in previous photographs. This cloud, which is higher than the rising ash clouds, is interpreted as a condensation layer resulting from a pressure pulse that rapidly drove a nearly water-saturated atmospheric layer below the dew point. The rapid development of this cloud is apparent from several different series of photographs. The initial dark condensation cloud was centered about 8 km north (Figure 10.4), and the disturbance that produced it was probably beneath its center. The fact that the earlier explosions at the head of the avalanche did not disturb the overlying atmosphere suggests that this later northern disturbance was of greater magnitude. Subse-

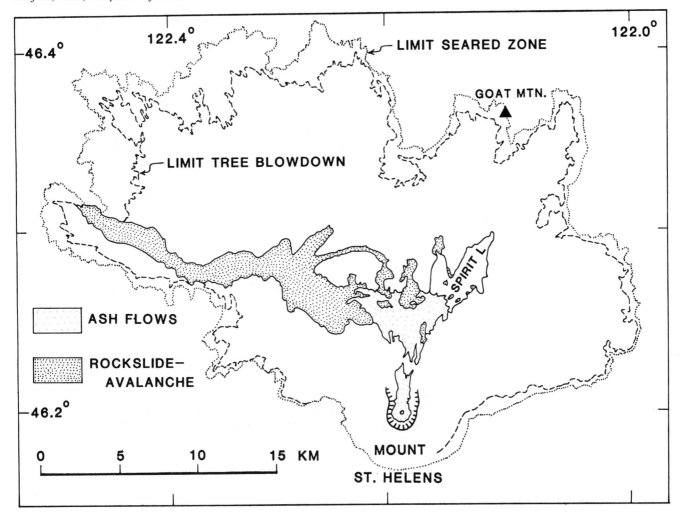

FIGURE 10.2   Map of Mount St. Helens showing deposits and effects of the eruption on May 18, 1980. Modified from Lipman (1981).

quent photographs show dramatic lateral and vertical growth of the atmospheric cloud. Within about 10 min it grew to a giant mushroom 25 km high and 70 km in diameter. The base of the mushroom cap remained at or slightly above the 6-km level of initial condensation, and the stalk, 20 km in diameter, rose from the entire devastated area.

The pressure disturbance that rapidly produced the horizontal atmospheric cloud is believed to have been generated by a northern explosion emanating from the moving landslide at about 15:34.3. Alternately, the horizontal cloud may have been produced by convective rise and condensation in a moisture-rich air column above the hot pyroclastic surge or by a negative pressure pulse rising above the moving landslide caused by displacement of air in front of the advancing avalanche and surge. The convection model is considered unlikely because of the rapid growth of the cloud and the northern position of the most dense part of the cloud when first observed. Convective rise would begin above the summit where the first explosions occurred. The negative pressure model is unlikely because the

horizontal cloud appeared so late, more than 1 min after the toe of the avalanche and surge arrived at the Toutle River valley (Figure 10.5).

Other evidence of a northern explosion was the appearance 12 to 13 km to the north of a rapidly rising ash plume (Figures 10.3E and 10.4) that rose more rapidly (70 m/sec) than any other one measured. Backward projection of this plume's trajectory indicates its origin from the ground 8 km north at about 15:34.3 (Figure 10.4). Finally, the acceleration of the pyroclastic surge front at 15:34.5 is attributed to this northern explosion (Figure 10.5). Satellites observed this acceleration directly, as is described in the next section.

The low-level white condensation clouds capping the eastern front of the pyroclastic surge, which first appeared at 15:34.5 (Figure 10.3C), may also have formed in response to a pressure pulse originating at the northern explosion. Alternately, they may have resulted from copious moisture picked up by the surge cloud when the avalanche toe plowed into Spirit Lake and several stream valleys at about 15:33.5 (Figure 10.5).

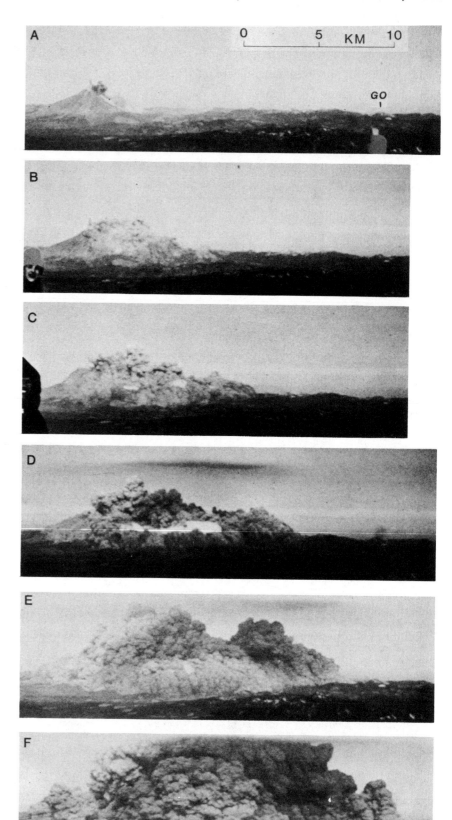

FIGURE 10.3 Photographs of developing pyroclastic surge as seen from Mount Adams, 53 km east of the summit of Mount St. Helens; the timing is described in the text. North is to the right. A, 15:32.9 UT: initial explosions flanking block II; light-colored block I is on the right. GO designates Goat Mountain, 20 km north of the preeruption summit. B, 15:33.9: the two explosion clouds have coalesced, and both the avalanche and the pyroclastic surge have reached Toutle River and Spirit Lake. C, 15:34.5: surge has begun descending the south slope, and two white condensation clouds appear in the eastern surge. The northern explosion occurred slightly before this time. D, 15:35.0: the southern surge has nearly reached the base of the mountain, the white clouds have coalesced, and the dark atmospheric cloud has appeared at 6 km elevation. E, 15:35.5: the rapidly rising dark ash cloud appears to the north (right), while the north surge has nearly reached Goat Mountain. F, 15:36.5: the top of the rising ash cloud is obscured by the atmospheric cloud. Copyrighted photographs courtesy of John Christiansen and Vince Larson.

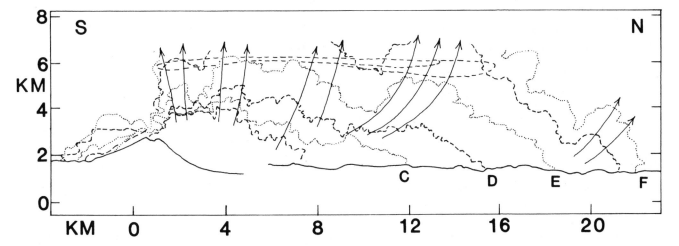

FIGURE 10.4 Outlines of the ash and surge clouds as viewed from Mount Adams. Dashed horizontal lines outline the first appearance of the atmospheric cloud as shown in Figure 10.3D. Arrows show trajectories of rising ash clouds, including the rapidly rising cloud (12 to 13 km north) that apparently originated at the northern explosion source 8 km north of Mount St. Helens. Letters designate cloud outlines of the appropriate photographs of Figure 10.3.

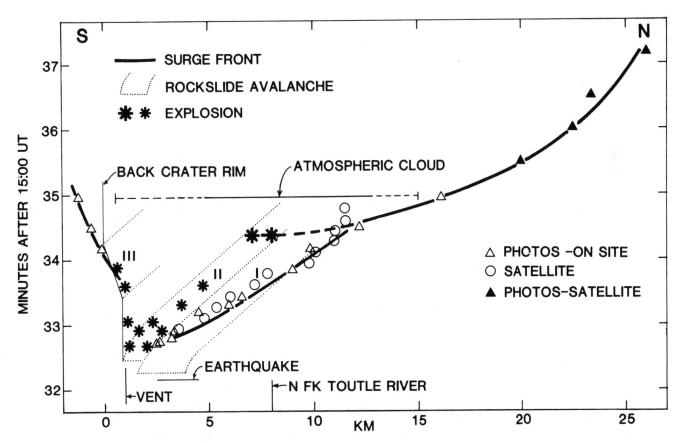

FIGURE 10.5 Time-travel plot showing north-south position of landslide blocks (I, II, and III), surge-front explosion sources, and the atmospheric cloud. Distance is measured north of the posteruption back crater rim.

SATELLITE MEASUREMENTS

Observations from infrared sensors aboard two U.S. Department of Defense satellites show the development of the pyroclastic surge and associated ash clouds from the initial sighting of the eruption plume at 15:32.9 UT (Rice, 1981). The satellites' view angles were similar to those of the GOES-East and GOES-West weather satellites. This dual monitoring permits triangulation on both the horizontal and the vertical development of ash clouds. Because the cloud is viewed from an angle relative to the local vertical, elevations must be known for correct planimetric location. The angle of view permitted viewing for some distance under the southern periphery of the developing mushroom cloud. Therefore, measurements of the southern part of the expanding pyroclastic surge are most reliable.

The satellite sensors, which use a short-wavelength infrared spectral band, could record scattered solar radiation from the rising ash clouds, provided that the cloud was not obscured by atmospheric attenuation. Solar scatter was the dominant source of the observed signal from relatively cool high-altitude clouds. However, early in the eruption, ash clouds near ground level were sufficiently hot to be seen in self-emission. At later times the sensors could view surface features when the high-altitude ash-cloud configuration was favorable. Estimation of cloud temperatures depends on assumptions of atmospheric attenuation as well as size, uniformity, and emissivity of the emitting mass.

The earliest eruption period was characterized by a complex sequence of emissions (see Figure 10.6). Before about 15:33.1, relatively cool material ($T < 400$ K) was emitted, part of which moved north at about 100 m/sec. Three intense emissions (explosions) of hot material ($T \geq 500$ K) occurred at 15:33.3, 15:33.6, and 15:34.4 (Figure 10.6). None of this material exceeded an altitude of 6 km. The timing and relative intensity of these three emissions are comparable for the two satellites, although one of the satellites recorded a lower intensity because of the greater signal attenuation of a much longer slant path through the atmosphere. The hot material ejected prior to 15:34.8 was confined primarily to the north face of the mountain and the Toutle River-Spirit Lake area.

The explosion producing the intensity peak at 15:34.4 (Figure 10.6) is correlated with that which generated the horizontal atmospheric cloud. Afterward all intense emissions declined, at least partly because of the development of the obscuring atmospheric condensation cloud. The earlier 15:33.6 explosion produced the largest-intensity peak, possibly because obscuring ash clouds were then at a lower level (Figures 10.3B and 10.C). At 15:34.8 a sudden, rapid lateral expansion of the previously evolved material occurred, and the surge appeared to split and be driven east and west at velocities of 150 m/sec or greater. This period of rapid expansion caused by the northern explosion lasted less than 0.5 min and was followed by lower velocities that carried the pyroclastic surge out to the edge of the tree-blowdown zone by about 15:37. Northward expansion could be monitored only indirectly because the explosion produced a thick obscuring cloud above about 6.5-km altitude (as noted in Figure 10.3D). This rapidly appearing cloud covered most of the region above the expanding surge; the southern

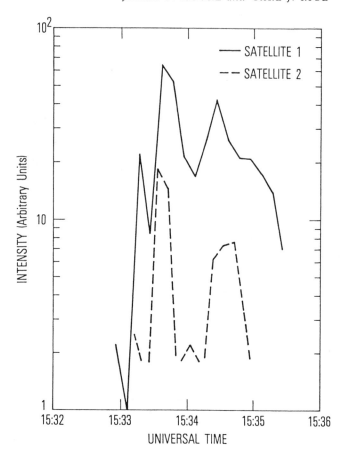

FIGURE 10.6   Intensity of infrared emission of low-level ash clouds as monitored by U.S. Department of Defense satellites during the early part of the eruption. Satellite I is north-looking, and satellite 2 is northwest-looking. Major explosion near Toutle River valley produced the third intensity peak at 15:34.4.

periphery could still be monitored, however, because of the favorable viewing angle.

The hot material near the ground caused a local lifting of constant pressure levels at higher elevations as the atmosphere attempted to reestablish equilibrium. This affected a general haze layer, which already existed at 10-km altitude (near the tropopause) as well as the expanding explosion-induced condensation cloud (Figure 10.3D), causing the haze layer to rise and become disturbed above the hotter surge and ash plumes. Along the southern edge, solar scattering from these localized upwelling "patterns" can be matched with direct emission from the individual ground-level plumes that produced them. These "pattern clouds" were about 9 km above their source, and, where the hot source in the ground-level surge was moving, the upper disturbance moved in response, lagging by about 30 sec, which implies that the disturbance in the constant pressure-level surface had traveled upward at the speed of sound.

Perhaps most important is that by monitoring the position of these pattern clouds, information can be obtained on the position of those parts of the ground-level surge obscured from direct satellite observation by the overlying cloud. A combination of photographic measurements and the direct and indirect satellite measurements was used to derive a map showing the low-level pyroclastic surge front at half-minute intervals (see Figure 10.7). The rapid expansion has been described by Rice (1981). At that time it was not recognized that part of the satellite image originated from the higher-level pattern clouds (about 10-km elevation), and appropriate corrections for projection effects were not made. Consequently, an excessive initial northward velocity and ground-level extension were originally ascribed to the low-level pyroclastic surge front.

The pyroclastic surge reached the eastern limit of the devastated zone by 15:36 and the western limit somewhat after 15:37 (Figure 10.7). The time at which the western toe of the avalanche reached its western limit is unknown but is estimated to be about 15:42 (Voight *et al.*, 1981). A great convecting column rose rapidly from the entire devastated zone, forming the mushroom cloud. Its altitude was 10 km at 15:39, 20 km at 15:44, and about 30 km at 16:00 (Rice, 1981), at which time it was fully developed with a diameter of 110 km. Just prior to 15:42, the satellite sensors detected hot material rising vertically from the crater vent, producing the plinian ash column that was the dominant feature of the eruption until late afternoon.

ORIGIN OF SECONDARY EXPLOSIONS

Previous studies (Rosenfeld, 1980; Kieffer, 1981; Moore and Albee, 1981) have assumed that a single, early explosive episode occurred near the volcano summit and produced most of the surge deposits and damage. Therefore, the recognition of several explosive episodes culminating in the northern explo-

sion at 15:34.3, nearly 2 min later than the first explosions near the summit, requires revisions of our concept of the Mount St. Helens eruption but better explains previously puzzling features. The first summit explosions are regarded as phreatic steam blasts similar to those that occurred prior to May 18; these were triggered by cracking and pressure release when landsliding began. The initial high moisture content of these clouds is indicated by their dark color and the presence of white condensation trails behind ejected blocks (probably partly ice) (Williams, 1980, p. 90).

The northern explosion was larger than those that preceded it. Only the 15:34.3 explosion disturbed the overlying atmospheric layer. This explosion caused most of the blast damage and fatalities and presumably produced the audible signal heard in Seattle (150 km north) and beyond (Reed, 1980). It appears to have emanated from that part of the landslide that contained block II (the graben block of Moore and Albee, 1981). This landslide block, unlike the northern block I, contained the major part of the dacitic cryptodome. During transport in the avalanche the hot dacite of the crypotodome was, no doubt, continuously broken and fragmented, which triggered secondary explosions. After nearly 2 min and several kilometers of northward transport, a major explosion was emitted from the avalanche, which produced a secondary higher-velocity pyroclastic surge that spread outward in every direction.

The time and place of this northern explosion can be estimated from a variety of criteria. Satellite measurements indicate that the last and most sustained maximum in the infrared intensity occurred between 15:34.4 and 15:34.7 (Figure 10.6), that the eastern and western surge lobes underwent rapid expansion at 15:34.8, and that the high-velocity northern surge overtook the previous surge at 15:34.5 (Figure 10.5). Photographs from Mount Adams indicate that the white condensation

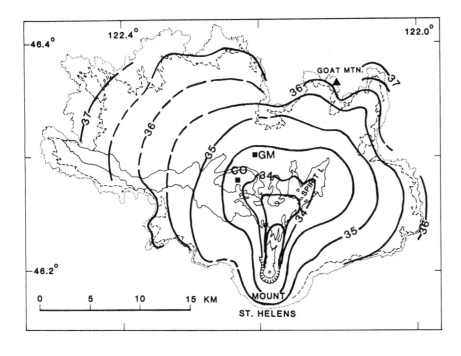

FIGURE 10.7 Map showing approximate position of the pyroclastic surge front at half-minute intervals. The lines are dashed where speculative. Squares indicate location of Coldwater II station (CO) and position of Gerry Martin (GM).

clouds first appeared about 15:34.5 on the east front of the surge cloud centered 5.5 km north, and the dark atmospheric cloud at 6.5-km elevation appeared between 15:34.5 and 15:35 centered 8 km north (Figures 10.3D and 10.5). Seismic measurements indicate that a second large earthquake and explosion occurred about 2 min after the first earthquake of 15:32.2 (Kanamori and Given, 1982). These data suggest that the major explosion occurred about 8 km to the north, near the valley of the North Fork Toutle River and Spirit Lake, at about 15:34.3.

Additional evidence for the major explosion source near the Toutle River valley is apparent in field measurements of blast damage and distribution of surge deposits (Moore and Sisson, 1981). The aerial distribution of these features tends to center near the North Fork Toutle River area rather than farther south near the crater. Maps of the thickness of the coarse basal surge deposit and of the thickness of wood removed (Figures 245 and 249 in Moore and Sisson, 1981) illustrate this tendency. A similar distribution is shown by the median grain size of the surge deposit (see Figure 10.8) and the percent of bark removed from small trees (see Figure 10.9). The distribution of tree-blowdown directions (Lipman, 1981) can also be reconciled with the northern explosion source, even though the earlier explosions nearer the crater area did blow down trees closer to the summit crater and radial to it. Tree-blowdown directions record only the initial fall directions; trees on the ground are unlikely to be rotated by later explosions, but the common occurrence of abrasion of the roots of downed trees indicates the occurrence of later explosions.

Block II, containing the cryptodome and its associated hydrothermal system, began moving and exploding at about 15:32.7. At an average velocity of 60 m/sec, it would have traveled 5.7 km by 15:34.3, placing its center about 8 km north of the posteruption south crater rim at the time of the northern explosion (Figure 10.5). This explosion may have occurred when

block II, containing the hot dacite of the cryptodome, collided with the north wall of the Toutle River valley. Alternately, the explosion may have resulted when block II overrode debris from block I (which was rich in moisture and glacial ice) as well as water-soaked ground in the Toutle River valley and, perhaps, Spirit Lake itself (see Figure 10.10). The initial entry of block I into Spirit Lake at about 15:33.5 (Figure 10.5), nearly a minute before the explosion, may have displaced lake water, sending a torrent down the Toutle valley, which mixed with the hot dacite carried in block II. The dacite at magmatic temperatures heated underlying and admixed water, causing it to flash to steam, producing the explosion. Whatever their origin, these later explosions accelerated the pyroclastic surge, which moved outward to the limit of the devastated zone, carrying with it a high proportion (about 50 percent according to Moore and Sisson, 1981) of hot dacite that served as the explosion heat source.

Similar, though much smaller, explosions continued for more than a month at the general site of the northern explosion in the valley of the North Fork Toutle River. On the afternoon of May 18, such explosions generated clouds that rose 3 km above the site of this activity (Williams, 1980, p. 105). At least 37 small explosion craters (Figure 10.9) extend 5 km downstream from Spirit Lake (Lipman, 1981). Some of these phreatic explosions were, no doubt, produced when the hot pumice flows, fed from the vertical eruption column, swept over the wet parts of the avalanche and the margin of Spirit Lake. However, many of the craters occur only in avalanche material, and others extend downward through the ash flows deep into avalanche rubble. These features suggest that the craters were produced by continued heating of Toutle River moisture by remaining fragments of hot cryptodome dacite buried in the avalanche deposits. The much larger craters that must have been produced by the major explosions at 15:34.3 were ap-

FIGURE 10.8   Median grain size of the surge deposit. Stars indicate the position of the major explosion.

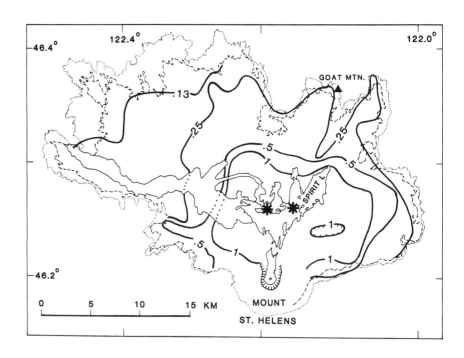

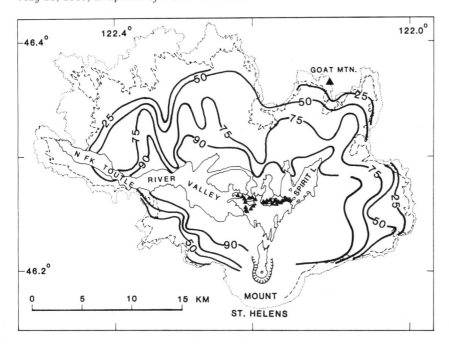

FIGURE 10.9 Percentage of bark removed from small trees. Triangles indicate the small phreatic explosion craters (after Lipman, 1981).

parently filled and covered by the tail of the avalanche, fed in part by the collapse of the south crater wall at 15:33.4 (Figure 10.5) and later by the hot pumice and ash flows. Pumice Pond, 2 km west of Spirit Lake in the Toutle River valley, may be a partially filled vestige of a crater produced by the northern explosions (Lipman, 1981).

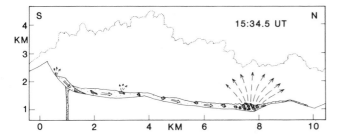

FIGURE 10.10 Section showing the development of the northern explosions when the hot fragmented cryptodome (stippled) of avalanche block II collided with the Toutle River valley wall or encountered water in the Toutle River drainage and Spirit Lake about 8 km north of the posteruption back crater rim.

ACKNOWLEDGMENTS

Permission by John Christiansen and Vince Larson to reproduce their remarkable photographs from Mount Adams (Figure 10.3) is appreciated. The videotape timing of the Mount Adams photographs is the result of detailed studies by Stephen Malone; we thank him for his considerable help. Reviews and comments by F. R. Boyd, P. W. Lipman, R. L. Christiansen, M. H. Beeson, and K. H. Wohltez are gratefully acknowledged.

This work was supported in part by the Aerospace-Sponsored Research Program and in part by the U.S. Air Force, Space Division, under contract F04701-81-C-0082.

This paper is dedicated to David Johnston and Gerry Martin, who lost their lives on May 18 while reporting observations on eruptive activity at Mount St. Helens.

REFERENCES

Kanamori, H., and J. W. Given (1982). Analysis of long-period seismic waves excited by the May 18, 1980, eruption of Mt. St. Helens—A terrestrial monopole? *EOS 63*, 239.

Kieffer, S. W. (1981). Blast dynamics at Mount St. Helens on 18 May 1980, *Nature 291*, 568-570.

Lipman, P. W. (1981). Geologic map of proximal deposits and features of the 1980 eruptions of Mount St. Helens, Washington, in *The 1980 Eruptions of Mount St. Helens, Washington*, P. W. Lipman and D. R. Mullineaux, eds., U.S. Geol. Surv. Prof. Pap. 1250, Plate 1.

Moore, J. G., and W. C. Albee (1981). Topographic and structural changes, March-July 1980 photogrammetric data, in *The 1980 Eruptions of Mount St. Helens, Washington*, P. W. Lipman and D. R. Mullineaux, eds., U.S. Geol. Surv. Prof. Pap. 1250, pp. 123-134.

Moore, J. G., and T. W. Sisson (1981). Deposits and effects of the

May 18 pyroclastic surge, in *The 1980 Eruptions of Mount St. Helens, Washington*, P. W. Lipman and D. R. Mullineaux, eds., U.S. Geol. Surv. Prof. Pap. 1250, pp. 421-433.

Reed, J. W. (1980). Air pressure waves from Mount St. Helens eruptions, *EOS 61*, 1136.

Rice, C. J. (1981). Satellite Observations of the Mount St. Helens Eruption of 18 May 1980, The Aerospace Corporation, Space Sciences Laboratory, Report SSL-81(6640)-1, 20 pp.

Rosenfeld, C. L. (1980). Observations on the Mount St. Helens eruption, *Am. Sci. 68*, 494-509.

Voight, B (1981). Time scale for the first moments of the May 18 eruption, in *The 1980 Eruptions of Mount St. Helens, Washington*, P. W. Lipman and D. R. Mullineaux, eds., U.S. Geol. Surv. Prof. Pap. 1250, pp. 69-86.

Voight, B., H. Glicken, R. J. Janda, and P. M. Douglass (1981). Catastrophic rockslide avalanche of May 18, in *The 1980 Eruptions of Mount St. Helens, Washington*, P. W. Lipman and D. R. Mullineaux, eds., U.S. Geol. Surv. Prof. Pap. 1250, pp. 347-377.

Williams, C. (1980). *Mount St. Helens, A Changing Landscape*, Graphic Arts Center Publishing Co., Portland, Oreg., 128 pp.

Factors Governing the Structure of Volcanic Jets

11

SUSAN WERNER KIEFFER
U.S. Geological Survey

ABSTRACT

Observable characteristics of volcanic jets, such as temperature, shape, and velocity, depend in a complex way on the subsurface features of the volcanic systems from which the jets emanate. The variety of sizes and shapes of jets associated with violent eruptions reflects differences primarily in the geometric shapes of the near-surface parts of the volcanic systems and, secondly, differences in thermodynamic properties and rheology of the erupting fluids. In this paper, differences in shape between plinian eruption columns and lateral blast flows on Earth and between umbrella-shaped and diffuse plumes on Jupiter's moon, Io, are discussed in terms of a generalized model for thermodynamic properties and geometry of a volcanic system. On Earth, decompression of the volcanic fluid to ambient pressure of 1 bar typically occurs within only a few tens of meters from the bottom of a crater. In contrast, ambient subsolar surface pressure on Io is about 10^{-12} bar (perhaps as high as 10^{-7} bar in the vicinity of an erupting volcano), and the pressure in a jet on Io will decay from the sonic conduit value to ambient within the crater only if the crater is more than 1 km deep. Therefore, on either body, two classes of jets may result from different crater depths: (1) "pressure-balanced" jets, if the jet pressure approximately equals ambient pressure as it emerges at the surface of the planet, or (2) "overpressured" (underexpanded) jets, if the jet pressure greatly exceeds atmospheric. The structure of pressure-balanced jets may vary from one planet to another, but on one planet the jets may tend to have rather similar structures from one volcano to another because they are equilibrated to atmospheric pressure. The structure of pressure-balanced jets is determined primarily by the initial thrust and, secondly, by external factors such as entrainment of air, winds, and mixture buoyancy. These factors also influence overpressured jets, but internal jet dynamics, including shock and expansion waves, also play a significant, sometimes dominant, role. Overpressured jets may show a much greater diversity in structure than pressure-balanced jets because the jet structure depends on the ratio of jet pressure to atmospheric pressure and typical volcanic eruption conditions allow a wide range of ratios. Plinian eruption columns are typical terrestrial pressure-balanced jets because they commonly emerge through craters and thus enter the atmosphere at atmospheric pressure. Plume 3 (Prometheus) on Io may be pressure matched to the Ionian atmosphere. The lateral blast of May 18, 1980 at Mount St. Helens was an overpressured jet, since it emerged directly at high pressure from the face of the mountain; it was able to expand toward atmospheric pressure only as it traveled tens of kilometers away from the mountain. On Io, plume 2 (Loki), which apparently emerges through a fissure, may have been overpressured when photographed during Voyager 1.

143

INTRODUCTION

The verb *to explode* is defined by *Webster's Third New International Dictionary* as "(1) to undergo rapid combustion with sudden release of energy in the form of heat that causes violent expansion of the gases formed and consequent production of great disruptive pressure and a loud noise, (2) to undergo an atomic nuclear reaction with similar but more violent results, and (3) to burst violently as a result of pressure from within." Although our understanding of the thermodynamic evolution of magma before and during an eruption has increased enormously over the last few decades (Graton, 1945; Verhoogen, 1946; Boyd, 1961; Sparks *et al.*, 1978; Self *et al.*, 1979; Wilson and Head, 1983), many misconceptions remain of processes occurring during explosive volcanism. This paper describes how jets from violently erupting volcanoes obtain their "explosive" velocities from the transformation of the initial enthalpy stored in magma into kinetic energy and how the transformation process depends on the interplay between the system geometry, fluid thermodynamic properties, reservoir initial conditions, and ambient atmospheric conditions. The concepts are applied to an interpretation of the structure of four jets of very different shapes: (1) the lateral blast that initiated the eruption on May 18, 1980, at Mount St. Helens; (2) the plinian plume that developed during the same eruption; (3) the umbrella-shaped plume of many volcanoes, such as Prometheus, on Jupiter's satellite, Io; and (4) the diffuse plume of Loki on Io. Further details are given in Kieffer (1981, 1982a, 1982b).

VOLCANIC JETS*

The fact that volcanic jets associated with violent eruptions come in a variety of sizes and shapes (see Figure 11.1) has important implications for understanding the relationships between jet structure, magma properties, and volcanic system geometries as well as for volcanic hazard modeling and comparative planetology. Two very different jet structures were evident during the course of the eruptions on May 18, 1980, at Mount St. Helens. The vertical jet that developed as the May 18 eruption sequence progressed (Figure 11.1a) is typical of jets from major plinian eruptions on Earth. The jet was a towering column of gas and pyroclastics and resembled a jet of water from a fire hydrant in that most of the flow was directed

---

*The gas-particulate emissions from volcanoes have commonly been referred to as *plumes*, which, by strict definition, resemble a feather in shape, appearance, or lightness. In the more rigorous fluid mechanics context, a plume is buoyancy dominated. Thus, a volcanic plume is generally thought of as a rising buoyant column of gas with entrained pyroclastic fragments. However, the emissions from many violent eruptions are unfeatherlike in shape, appearance, or lightness, e.g., nuées ardentes, lateral blasts, and other such dense flows. The word *jet* is used in fluid mechanics to describe a general flow from an orifice and, more specifically, refers to a flow that is momentum dominated. In this paper, I use the word *jet* in a more general context than the word *plume*.

upward and was approximately parallel to the presumed vertical center line of the conduit from which it erupted. The columnar form occurred because material in the column maintained the vertical direction of motion imparted by accelerations in the volcanic system. Convective mixing of air into the eruption column caused the column to grow slowly wider with height, but not until the plume reached altitudes at which it became neutrally buoyant did interactions with major atmospheric layers cause it to deviate from the columnar form.

In contrast, the jet of the lateral blast that initiated the May 18 eruption was very different in shape. The shape of the lateral blast is reflected in the shape of the devastated area north of the volcano where the blast was sufficiently dense to remain near the ground; at a distance of about 25 km it became buoyant enough to lift from the land. Photographs of the inception of the blast show that initial particle motions were largely directed northward. However, over much of the region covered by the jet, the streamlines and the flow boundaries were *not* directed northward. In contrast to the way that the material in the plinian column maintained its initial direction of motion, the material in the lateral blast departed from the initial direction as soon as it left the face of the mountain. The flow initially spread out over a broad fan-shaped sector diverging around the mountain as far as southeast and southwest (see Figure 11.1b). The overall shape differences between the columnated vertical plume and the fanning lateral blast cannot be explained by differences in the orientation (vertical versus subhorizontal) of the jets alone, and the intriguing question of why the shapes of the two jets at Mount St. Helens were so different remains.

Observations by Voyager spacecraft show that jets also come in a variety of shapes on Io, the only other planet in the solar system where current volcanic activity exists (Strom *et al.*, 1979; Strom and Schneider, 1982; McEwen and Soderblom, 1983). The volcanoes on Io seem to have two types of behavior. One group of volcanoes (Pele, Surt, Aten Patera) have short-lived eruptions (days to weeks probably), eject plumes to heights of ~300 km, and deposit ejecta of sulfur in discrete rings at distances up to 1400 km from the vent. Surface temperatures near the source regions are in the range of 500 to 700 K. The other group of volcanoes (Prometheus, Volund, Amirani, Maui, Marduk, Masubi, and two plumes from Loki) are long lived (years), eject plumes to heights on the order of only 100 ± 40 km, and deposit white ejecta (SO_2?) in discrete rings of only 200 to 300 km diameter or in diffuse rings. They have surface temperatures near the source regions that are less than 400 K. McEwen and Soderblom (1983) suggested that these two differences in class represent sulfur- versus sulfur-dioxide-driven plumes. Deposits from Loki suggest that it varies between the two classes in behavior, but for the purposes of this discussion its behavior while it appeared to be in the second group during the Voyager 1 encounter is analyzed.

Of particular interest here is the fact that most of the plumes have beautiful umbrella shapes and deposit ejecta in discrete rings (e.g., Pele, Surt, Aten Patera, Maui, Prometheus) (see Figure 11.1c). However, Loki (plume 2, Figure 11.1d) and Volund, as observed by Voyager 1, have more diffuse plumes and more irregular deposits.

FIGURE 11.1   Eruption jets on Earth and on Io. (a) The plinian eruption column at Mount St. Helens, May 18, 1980. The crater from which the plume emerges is approximately 1 km in diameter. Photograph by Austin Post, U.S. Geological Survey. (b) Satellite photo of the area devastated by the lateral blast on May 18, 1980, at Mount St. Helens (EROS Data Center, Goddard Space Flight Center). The distance from the mountain to the farthest northwest and northeast boundaries of the devastated region is about 25 km. The devastated area is assumed to mimic the shape of the jet from the lateral blast over the area where the jet remained close to the ground. The jet actually extended beyond these boundaries but lifted from the ground when it expanded to ambient atmospheric density (Kieffer, 1982a). (c) Umbrella-shaped plume 3 from Prometheus on Io. The jet is about 70 km high (image by NASA). (d) Diffuse plume 2 from Loki on Io. The jet is 165 km high (image by NASA).

VOLCANIC SYSTEM GEOMETRY

The structure of volcanic jets cannot be isolated from the behavior of magmatic fluid in subsurface parts of a volcanic system. To analyze jet structure, the fluid composition, the mechanical behavior, and flow conservation laws through all parts of the volcanic system must be considered. To formalize these assumptions, I have divided a volcanic system into five regions that are separately susceptible to analysis of fluid dynamic and thermodynamic behavior: *supply region*, in which magma is generated; *reservoir*, in which it collects; *conduit* and *crater*, in which accelerations to high velocities can occur; and *jet or*

FIGURE 11.2 Schematic geometry of volcanic systems used for models discussed in text. (a) Sketch illustrating the concept of the five parts of a volcanic system discussed in the text. (b) Model for simple, direct discharge of reservoir into jet at Mount St. Helens for lateral blast. Blocks 1 and 2 represent the two major blocks involved in the landslide (after Moore and Albee, 1982). The stippled area surrounded by hachuring represents the combined magmatic-hydrothermal system assumed to have underlain the north slope of the mountain; configuration is unknown. The complex reservoir geometry that fed the blast is simplified by assuming that the material discharged from a tabular reservoir that was 1 km in east-west dimension, 0.25 km in vertical height, and 0.5 km in north-south depth. It is assumed that the flow of the dacite from the deep reservoir (as illustrated in sketch c) was not involved in the lateral blast. (c) Discharge of deep reservoir through conduit and, perhaps, the shallow surface crater into jet to form plinian eruption column at Mount St. Helens. (d) Discharge of reservoir through conduit with negligible surface crater on Io (called a *fissure system*). (e) Discharge of reservoir through conduit with substantial surface crater on Io (called a *crater system*).

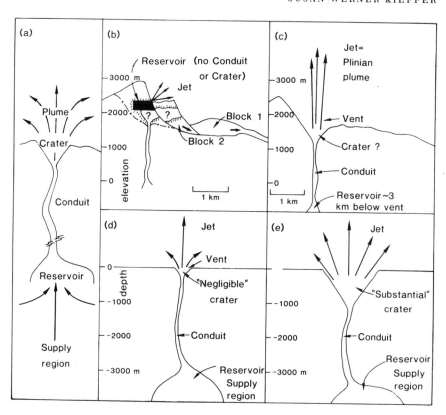

plume, the visible ejecta above the planetary surface (see Figure 11.2a). The geometry of a volcanic system is unique to each volcano and, perhaps, even to each eruption at a given volcano.

Consider nominal system geometries for the four examples under discussion. The lateral blast at Mount St. Helens is modeled as the discharge of a simple, shallow, tabular reservoir exposed suddenly to atmospheric pressure by the landslide that initiated the eruption at 8:32 A.M. (Figure 11.2b). The tabular shape is compatible with observations that the discharge occurred over a surface area at least 1 km in lateral dimension and 0.25 km in height and that about 0.25×10^{15} g of material were ejected. However, it is a much oversimplified representation of a reservoir that was certainly complex in geometry and in temporal evolution. The plinian eruption at Mount St. Helens is modeled as the discharge of a deeper (~3 to 5 km) reservoir through a conduit, with perhaps a crater (Figure 11.2c). This model is consistent with seismic observations of the depth of emplacement of magma by May 18.

The Ionian eruptions are modeled as the discharge of shallow crustal reservoirs through cylindrical or tabular conduits several kilometers deep, similar to the plinian system assumed for Mount St. Helens. One system, which for convenience I call a fissure system, is assumed to have no substantial surface crater (Figure 11.2d). The other system, which I call a crater system, has a substantial surface crater (Figure 11.2e). (A substantial surface crater will be defined later in this paper.) I will demonstrate that the geometry of a volcanic system in the near-surface region has a larger influence on the structure of a volcanic jet than do plausible variations in thermodynamic properties of the magma and initial pressure conditions.

THERMODYNAMIC PROPERTIES OF MAGMATIC FLUIDS

In addition to simplifying the geometry of a volcanic system, it is also necessary to simplify the rheologic behavior of its magmatic fluid in order to obtain an equation of state. Even though magmatic ascent probably involves many nonequilibrium processes, representative thermodynamic paths of ascent on phase diagrams of magmas would show useful reference equilibrium conditions. In practice, even this limited goal of showing equilibrium thermodynamic paths has not yet been done for any magma. Thermodynamic analysis is still limited to highly idealized fluids by two constraints: (1) the properties of magmas are so complex that complete phase diagrams have not yet been formulated, and (2) the equations describing fluid flow can be solved only for the simplest equations of state and thus cannot accommodate the complexities that are necessary to describe magma flow, e.g., phase changes, gas-liquid-particle interactions, radiative heat transfer.

Magma rheology is conveniently simplified by recognition of three types of fluid behavior corresponding to decreasing depth in the volcanic system (Wilson *et al.*, 1980): (1) the *lower zone*, where volatiles are dissolved in the magma; (2) the *middle zone*, where the volatiles are exsolved within a liquid phase; and (3) the *upper zone*, where the volatiles are the dominant phase by volume and the magma is a gas with entrained liquid droplets and solid fragments. These three rheologic zones need not, of course, correspond to any idealized geometric zones in a volcano. For example, fragmentation (upper-zone behavior) may occur deep in a conduit, or even within a reservoir, in a deep-seated kimberlite or caldera eruption, whereas liquid magma with little exsolved gas may appear in surface craters in Hawaiian eruptions. In this paper I discuss systems in which the rheology is gas dominated at the conduit level.

Much can be illustrated about the thermodynamics of upper-zone magmatic behavior by considering the phase diagrams of the volatiles that propel the ascent of magma, because most of the velocity during ascent is obtained after the magma has fragmented into a gas-pyroclast mixture (see Wilson *et al.*, 1980). Therefore, if hypothetical initial states for the volatile components of a magma can be specified from knowledge of pressure-temperature relations in volcanic regions (geotherms

on the Earth, iotherms on Io), the thermodynamic history of the volatiles can be outlined and the thermodynamic history of the magma approximated. The pressure-temperature phase diagrams of H_2O, CO_2, SO_2, and S are shown in Figure 11.3 with superposed geotherms and iotherms. These diagrams provide the initial pressures and temperatures for solution of the fluid flow equations. The thermodynamic history of the fluids during ascent and eruption, however, is more easily analyzed on temperature-entropy (T-S) phase diagrams, shown in Figure 11.4, because many eruptions are quasi-isentropic and analysis of the fluid flow is often restricted by the assumption of isentropic processes. Even if the expansion of the erupting volatile

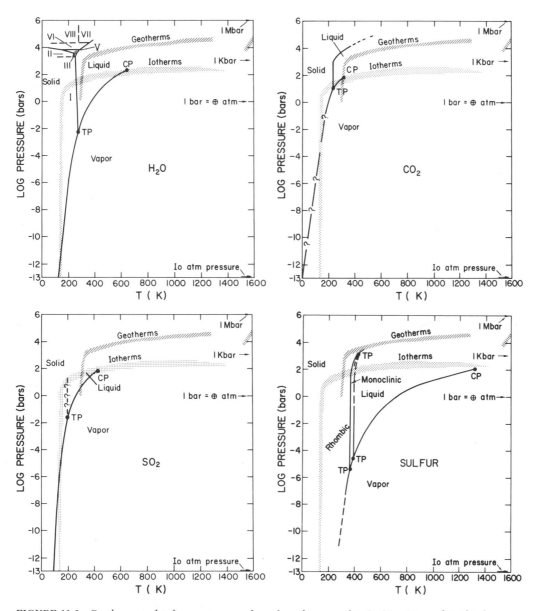

FIGURE 11.3 Geotherms and iotherms superposed on phase diagrams of H_2O, CO_2, SO_2, and S. This figure reproduced from Kieffer (1982b, with permission of the the University of Arizona Press), where data sources for the phase diagrams and phase diagrams can be found in the caption to Figure 18.3.

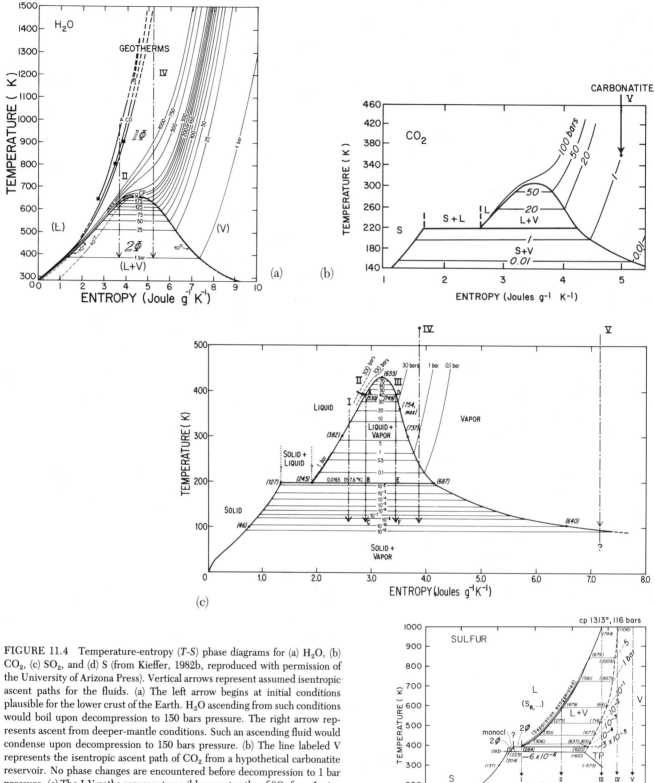

FIGURE 11.4 Temperature-entropy (*T-S*) phase diagrams for (a) H_2O, (b) CO_2, (c) SO_2, and (d) S (from Kieffer, 1982b, reproduced with permission of the University of Arizona Press). Vertical arrows represent assumed isentropic ascent paths for the fluids. (a) The left arrow begins at initial conditions plausible for the lower crust of the Earth. H_2O ascending from such conditions would boil upon decompression to 150 bars pressure. The right arrow represents ascent from deeper-mantle conditions. Such an ascending fluid would condense upon decompression to 150 bars pressure. (b) The line labeled V represents the isentropic ascent path of CO_2 from a hypothetical carbonatite reservoir. No phase changes are encountered before decompression to 1 bar pressure. (c) The I-V paths represent possible ascent paths of SO_2 from Ionian and/or terrestrial reservoirs (initial conditions given in detail by Kieffer, 1982a). Note that all paths, except V, enter multiple phase fields. (d) I-V paths represent possible paths of ascent of sulfur on Io, as discussed by Kieffer (1982b). Irreversible processes would swing the paths counterclockwise.

phase is not isentropic, as will be discussed below, the *T-S* diagrams provide a convenient reference for determination of paths of ascent and stable phases during different parts of the flow.

Consider first the ascent of the pure fluids illustrated. On the phase diagrams of Figure 11.4, possible initial conditions in different types of terrestrial and Ionian reservoirs are illustrated on the basis of estimated geotherms in different tectonic settings on the two planets (Figure 11.3). All plausible initial conditions give reservoir fluids that are highly compressed, liquid, or supercritical. As the fluids ascend from the supply regions or reservoirs in which they are initially at rest, their velocities are low because the volume changes associated with decompression are small as long as the fluids are in this compressed state. As they enter higher levels in the volcanic systems the fluids begin to expand, but two thermodynamically different processes may occur that would complicate the expansion (compare the left and right sides of the phase diagrams in Figure 11.4): (1) transformation (boiling) of a liquid (low-entropy) phase to a liquid-plus-vapor mixture and (2) transformation (condensation) of a vapor (high-entropy) phase to a vapor-plus-liquid mixture. Only the expansion of CO_2 from mantle conditions on the Earth (Figure 11.4b), or of SO_2 or S from a very high temperature reservoir on Io (Figures 11.4c and d), to ambient atmospheric pressure could occur without complications in the fluid flow due to phase changes. The effects of phase changes on erupting fluids have been qualitatively discussed by Kieffer (1982b).

The thermodynamic history and fluid-mechanical properties of a volatile phase are altered by the presence of entrained pyroclastic fragments. The entrained fragments increase the bulk density and effective molecular weight of the erupting fluid. They can also alter the expansion of the volatile phase by transferring heat to that phase. Two limiting cases are commonly considered: (1) if the entrained fragments are small, heat can be transferred from the solid fragments into the gas, and the gas will expand nearly isothermally, and (2) if the entrained fragments are large, heat cannot be effectively transferred to the gas and the gas will expand nearly adiabatically. If the process is reversible, the expansion is also isentropic. For many volcanic flows the ascent time is 10 to 100 sec, and particles less than a few millimeters in radius can be considered small in the sense of heat-transfer properties.

These considerations can be quantified in a simple way by assuming that the vapor-pyroclastic fragment mixture can be modeled as a pseudogas. The thermodynamic history of the solid and gas phases can be calculated separately (details are given in Appendix A of Kieffer, 1982b). A typical result for H_2O is shown in Figure 11.5, where the ascent path of H_2O alone from a hypothetical kimberlite reservoir is compared with the path of the H_2O if it contains a weight ratio m of pyroclastic fragments while it ascends. In many magmas, m lies in the range of 25 to 100. By these standards a relatively small weight fraction of entrained pyroclastic fragments ($m \geq 10$) can transfer sufficient heat to the vapor phase such that phase transformations do not occur until the fluid has decompressed to 1-bar pressure, when processes other than isentropic ascent influence the cooling history. However, the complications that

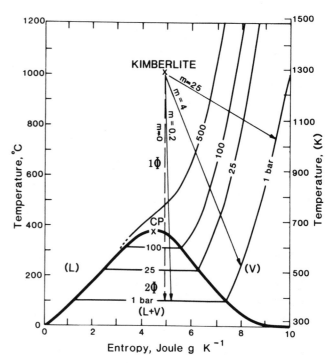

FIGURE 11.5 Temperature-entropy phase diagram for H_2O comparing the thermodynamic history of the H_2O during isentropic ascent of the fluid alone ($m = 0$ path) with isentropic ascent of a mixture containing weight ratio, m, of solids to vapor.

phase changes introduce into the fluid flow are avoided only at the expense of the complexities introduced into the flow by other processes, such as particle drag and heat transfer. I focus here on flows that contain sufficient entrained material such that phase changes do not complicate the flow.

The gross effects of pyroclastic mass and solid-gas heat transfer can be accounted for by the simple pseudogas model mentioned above, although particle drag is not considered. To obtain an equation of state for the erupting fluid suitable for use in the fluid flow equations, the erupting fluid is assumed to be a mixture of vapor and pyroclastic fragments through its entire ascent. Thus, in the following sections, reference to the properties of a "reservoir fluid" is to the properties of this vapor-particulate mixture in a hypothetical rest state not to the properties of the magma that physically exists there. The equation of state of a pseudogas-gas has the algebraic form of the perfect gas law:

$$PV = R_m T \tag{11.1a}$$

and

$$PV^{\gamma_m} = \text{constant.} \tag{11.1b}$$

In these equations, P is pressure, V is the volume of the mixture, T is absolute temperature, R_m is the gas constant, and γ_m is the isentropic exponent. Two modifications reflecting the *mixture*, rather than *vapor*, properties are that (1) the gas constant R_m is modified from the value appropriate to the vapor alone, so that the mixture is modeled as a gas of high molecular weight, and (2) the value of γ_m is modified to account for heat-transfer processes within the expanding mixture. In the case of the flow duration being short and/or the entrained solids being so large that their heat cannot be appreciably transferred from the solids to the vapor, the isentropic exponent is left as that of the vapor phase alone, and the mixture expands adiabatically with the isentropic exponent of the vapor phase. In this case, γ_m is typically about 1.3. In the second case of the flow duration being long and/or the entrained solids being small (which allows appreciable heat transfer to occur rapidly), the isentropic exponent of the gas is modified to account for this heat flow. The *mixture* again expands adiabatically and isentropically, but the isentropic exponent for the mixture is close to unity, indicating that the *vapor* expansion is nearly isothermal. In most cases the latter assumption of flow with heat transfer is the most relevant one, and the flow expands with a very small value of γ_m. The specific volume, V, and the modified γ_m and R_m depend on the mass ratio, m, of solids to vapor (see equations in Kieffer, 1982b).

INITIAL THERMODYNAMIC CONDITIONS ASSUMED FOR THE RESERVOIRS

Initial conditions for the flow are determined in the supply and/or reservoir regions. It is here that differences in average temperature, pressure, and fluid composition (manifested in the weight ratio, m, and the gas constant, R, of the pseudogas equations) arise. Most current thermodynamic models for eruptions are similar to the one described above. In these models the thermodynamic behavior is independent of assumptions about the origin of the volatile phase (magmatic versus meteoric), except as the origin is reflected in the chosen values of P, T and m; as a result, the fluid-dynamic conclusions also are independent of these assumptions and cannot be used to test the classic problem of magmatic, phreatic, or phreatomagmatic origin of an eruption.

The differences in initial temperatures and compositions, however, are of secondary importance to the geometric factors in this discussion. The most important reservoir variable influencing overall jet structure is the ratio of initial reservoir pressure to atmospheric pressure. On Earth or on Io the range of tens of bars to perhaps a thousand bars brackets most initial reservoir pressures; for rare kimberlite eruptions on Earth, initial reservoir pressures may even be on the order of tens of thousands of bars. On Earth atmospheric pressure is approximately 1 bar; on Io ambient subsolar atmospheric pressure is between 10^{-12} and 10^{-7} bars; the pressure on the dark side could be as low as 10^{-17} bars. Hence, for different combinations of reservoir and atmospheric conditions, the ratio of reservoir to atmospheric pressure on Earth may be between ~10 and 10^5, and on Io between 10^{20} and 10^8.

The following reservoir conditions appropriate to the reservoirs shown in Figure 11.2 are assumed in order to focus the discussion, but the results are virtually insensitive to the values assumed:

1. For the Mount St. Helens lateral blast the reservoir fluid is assumed to have been a mixture of rock and H_2O vapor, with a mass ratio of solids to vapor of 25:1 (see Kieffer, 1982a, for details). On the basis of an average depth at which the blast material emanated from the mountain, the average initial pressure is assumed to have been 125 bars (appropriate to 650 m of lithostatic overburden) and the initial temperature to have been 600 K. This is the saturation temperature of pure H_2O at 125 bars and is perhaps a reasonable number to pick *a priori* in the absence of any information about the nature of the reservoir. This temperature may also be thought of as an averaged temperature of an erupting fluid that, after only a short distance of travel, was a thoroughly mixed fluid of hot dacite, hydrothermally altered rock and water, and glacial ice and organic debris. It therefore does not imply that the original reservoir nor the eruption that followed was strictly "hydrothermal." The median grain size in the blast material was less than 1 mm (Moore and Sisson, 1982), so there was good heat transfer between the particles and the vapor. Assuming continuous thermal equilibration of the particles and vapor, the value of γ_m for this mixture is 1.04.

2. For the Mount St. Helens plinian eruption, an initial pressure of 1000 bars (appropriate to a reservoir with an average depth of 3 km) and an initial temperature of 1200 K (approximately the liquidus temperature for the Mount St. Helens dacite) were assumed. The same mass ratio of solids to vapor of 25:1 as for the lateral blast was assumed for simplicity, so that the effective value of γ_m is 1.04. Variations of the value of γ_m to more nearly match the water content of the dacite would not significantly alter γ_m.

3. For Ionian volcanoes a variety of reservoir conditions and reservoir fluids, SO_2 and S with and without entrained pyroclastic fragments, were examined (details are given in Kieffer, 1982b). For discussion here I will assume an initial reservoir pressure of 40 bars and an initial temperature of 1000 K— higher than has yet been observed on Io but chosen specifically to provide an upper limit comparable to the Mount St. Helens case. A value of $\gamma_m = 1.1$ can be used to model either pure S_6 vapor or SO_2 vapor with a pyroclastic mass loading ratio of solids to vapor of $m = 1:1$.

FLUID-DYNAMIC CONDITIONS

Equations

The supply and reservoir regions have been defined above as places where the fluid is at, or nearly at, rest and the conduit as the place within the volcanic system where the fluid obtains an appreciable velocity. Therefore, the first stage in calculation of fluid motion is calculation of flow conditions in the conduit.

Flow is governed by the laws of conservation of mass, momentum, energy, and entropy. For simplicity, quasi-one-dimensional steady flow is examined. The general equations of motion are as follows:

$$\frac{1}{\rho}\frac{d\rho}{dz} + \frac{1}{u}\frac{du}{dz} + \frac{1}{A}\frac{dA}{dz} = 0 \qquad \text{(continuity)} \quad (11.2)$$

$$\rho u \frac{du}{dz} + \frac{dP}{dz} \pm G + \frac{2}{D} f \rho u^2 = 0 \quad \text{(momentum) (11.3)}$$

$$\frac{d}{dz}\left(h + \frac{u^2}{2} + \psi\right) = 0 \qquad \text{(energy) (11.4)}$$

$$dS \geq dQ/T. \qquad\qquad \text{(entropy) (11.5)}$$

In these equations, z is the vertical coordinate, ρ is density, u is velocity, A is cross-sectional area, G is a body force (gravity), h is enthalpy, ψ is a field potential ($\psi_{,k} = -G_k$), D is the conduit diameter, and f is the Fanning friction factor ($f = 2\tau/\rho u^2$, where τ is the shear stress at the boundary of the fluid). S is entropy and Q is heat added to the system. The flow is assumed to be adiabatic and isentropic. Because field observations suggest that conduits are approximately straight, flow in the conduit is assumed to be strictly one-dimensional ($dA/dz = 0$); all significant area changes are assumed to occur in the crater. For this discussion the roles of gravity and friction during flow in the conduit and crater are ignored.

The integrated form of the conservation of energy Eq. (11.4) gives the well-known Bernoulli equation, which holds for thermodynamic states along any streamline:

$$h_0 + u^2/2 + \psi_0 = h_1 + u^2/2 + \psi_1, \qquad (11.6)$$

where the subscripts 0 and 1 refer, respectively, to reservoir and flow conditions. For purposes of simple illustration, choose $u_0 = 0$ and ignore the gravity terms. Then

$$h_0 = h_1 + u^2/2. \qquad\qquad (11.7)$$

This equation shows that the velocities obtained by fluids in explosive volcanism result from the simple conversion of the initial enthalpy of the fluid into kinetic energy (enthalpy = internal energy + energy stored in P-V compression, where P is pressure and V is volume). Latent heat of phase changes, exsolution of gases, or chemical reactions would be energetically similar to combustion terms in the above equations and must, of course, be included in a generalized form of this equation, but as will be shown below, their heat is not required to obtain large velocities of eruption and their effect is generally believed to be small during actual eruptions.

Flow in Conduits

The acceleration of the fluid as it moves from the reservoir through the volcanic system depends primarily on the pressure difference between the reservoir and the atmosphere into which the volcano is erupting and, secondly, on frictional and gravitational forces and on details of fluid properties. For the examples of interest on Earth and Io, atmospheric pressure is sufficiently less than the reservoir pressure, so that very large accelerations from initial to final pressure are possible. However, *accelerations within the conduit are limited by the geometry*. This limitation is best described with reference to the local fluid sound speed, a thermodynamic property that depends primarily on the fluid composition and temperature. For the pseudogas model used here, the reservoir sound speed is

$$c_0 = (\gamma_m R_m T)^{1/2}. \qquad\qquad (11.8)$$

For the Mount St. Helens lateral blast fluid, c_0 was 105 m/sec; for the plinian fluid it was 150 m/sec; and for the Ionian reservoir fluid, it is about 220 m/sec. As the fluid accelerates up the conduit the flow velocity increases, but the sound velocity decreases because the temperature drops as the fluid expands. The fluid can accelerate within the conduit *only until the local flow velocity, u, equals the local sound velocity,* $c^* = [2/(\gamma_m + 1)]^{1/2} c_0$.

The conservation equations show that the fluid cannot accelerate above this velocity unless the conduit diverges. In my conceptualized volcanic system all significant divergences occur in the near-surface region defined as the surface crater. Thus, in a narrow, nearly straight conduit the flow velocity is approximately limited to the sonic velocity, a condition known as *choked flow*. Pressure, temperature, and density decrease until they reach the sonic conditions; to an order of magnitude, frictional effects do not change the results of this discussion. Typically, for vapor-particulate mixtures, sonic flow velocities are 100 to a few hundred m/sec, depending on the temperature and the solid-loading factor, m. For the examples here the sonic flow velocities are very close to the original reservoir sound speeds: Mount St. Helens blast, 104 m/sec; Mount St. Helens plinian column, 148 m/sec; and Ionian jets, 213 m/sec.

The pressure obtained at sonic conditions is

$$P^* = P_0 \left[\frac{2}{\gamma_m + 1}\right]^{\gamma_m/(\gamma_m - 1)} \qquad (11.9)$$

For the Mount St. Helens lateral blast model, $P_0 = 125$ bars and $P^* = 75$ bars. The fluid volume increases by a factor of 1.64 from reservoir to sonic conditions. At sonic conditions, $T^*/T_0 = 2/(\gamma_m + 1) = 0.98$ for the proposed mixture. These pressure, volume, and temperature changes show that the initial sonic velocity of 104 m/sec is obtained by decompression of the reservoir fluid to 0.60 of the initial pressure, with accompanying volume expansion of a factor of 1.64 and conversion of only 2 percent of the internal energy into kinetic energy.

Sonic pressures are typically about half of the initial pressure for the fluids considered here, i.e., they are typically a few tens to a few hundreds of bars. These pressures are higher than the atmospheric pressure on Earth by factors of tens to hundreds and are higher than atmospheric pressure on Io by at least 8, and possibly 14, orders of magnitude! Thus, if the fluid emerged directly from the conduit into the atmosphere, it would be at a higher pressure than atmospheric, a condition discussed further below.

The high fluid pressures that exist in fluid erupting directly from a straight conduit or fracture cause surface craters to form, either by direct erosion of the sharp lips of such features or indirectly by slumping of unstable near-surface material into the flow (which has a great capacity for material transport in this relatively high pressure state; see discussion in Kieffer, 1982b). Surface craters are ubiquitous over most volcanic conduits associated with high-velocity jets on Earth. What effect do they have on the properties of the erupting fluid and, in particular, on pressure?

Flow in Craters

The area across which the fluid flows increases rapidly if it flows through a surface crater. For example, in Figure 11.6 a crater with internal slopes at angle of repose (30°) is shown overlying a conduit with a 5-m radius. As shown by the four columns on the right, the ratio of crater area to conduit area, A/A^*, increases dramatically with height above the crater floor. At a height of 200 m above the conduit, the crater has 5000 times the area of the conduit. This increasing area affects all of the flow variables. Application of Eqs. (11.1-11.5) allows all flow variables to be calculated within the framework of the quasi-one-dimensional approximation. The flow picks up speed, accelerating from Mach 1 at sonic conditions in the conduit to very high Mach numbers in the crater, and the temperature and density of the fluid decrease. Most importantly, for the discussion here, the pressure decreases substantially (Figure 11.6). There is about an order of magnitude difference in the rate of pressure decrease depending on the properties of the vapor phase and/or the amount of heat transfer, as manifested in different values of γ_m (values for γ_m of 1.3 and 1.04 are shown in Figure 11.6). However, compared with a pressure drop of only about 50 percent obtained during flow between the reservoir and the top of the conduit, the decrease in pressure by many orders of magnitude within even short distances in the crater is dramatic, e.g., it decreases nearly five orders of magnitude in the first 200 m.

Within a crater, then, the fluid pressure will tend to drop toward ambient atmospheric pressure simply because of the area change. On Earth, a pressure drop by a factor of 100 to 1000 would be sufficient to reduce the fluid pressure to ambient atmospheric pressure. This can happen within a shallow surface crater—one with a depth of tens of meters. (In fact, the effect of friction alone in a very long conduit could reduce the fluid pressure to ambient on Earth without a surface crater if initial reservoir pressures were on the low side of normal reservoir values.) Thus, *in most cases of terrestrial volcanism where there are even minor surface craters, jets enter the atmosphere with a fluid pressure approximately equal to ambient atmospheric pressure,* a condition that I will call *pressure balanced.* If they enter the atmosphere at a pressure significantly above ambient, I call them *overpressured,* although the commonly used description in fluid mechanics is *underexpanded.*

On Io, ambient atmospheric pressure is about 10^{-12} bars but could be higher (10^{-7} bars?) around volcanic eruption sites. With typical reservoir pressures on the order of tens of bars, a pressure drop of about 8 to 15 orders of magnitude is required to reduce the volcanic fluid pressure to ambient pressure. A crater on the order of 1000 m or deeper would be required for this depressurization. Thus, there is the possibility of pressure-balanced jets occurring on Io *if and only if* they erupt from volcanic systems with craters more than 1 km deep. The jets will be overpressured if the eruptions occur through shallow craters or fissures without craters.

These conclusions depend on the assumed conduit-crater geometry. The amount of depressurization is directly proportional to the ratio of crater to conduit area, as can be seen from Figure 11.6. The conduit dimension chosen, 5-m radius, is representative of terrestrial dikes that might be interpreted as conduits of ancient volcanoes. Therefore, the minimum crater depth of about 1 km referred to above as necessary for creation of a pressure-balanced jet on Io applies *if* the geometry of Ionian volcanic systems is similar to those postulated for Earth. However, if the conduit-crater geometry were drastically different,

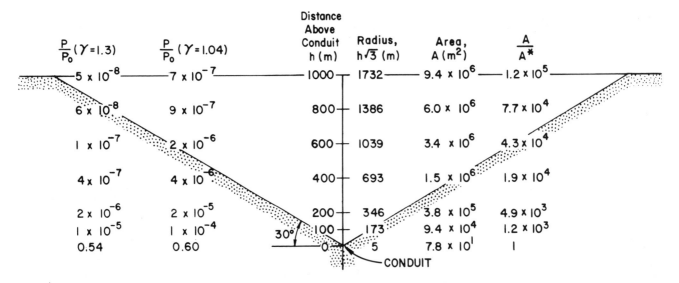

FIGURE 11.6 Variation of geometric and flow variables in crater. For this nominal geometry the conduit radius has been taken as 5 m, and the crater has been taken with the internal slopes at angle of repose, 30° from horizontal. The two central columns give height above the crater floor (in meters) and the ratio of crater area, A, to conduit area, A^*. The columns on the far right and far left give the ratio of fluid pressure, P, to reservoir pressure, P_0, for the two values of γ_m shown (from Kieffer, 1982b, modified and reproduced with permission of the University of Arizona Press).

e.g., if a conduit had a radius of 500 m and the surface crater were as postulated above, the inferred depth at which pressure equilibration would occur would need to be scaled to the place in the crater where comparable area ratios, A/A^*, were obtained. Similarly, if the crater were simply a gradual steep-walled enlargement of the conduit, as proposed by Hawthorne (1975) for kimberlite geometries, the pressure equalization distances would also need to be scaled. In Hawthorne's model of a kimberlite pipe, the conduit enlarges by a factor of 100 over the top 2.5 km of its profile. For a fluid with γ_m in the range of 1.0 to 1.1, the pressure at the exit plane would be 0.1 percent of the stagnation pressure (assumed to be the reservoir pressure in this model that ignores friction). If the initial reservoir pressure were 1 kbar, the jet could become pressure-balanced traversing 2.5 km through this steep-walled geometry. At higher pressures plausible for kimberlite reservoirs, passage through a conduit of such a geometry would not (in the absence of friction) reduce the jet pressure to ambient, and it would exit as an overpressured jet.

IMPLICATIONS FOR JET STRUCTURE

Pressure-Balanced Jets

Pressure-balanced and overpressured jets have different structures because of the respective absence and presence of internal-pressure gradients. Their structures are summarized in Figures 11.7 and 11.8.

In a pressure-balanced jet under terrestrial conditions, the flow maintains a direction approximately parallel to the centerline of the conduit in which the fluid accelerates, because internal pressure gradients that could cause it to expand laterally are dissipated by passage through a small crater (Figure 11.7a). Thus, flow is unidirectional in terrestrial pressure-balanced jets, the initial velocity vector is parallel to the centerline of the volcanic conduit, and the jets have no internal wave structures because of pressure gradients. Plinian jets on Earth (Figure 11.1a) are usually pressure balanced because they rise through long conduits (in which friction helps reduce the pressure toward ambient) and/or shallow surface craters. In the case of the plinian column on May 18 at Mount St. Helens, the jet probably rose through a conduit on the order of 3 km long and through a small crater in the floor of the amphitheater created by the lateral blast (inferred from the morphology of the crater floor after cessation of the eruption). If this crater had been only a few tens of meters deep, it would have been sufficiently deep to allow decompression to 1-bar pressure. Such plinian jets rise upward in the atmosphere *not* because of internal pressure of the gas but through a combination of initial momentum combined with increasing buoyancy as air is entrained into the jet (these are the jets carefully studied and modeled by Wilson, 1976; Sparks *et al.*, 1978; Wilson *et al.*, 1980; see also Wilson and Head, 1983).

The structure of pressure-balanced jets on Io might be expected to be quite different from those on Earth, not only because of the effects of different gravity and atmospheric pressure but also because of the much deeper crater required to

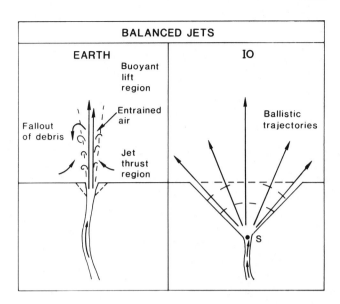

FIGURE 11.7   Schematic drawings of the structure of pressure-balanced jets on Earth and on Io.

produce a pressure-balanced jet on Io (Figure 11.7). The high internal pressure (relative to the ambient pressure) that causes the flow to expand laterally will be maintained through a fairly large crater traverse, and the flow streamlines will diverge to become parallel to the crater walls. As the flow traverses the surface crater, fluid pressures drop from several bars to atmospheric pressure. The velocity vectors tend to depart more from the initial vertical direction imparted by accelerations in the restricted conduit. Within the crater, flow streamlines will not be parallel to the centerline of the conduit nor perpendicular to planes across the crater; rather they diverge as the flow follows the contours of the crater. A rigorous model of the flow would require two- or three-dimensional flow equations in the crater. However, to a first approximation the spreading streamlines diverge from an apparent spherical center of symmetry at the bottom of the crater. Surfaces of constant flow properties are, then, hemispheres rather than planes. Strom and Schneider (1982) showed that the shape and optical characteristics of the umbrella jet at plume 3, Prometheus, can be well modeled by using a point-source ballistic model in which particles radiate with a uniform velocity through angles from 0 (vertical) to 55 (about angle of repose). I suggest that this is because Prometheus (Figure 11.1c) and, by analogy, the umbrella jets are pressure-balanced jets, in which internal-pressure gradients are negligible. After passage through the surface crater, the gas and the particles coupled to the gas follow simple ballistic trajectories in the near-vacuum environment of the Ionian surface. These ballistic trajectories are determined as the streamlines of the flow diverge to follow the crater walls. From this analysis it can be inferred that the jet of Prometheus emerges from a rather deep surface crater ($> \sim 1$ km). [Note that the surface crater discussed here is several kilometers in diameter (see Figure 11.6), which is below the limits of resolution of Voyager images.]

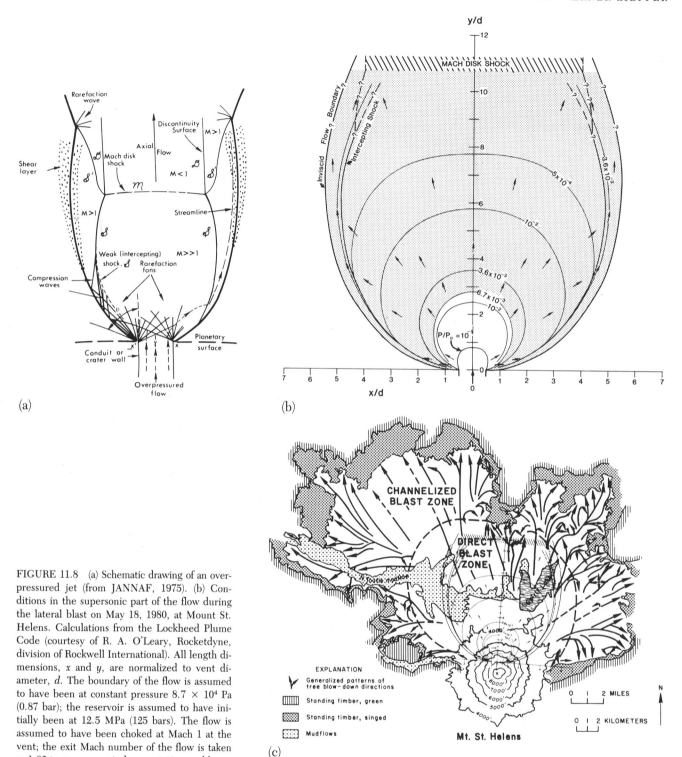

(a)

(b)

(c)

FIGURE 11.8 (a) Schematic drawing of an over-pressured jet (from JANNAF, 1975). (b) Conditions in the supersonic part of the flow during the lateral blast on May 18, 1980, at Mount St. Helens. Calculations from the Lockheed Plume Code (courtesy of R. A. O'Leary, Rocketdyne, division of Rockwell International). All length dimensions, x and y, are normalized to vent diameter, d. The boundary of the flow is assumed to have been at constant pressure 8.7×10^4 Pa (0.87 bar); the reservoir is assumed to have initially been at 12.5 MPa (125 bars). The flow is assumed to have been choked at Mach 1 at the vent; the exit Mach number of the flow is taken as 1.02 to ease numerical computation problems. Contours of constant thermodynamic properties are shown, as summarized in Table 11.1. Small arrows indicate the local flow direction. The flow diverges through rarefaction waves (not shown; see Figure 11.8a for schematic representation) that reflect from the flow boundaries to form the intercepting shocks shown. Across these shocks the flow velocities decrease, the pressure increases, and the flow changes direction. The intercepting shocks coalesce across the flow to form the strong Mach disk shock. In most of the flow the pressure is subatmospheric (stippled zone). It rises back toward atmospheric across the Mach disk and intercepting shocks. The location of the Mach disk shock is not predicted by the computer code; it is estimated from the JANNAF (1975) equations. (c) Figure 11.8b, scaled to and superposed on the devastated area at Mount St. Helens (same orientation and placement as in Kieffer, 1982a). Note approximate coincidence of supersonic jet and direct blast zone.

TABLE 11.1   Thermodynamic Conditions on the Contours in Figure 11.8B

P/P_0	M	T/T_0	ρ/ρ_0
1.0×10^{-1}	2.1	0.92	1.9×10^{-1}
1.0×10^{-2}	3.1	0.84	1.8×10^{-1}
6.7×10^{-3} (ambient)	3.3	0.82	1.4×10^{-2}
3.6×10^{-3}	3.5	0.80	6.9×10^{-3}
1.0×10^{-3}	3.9	0.76	2.0×10^{-3}
5.0×10^{-4}	4.1	0.75	1.1×10^{-3}

Overpressured Jets

The structure of overpressured jets, shown in Figure 11.8a, is much more complex than that of pressure-balanced jets because there are initially large pressure gradients within the jet. These pressure gradients cause the jet to expand through rarefaction or expansion waves that emanate from the sides of the vent. The fluid thus diverges from its initial centerline direction (flares strongly) near the vent, and streamlines of the flow diverge from the simple paths determined in the conduit. If the fluid is at substantially higher pressure than ambient, it does not simply expand to ambient pressure as it emerges but actually overexpands, developing a low-pressure zone in the center of the flow. The atmosphere then presses back in on the fluid, causing the flow boundaries to curve around to become more subparallel to the initial direction of flow. A complex system of internal expansion, compression, and shock waves is set up in response to the decompression. These internal waves, as well as gravity, turbulence, air entrainment, buoyancy, and atmospheric structure, must be considered in models of jet structure.

Equations of motion for an axisymmetric jet, in the absence of viscous forces and of gravitational accelerations, are as follows:

$$\overline{\nabla} \cdot (\rho \overline{u}) = 0 \qquad \text{(continuity)} \quad (11.10)$$

$$\overline{\nabla} \times \overline{u} = 0 \qquad \text{(irrotationality)} \quad (11.11)$$

$$\overline{\nabla} \left(\frac{u^2}{2} \right) + \frac{1}{\rho} \overline{\nabla} P = 0. \qquad \text{(momentum)} \quad (11.12)$$

In these equations, ρ is the density, $\overline{u}$ is the vector velocity, P is the pressure, and $\overline{\nabla}$ is the spatial differential operator. The equations are nonlinear and not solvable analytically. Detailed structure of overpressured jets is difficult to calculate because internal shock waves are difficult to handle mathematically, even with the aid of large computers. Within regions of flow where there are no strong shock waves, solutions can be obtained by the method of characteristics. A solution for the initial conditions proposed for the Mount St. Helens lateral blast obtained by hand calculations was given by Kieffer (1982a); a refined computer-generated version is given here in Figure 11.8b. In such solutions the position of strong shocks must be determined independently, e.g., from compilations of laboratory data, such as JANNAF (1975) or Ashkenas and Sherman (1966). The major features of the two models are in good agreement. However, a significant difference is the curvature of the

boundary of the supersonic region. In the model given by Kieffer (1982a) the boundary was determined by use of the JANNAF (1975) construction formula; its use for the pressure ratios assumed for the blast represented an extrapolation beyond limits validated by laboratory experiment. As a result, the new theoretical boundary is preferable even though the discrepancy with the mapped boundary between the direct and channelized blast zones is increased.

Because the computer-generated flow field presented is so similar to the model given by Kieffer (1982a), detailed discussion of the flow-field properties will not be repeated here. However, several points are worth reemphasizing because of their relevance to jet structure, and several new quantitative statements can be made.

First, the flow originally emanating northward from the vent is immediately deflected through a range of angles, up to 96°. Thus, highly overpressured jets do not form simple fire-hydrant-shaped jets; instead they form fan-shaped ones. The flow is turned back more nearly parallel to the initial flow direction as it passes through the intercepting shocks (see Figure 11.8b, small arrows).

Second, much of the region within the flow is at subatmospheric pressure (the stippled region in Figure 11.8b). The atmosphere responds to the formation of this low-pressure zone by pressing inward, causing the intercepting and Mach disk shocks to form. The pressure drops to 4 percent of atmospheric pressure immediately in front of the Mach disk shock. When the fluid passes through the lateral intercepting shocks and the Mach disk shock, the pressure is brought back up toward atmospheric.

Third, everywhere in the flow, except possibly for a small region right in front of the Mach disk shock, the density of the fluid remains greater than the atmospheric density. (Uncertainties in the computer handling of the convergence of the lateral intercepting shocks preclude a statement on whether there is actually a small region that decompresses to or below atmospheric density.) Thus, with the possible exception of a small "chimney" in front of the Mach disk shock, the flow is negatively buoyant within the supersonic zone. At the inviscid boundary of the supersonic plume, the density is 0.014 g/cm³, about 16 times atmospheric density. In front of the Mach disk shock it is approximately atmospheric density or slightly less; upon passing through the Mach disk shock the density also increases. In my original work (1982a) I assumed that the Mach disk shock was a weak shock; thus, I extrapolated the densities through the shock. An alternative and better-founded assumption (Ashkenas and Sherman, 1966) is that the Mach disk shock is a strong shock that brings the flow pressure back to atmospheric; in this case the density of the fluid increases to about 10 times atmospheric. Thus, at the margins of the supersonic flow region, the fluid is denser everywhere than in the atmosphere. The temperature of the fluid also increases dramatically as it passes through the Mach disk shock. Thus, the shock could be a source of both major flow disturbances and atmospheric phenomena because of the pressure, density, and temperature changes (see Chapter 10 of this volume).

Fourth, the "boundary" of the plume shown in Figure 11.8b is a boundary calculated by assuming inviscid flow. No estimates of the conditions in the shear layer that exists on the flow boundary (shown in Figure 11.8a) are included, which means that the calculated flow velocities at the boundary are not realistic. The calculated inviscid flow velocities are about 300 m/sec. Entrainment and mixing would substantially reduce these velocities and would alter the density of the flow. The model strongly suggests, however, that around the perimeter of the supersonic zone the flow transforms from a supersonic jet to a dense, largely subsonic flow. Gravity would play a major role in the flow motion beyond the supersonic flow boundary.

When scaled to and superposed on a map of the devastated area at Mount St. Helens (Figure 11.8c), the overpressured jet mimics the shape of the direct blast zone, leading me to conclude that the direct blast zone was a zone dominanted by supersonic fluid flow and that the surrounding channelized blast zone was the zone of denser subsonic flow (Kieffer, 1982a). The model explains a number of features related to flow properties, such as velocity, density, downed-tree pattern, and temperatures throughout the devastated area. For example, internal rarefaction waves at the vent deflected the flow through more than a 90° angle initially, causing initial flow directions of some of the fluid to be approximately east-west. However, because the rarefaction waves were reflected from the boundary between the flow and the atmosphere as compressive waves, coalescing to form the Mach disk shock, the flow turned back to a more northerly direction as it diverged from the vent. Streamlines of overpressured flows are extremely complex, and all of the internal waves affect the velocities and direction of flow that crosses them. In fact, if streamlines from the lateral parts of the flow field near the source are extrapolated in a simple linear manner back toward the vent, they do not converge at the vent but rather at an apparent source at least 1 km in front of the actual vent. No actual physical source in such a location is required to explain the nature of such streamlines.

What would be the characteristics of an overpressured jet on Io? Internal-wave dynamics within the jet would destroy the smooth ballistic trajectories that give the umbrella shape to pressure-balanced jets on Io. Perhaps in low-resolution photos such jets would appear to be rather structureless blobs like the jet from Loki (Figure 11.1d) or, on a smaller scale, their deposits would be irregular like the white frost deposits described by McCauley et al. (1979). If diffuse jets do exist on Io and are indeed overpressured, it can be inferred that the internal fluid pressure at the vent is greater than atmospheric and that the jets originate from conduits or fissures that have only shallow surface craters.

It is at least consistent with the idea that the regolith of Io consists of interbedded pyroclastic material, sulfur lava flows, and perhaps some silicate ash, i.e., weakly cohesive material in which kilometer-deep volcanic craters could easily be excavated, that most of the plumes on Io are umbrella shaped. The two plumes that have at least at times been diffuse are Loki (plume 2) and Amirani (plume 5). These plumes emanate from fissures that are perhaps in sufficiently strong surface materials that deep craters are not easily eroded, so that overpressured plumes can at least be maintained temporarily. An intriguing implication of this model is that plume structure on Io could change substantially as atmospheric pressure changes, e.g., from day to night.

SUMMARY

The structure of pressure-balanced jets may vary from one planet to another, but on one planet the jets may tend to have rather similar structures from one volcano to another because they are equilibrated to atmospheric pressure. This is perhaps the reason that plinian jets around Earth look rather similar. However, overpressured jets may show a much greater diversity in structure because pressure imbalances may range from slight ($P > P_a$) to huge ($P >>> P_a$), and jet structure depends on the degree of overpressurization. Thus, either as a class, or as individuals like the Mount St. Helens blast, overpressured jets are harder to categorize, analyze, and model. This paper describes the role of compressibility, without considering the influences of other important effects, such as gravity; buoyancy; viscous effects at boundary layers; and, for lateral jets, topography. Accurate prediction of hazards from jets during explosive eruptions must account simultaneously for the influence of all of these effects.

REFERENCES

Ashkenas, H., and F. S. Sherman (1966). The structure and utilization of supersonic free jets in low density wind tunnels, in *Rarified Gas Dynamics*, Vol. 2, Academic Press, New York, pp. 84-105.

Boyd, F. R. (1961). Welded tuffs and flows in the rhyolite plateau of Yellowstone National Park, Wyoming, *Geol. Soc. Am. Bull.* 72, 387-426.

Graton, L. C. (1945). Conjectures regarding volcanic heat, *Am. J. Sci.* 243A, 135-259.

Hawthorne, J. B. (1975). Model of a kimberlite pipe, *Phys. Chem. Earth 9*, 1-16.

JANNAF [Joint Army, Navy, NASA, Air Force] (1975). *Handbook of Rocket Exhaust Plume Technology*.

Kieffer, S. W. (1981). Blast dynamics at Mount St. Helens on 18 May 1980, *Nature 291*, 568-570.

Kieffer, S. W. (1982a). Fluid dynamics of the May 18 blast at Mount St. Helens, in *The 1980 Eruptions of Mount St. Helens, Washington*, P. W. Lipman and D. R. Mullineaux, eds., U.S. Geol. Surv. Prof. Pap. 1250, pp. 379-400.

Kieffer, S. W. (1982b). Ionian volcanism, in *Satellites of Jupiter*, D. Morrison, ed., U. Arizona Press, Tucson, pp. 647-723.

McCauley, J. F., B. A. Smith, and L. A. Soderblom (1979). Erosional scarps on Io, *Nature 280*, 736-738.

McEwen, A. S., and L. A. Soderblom (1983). Two classes of volcanic plumes on Io, *Icarus 55*, 191-217.

Moore, J. A., and W. C. Albee (1982). Topographic and structural changes March-July 1980—photogrammetric data, in *The 1980 Eruptions of Mount St. Helens, Washington*, P. W. Lipman and D. R. Mullineaux, eds., U.S. Geol. Surv. Prof. Pap. 1250, pp. 123-134.

Moore, J. A., and T. W. Sisson (1982). Deposits of the May 18 pyroclastic surge, in *The 1980 Eruptions of Mount St. Helens, Washington*, P. W. Lipman and D. R. Mullineaux, eds., U.S. Geol. Surv. Prof. Pap. 1250, pp. 421-438.

Self, S., L. Wilson, and I. A. Nairn (1979). Vulcanian eruption mechanisms, *Nature 277*, 440-443.

Sparks, R. S. J. (1978). The dynamics of bubble formation and growth in magmas: A review and analysis, *J. Volcanol. Geotherm. Res. 3*, 1-37.

Sparks, R. S. J., L. Wilson, and G. Hulme (1978). Theoretical modeling of the generation, movement, and emplacement of pyroclastic flows by column collapse, *J. Geophys. Res. 83*, 1727-1739.

Strom, R. G., and N. M. Schneider (1982). Volcanic eruption plumes on Io, in *Satellites of Jupiter*, D. Morrison, ed., U. Arizona Press, Tucson, pp. 598-633.

Strom, R. G., R. J. Terrile, H. Masursky, and C. Hansen (1979). Volcanic eruption plumes on Io, *Nature 280*, 733.

Verhoogen, J. (1946). Volcanic heat, *Am. J. Sci. 244*, 745-770.

Wilson, L. (1976). Explosive volcanic eruptions—III. Plinian eruption columns, *Geophys. J. R. Astron. Soc. 45*, 543-556.

Wilson, L., and J. W. Head III (1983). A comparison of volcanic eruption processes on Earth, Moon, Mars, Io and Venus, *Nature 302*, 663-669.

Wilson, L., R. S. J. Sparks, and G. P. L. Walker (1980). Explosive volcanic eruptions—IV. The control of magma properties and conduit geometry on eruption column behavior. *Geophys. J. R. Astron. Soc. 63*, 117-148.

Experimental Studies of Hydromagmatic Volcanism

12 ———————————————————

KENNETH H. WOHLETZ *and* ROBERT G. McQUEEN
Los Alamos National Laboratory

ABSTRACT

Hydromagmatic volcanism was modeled in experiments in which thermite melt (Fe + Al_2O_3) explosively interacted with water. Several designs were explored using different contact geometries, water-melt ratios, and confinement pressures. The explosions featured ejection of steam and fragmented melt. The modeled volcanic phenomena includes melt fountains (Strombolian), dry and wet vapor explosions (Surtseyan), and passive chilling of flows (submarine pillow formation). The pertinent experimental parameters are: (1) ejection velocities of 20 to 100 m/sec, (2) confining pressures of 10 to 40 MPa, (3) melt ejecta sizes of microns to centimeters in diameter, (4) steam production at temperatures of 100°C to high levels of superheating (300 to 500°C), and (5) ejection modes that are both ballistic and surging flow in a turbulent expanding cloud of vapor and fragments. The results indicate that explosive efficiency is strongly controlled by water-melt mass ratio and confining pressure. Optimum thermodynamic efficiency measured as the ratio of mechanical to thermal energy occurs at water-melt ratios between 0.3 and 1.0. Fragmentation increases with explosive energy and degree of water superheating.

INTRODUCTION

The understanding of explosive volcanism has been limited by the difficulty of documenting many of the physical parameters involved. Visual assessments of the eruption products and their movement, like those detailed by Moore and Rice (see Chapter 10 of this volume), are necessarily distant observations. Only qualitative inferences can be made about the explosion process. Experimental modeling can be used to evaluate the energetics as well as boundary conditions of an eruption in that important parameters such as pressure, temperature, velocity, and density can be measured.

Volcanic explosions are the result of explosive expansion of volatile materials. Two end-member processes account for the origin of these volatiles: (1) *hydromagmatic* (hydrovolcanic) processes in which hot magma interacts with external water at or near the surface of the Earth, producing vapor explosions (see Figure 12.1), and (2) *magmatic* processes in which volatiles in the melt (dominantly H_2O) exsolve and explosively fragment

the magma by rapid decompression. These two processes may operate simultaneously during an eruption if the magma composition and environmental factors permit.

Experiments (Wohletz and Sheridan, 1981, 1982; McQueen and Wohletz, in preparation) reviewed in this paper provide information on hydromagmatic processes. Although experimentation on hydromagmatic volcanism has just begun, considerable work on vapor explosions has been done in the field of nuclear reactor safety (Sandia Laboratories, 1975). These studies build on previous work aimed at understanding disastrous industrial explosions in which molten metal has come into accidental contact with water (Lipsett, 1966). Such explosions are termed *fuel-coolant interactions* (FCIs), and their behavior has been studied in small-scale laboratory experiments (Board *et al.*, 1974; Peckover *et al.*, 1973). Recent work (Buxton and Benedict, 1979; Corradini, 1980; Nelson and Duda, 1981) dealing with theoretical and laboratory investigations of nuclear reactor systems also lends considerable insight into volcanic processes.

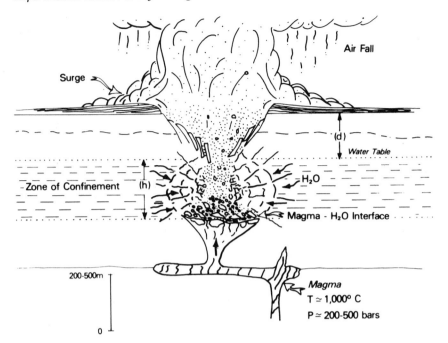

FIGURE 12.1 Diagrammatic cross section of a hydrovolcanic eruption in which magma interacts with near-surface groundwater. The interaction fragments the magma and country rock, vaporizes the water, and produces steam explosions that excavate a crater and eject tephra. Scale shown is for maar volcanoes of typical size.

GEOLOGIC BACKGROUND

Thomas Jaggar (1949) was an early advocate of the idea that many volcanic eruptions are the result of hydromagmatism. The importance of this mechanism to explosive volcanism was not totally appreciated, however, until the pyroclastic surge deposits from these eruptions were recognized for their widespread occurrence (Moore, 1967; Fisher and Waters, 1970; Waters and Fisher, 1971). Classical terms for eruptive styles of hydromagmatic volcanoes include (1) *Surtseyan* (Thorarinsson, 1966), exemplified in the explosive birth of the island Surtsey near Iceland, where repeated blasts deposited numerous layers of ash from base surges and ash falls (Figure 12.1); (2) *Strombolian* (Walker and Croasdale, 1971; Chouet *et al.*, 1974), typified in the growth of the Stromboli Volcano in the Mediterranean, where limited interaction of seawater with magma and magmatic gases produced low-energy bursts forming cinder cones; and (3) *submarine* eruptions where lava is extruded passively on the ocean floor with little or no explosive interaction with water (see Figure 12.2).

EXPERIMENTAL APPROACH

In the experiments described here the volcanic environment was simulated by bringing molten thermite into contact with water. The objective was to determine the parameters that control explosivity. Explosions occur when water vaporization at extreme *P* and *T* results in rapid conversion of the melt's thermal energy into mechanical energy.

Thermite

The process of thermite converting to melt arises through a highly exothermic oxidation-reduction reaction of a metal and metal oxide. The process has many applications, the most familiar of which relate to welding. The following physical properties of thermite resemble those of basaltic magmas and thus make it useful for our modeling:

1. the reaction temperature is greater than 1000°C,
2. viscosity is in the range of 10^3 poise,
3. density is 3 to 5 g/cm³,
4. the melting yields a mixture of crystals and liquid within the *T* and *P* of investigation, and

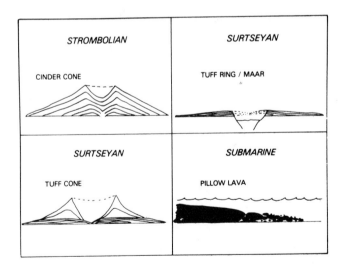

FIGURE 12.2 Land forms produced by four principal modes of hydrovolcanic activity. The layering of pyroclastic material illustrated results from periodic eruption pulses, avalanching of debris, and base surge and emplacement. Surtseyan tuff rings result from more highly explosive eruptions than do Surtseyan tuff cones, hence their lower profiles.

5. the mixture flows over a surface in a manner similar to basalt.

There are limitations, however, that must be considered: the enthalpy of molten thermite ($\Delta H = 877$ cal/g) is about three times that of basalt; the melting temperature may exceed 2000°C, whereas basalt typically melts at 1200°C; and the chemical composition of the thermite is unlike that of a basalt. Because of large enthalpy of thermite, addition of oxides such as silica to the system produces a melt chemistry that more closely matches that of a basalt. The thermite reaction is

$$Fe_3O_4 + 8/3\ Al = 4/3\ Al_2O_3 + 3\ Fe + \Delta H. \quad (12.1)$$

After the melt contacts water, additional enthalpy results from the spontaneous oxidation of Fe, dissociation of H_2O, and subsequent burning of H_2. This effect is minor, however, and occurs after the initial explosive interaction. Because the thermite reaction is essentially adiabatic and occurs at low pressures, the total internal energy is equal to the system's enthalpy.

1. Sand Burial

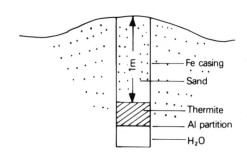

2. Confinement

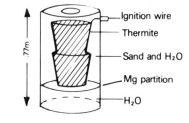

3. Water Box

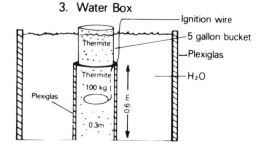

FIGURE 12.3   Three experiments for simulating volcanic eruptions.

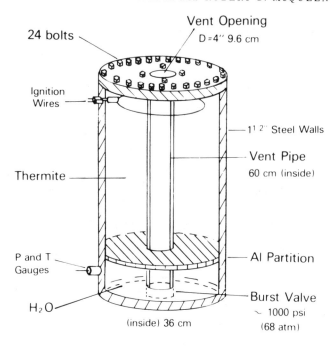

REACTION

$$Fe_3\ O_4 + 8/3\ Al \rightleftharpoons 4/3\ Al_2O_3 + 3Fe + Heat$$

FIGURE 12.4   A fourth experimental apparatus designed to measure P and T during eruption.

Design

Four experimental setups have been used to simulate different volcanic environments (see Figure 12.3). In the first configuration, the thermite was buried in sand to produce an explosive magma model. A second design gave further confinement by enclosing thermite and water in a sealed steel cylinder with a burst valve at the top. In the third design a large (1 m³) box with a Plexiglas side was filled with water and enclosed another plexiglas tube filled with thermite. This design allowed some visual assessment of the melt–water interaction. The fourth model, illustrated in Figure 12.4, allowed more precise documentation and control of experimental variables. The aims of this design were as follows:

1. To ignite the thermite from the top down so as to ensure a completely molten state prior to contact with water. In the setup an aluminum partition separating the thermite and the water must be melted for contact to occur.

2. To make the hydrostatic head of the melt force contact with water in the same way that basalt rising under a pressure head contacts surface water.

3. To simulate near-surface conditions by setting the burst valve at about 1000 psi (7 MPa).

4. To design the vent such that all of the melt must move through the water compartment during venting. Otherwise, trapped water would expel large amounts of melt above it without the chance for efficient heat exchange. In this way thermite is modeled as magma that had to move through near-surface water during extrusion.

5. To incorporate *T* and *P* recorders in the water compartment.

Documentation

High-speed cinematography was used as the primary means of documenting the experiments. Through photography it is possible both to compare the model qualitatively with volcanic eruptions and to measure quantitative features such as velocity, event timing, and ejecta trajectories by frame-to-frame analysis (Figures 12.5 and 12.6).

Pressure measurements made from gauges connected through the steel confinement casing were recorded on strip charts. Three gauges were used—two with a dynamic range from 0 to 5000 psi and one for recording nonpressure-related events. Records show timing for the initiation, duration, and intensity of the melt-water interaction.

Thermocouple records and postexplosion inspection of the confinement chamber provided only limited information. Thermocouple probes inserted into the water compartment measured minimum-temperature rise times and amplitudes for conditions in the zone of interaction. In the postexperiment inspection of the confinement chamber, the degree of melt-water mixing was assessed from the completeness of the chamber evacuation and the size and distribution of the ejecta.

FIGURE 12.5 Experiment 1 (Table 12.1) showing vertical ejection column and horizontally expanding base surge.

FIGURE 12.6 Water-box experiment in Strombolian activity.

EXPERIMENTAL RESULTS

This paper reviews the first stages of our experimental program; only tentative analyses are reported. The results are divided into three categories: venting phenomena, ejecta characteristics, and pressure histories.

Venting Phenomena

Descriptions of the system and the observed venting phenomena are given in Table 12.1 and Figure 12.7. These are characterized by three independent parameters: degree of confinement, water-melt mass ratio (R), and contact geometry.

The experimental eruption phenomena illustrated in Figure 12.7 are necessarily generalized. Also, no experiment shows just one phenomenon. Their classification is based on the dominant behavior. The most important requirement in this type of modeling is that the mechanism of the experiment be as close as possible to that operating in nature. As discussed earlier, the mechanism is basically the same as that in nature. The experimental explosions bear a remarkable resemblance to volcanic eruptions despite the following limitations: (1) use of a rigid cylindrical vent that directs ejecta vertically; (2) the high temperature of the melt, which limits the degree of melt-fragment quenching; and (3) the high superheat of the water, which on some occasions makes the steam released optically transparent instead of opaque, as is condensed or saturated volcanic steam from Surtseyan eruptions (Thorarinsson, 1966).

Rigorous scaling of the model's eruptive activity has not been attempted. Our efforts have thus far been directed at studying the feasibility of simulating volcanic explosions by this technique. Numerous experimental studies, including Board *et al.* (1975), Dullforce *et al.* (1976), Fröhlich *et al.* (1976), and Nelson and Duda (1981), have demonstrated the complexity of FCIs

and have shown that explosive efficiencies may increase with system size. Therefore, scaling formulas have yet to be determined. To scale time for the model, we can compare the length dimension to that of Mount St. Helens to give a geometric scale, n. Approximately 0.5 km^3 of new lava erupted from Mount St. Helens on May 18, 1980. Our model uses about 5 $\times$ 10^{-2} m^3 of thermite that results in $n = 10^4$. Model times approximately scale by a factor of $n^{1/2}$ or about 10^2.

Four types of volcanic activity have been modeled and are described in Table 12.1. Strombolian activity results in fountaining of ejecta in ballistic trajectories over periods of minutes. Strong Surtseyan activity is short lived (less than 1 to 2 sec). Material is ejected in unsteady surging flow, and ballistic behavior of ejecta is minor or entirely absent. Weak Surtseyan activity is more steady, lasting for periods of 10 or more sec. Characteristics of both ballistic and surge movement of ejecta are observed. Finally, the fourth type of activity is passive quenching of the melt into lumps with little or no ejection into the atmosphere. Vulcanian activity (Mercali and Silvestri, 1891) can be compared with strong Surtseyan blasts that are very short lived or cannon-like. The hydromagmatic origin of Vulcanian activity has yet to be conclusively demonstrated. However, recent studies at Vulcano, Italy (Frazzetta *et al.*, 1983), and at Ngauruhoe, New Zealand (Nairn and Self, 1978; Self *et al.*, 1979), suggest the strong likelihood of meteoric water interaction in this eruptive style.

Ejecta Characterization

Observations of ejecta grain sizes show that the model's melt fragments become finer with increasing energy and violence of ejection. A quantitative analysis of ejecta sizes has not yet been completed, but photo documentation and recovery of

FIGURE 12.7 Summary of experimental results. Unsteady and steady Surtseyan activities are thought to produce tuff rings and tuff cones, respectively, as shown in Figure 12.2.

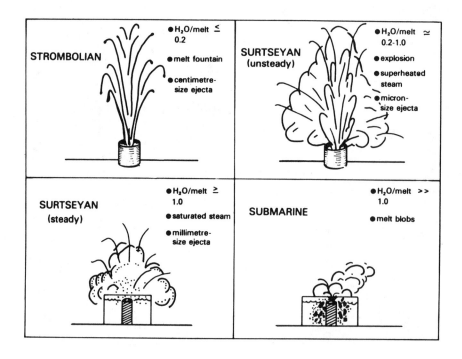

TABLE 12.1   Experimental Results

Experiment	R	Confinement	Description
Sand Burial			
1	0.23	0.5 m sand	Melt fragments of 1 to 2 mm in size fountained 2 to 3 m into the air for 4 to 5 sec; weak Strombolian.
2	0.13	0.5 m sand	Melt fragments of 1 cm in size fountained 7 to 10 m into the air for 2.5 sec; restricted neck vent; strong Strombolian.
Buried Confinement			
1	0.45	900 psi	Weak Surtseyan ejection of melt in less than 1 sec; eruption column 24 m high; horizontally moving base surge; 1-mm fragments; formed a tuff cone 2.5 m in diameter, 0.5 m high; continuous ejecta 5 m from rim.
2	0.22	900 psi	Strong Surtseyan blast lasting less than 1 sec; eruption column 40 m high; surge reaching 6 m from vent; submillimeter ejecta and accretionary lapilli; formed tuff ring 2.1 m in diameter, 0.27 m high; continuous ejecta 6 to 7 m from rim.
Water Box			
1	0.10-10.0	none	Variable access of H_2O to melt: Strombolian for several minutes with brief Surtseyan blasts (<1 sec), centimeter-size fragments ejected ballistically to a height of 10 m; Surtseyan blasts are brilliant hemispherically expanding clouds of micron-size particles in superheated steam; Surtseyan blast (<1 sec) destroying device, micron-size particles in horizontally directed surge.
2	0.10-10.0	none	Variable access of H_2O to melt: 50-sec Strombolian ballistic ejection of centi-meter-size fragments with brief (<1 sec) Surtseyan sprays of submillimeter size particles; 60-sec, weak Surtseyan ejection of millimeter-size quenched melt fragments moving ballistically and in surges, saturated steam; subma-rine quenching of melt into millimeter- or centimeter-size lumps (pillows).
77-1	0.31	900 psi	Violent Surtseyan blast lasting less than 1 sec; eruption column 50 m high; micron-size dust carried away by wind.
77-2	0.23	900 psi	Surtseyan blast lasting 16 sec, 35 m in height, micron-size fragments, super-heated steam; strong Strombolian eruption 90 sec long with two short Surt-seyan blasts, centimeter-size fragments ejected ballistically; weak Strombol-ian fountain 10 m in height.
78-1	0.26	900 psi	Surtseyan 5-sec ejection of incandescent gas and micron-size particles 50 m high; pulsating Strombolian lasting 5 sec and weak Strombolian also of 5-sec duration; fountains up to 20 m high.
78-2	0.22	none	Surtseyan blast lasting 2 sec, micron-size fragments, and superheated steam jetted 40 m high; restricted vent.

fragmental debris permit a qualitative assessment. For Strom-bolian-type bursts, centimeter-size melt fragments are ballis-tically ejected and fall back around the vent. After strong Surt-seyan explosions, little ejecta is noticeable around the volcano. Most of the ejecta is carried away as fine micron-sized dust (fine volcanic ash). Weak Surtseyan blasts are more steady in their ejection of material, as in the water-box experiments. Melt fragments are of millimeter size—roughly equivalent to coarse volcanic ash and lapilli. Passive quenching of the melt, analogous to formation of pillow lava, results in lumps or blobs of melt that range in size from millimeters to several centi-meters.

Scanning electron microscopy (SEM) analyses of the explo-sive products (see Figure 12.8) reveal blocky-equant shapes, spheroidal and drop-like shapes, as well as irregular moss-like shapes (Wohletz, 1983). Thermite debris recovered from strong Surtseyan blasts ranges in median diameter from 1 to 40 μm, while less violent explosions produce debris with median di-ameters in the range of 100 to 200 μm. The size and shape of

the melt fragments strongly resemble that of basaltic hydro-magmatic ash studied by Heiken (1972) and Wohletz and Krin-sley (1983).

Pressure Histories

Experimental pressure histories, presented in Figure 12.9, have been evaluated based on the magnitude and duration of pres-sure pulses. Strong Surtseyan activity generally is shown by off-scale (greater than 5000 psi or 350 MPa) response and short durations (several seconds or less). Weak Surtseyan activity also sends records off the scale, but durations may be up to 10 sec or more. Strombolian activity results in oscillating pressures from a few hundred psi to 1000-2000 psi. These oscillations are harmonic, with as many as 15 oscillations per second. If these oscillations are due to acoustic resonance, then sound speed for the system is about 10 m/sec, a value that agrees well with the values for liquid-gas mixtures calculated by Kieffer (1977). The Surtseyan blast of experiment 78-2 (Table 12.1) shows a

FIGURE 12.8 Scanning electron micrograph of some experimental explosion debris. Note the blocky equant grains and spherical shapes that strongly resemble basaltic volcanic ash particles. The debris is glassy and microcrystalline (undetermined mineralogy) and consists of dominantly Fe and Al oxides with subordinant amounts of SiO_2, TiO_2, MgO, MnO, Na_2O, CaO, and K_2O. The scale bar shown at the bottom of the photo is 10 μm, the average diameter of most grains pictured.

pressure buildup of only 400 psi, but the melt was completely ejected in less than 4 sec. Because no confinement valve was used, the low-pressure peak may be explained by lack of pressure buildup prior to ejection and an effectively larger orifice. The static pressure transducer used in experiment 77-2 (Table 12.1) recorded shock events characteristic of Surtseyan blasts. This record supports a theory developed by Board *et al.* (1975) of shock propagation during interaction of melt and water as well as observations of volcanic eruptions by Nairn (1976) and Livshits and Bolkhoritinov (1977).

DISCUSSION

The explosive energy of melt-water interaction is primarily manifested by the violence of melt ejection. The efficiency of an explosion is governed by several parameters. Four important controls are water-melt mass ratio, confining pressure, the amount of water superheating, and the degree of water-melt mixing.

Energy Calculations

Rough estimates of the mechanical energy developed in an explosive reaction can be obtained by analyzing the photographic record. Knowledge of the total mass of ejecta and the size of the vent orifice, coupled with estimates of the ejecta velocity history, can be used to determine a value for the system's kinetic energy. Pressure-time records are of a tremendous aid in this analysis. The ratio of kinetic energy to the initial thermal energy is the system's efficiency and a means by which to compare the relative explosivity of experiments.

A simpler way of estimating the thermal efficiency of these systems can be done by utilizing the known thermodynamics and a few simplifying assumptions. The first assumption required is that the actual method of heat exchange between the melt and the water is not important. Moreover, the melt and water need not be in thermal equilibrium; however, if equilibrium is not obtained, only minimum values of thermal efficiency are calculated. Known quantities are the specific volume, V, of the hot H_2O prior to ejection and the pressure, P, before expansion takes place. Recalling that the efficiency, E, of the system is

$$E = \frac{\text{work output}}{\text{heat input}} \quad (12.2)$$

and the heat, Q, equals the internal energy, U, plus the work, W

$$Q = U_f - U_i + W, \quad (12.3)$$

where f and i denote the final and initial states of the system. For thermodynamic systems where S denotes the entropy and T the temperature

$$dQ = dU + PdV \quad (12.4)$$

and

$$dU = TdS - PdV. \quad (12.5)$$

If the hot H_2O (steam) expands adiabatically to $P = 1$ atm, the work done on the steam is just

$$-W = U_f - U_i. \quad (12.6)$$

This calculation does not consider possible heat transfer to the steam during the expansion process, which would produce a greater thermal efficiency.

For the four experiments where pressure records were ob-

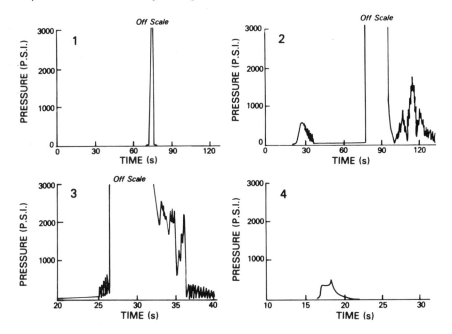

FIGURE 12.9 Pressure records for experiments 77-1, 77-2, 78-1, and 78-2 (Table 12.1), from which explosive efficiencies are calculated.

tained, 90 kg of thermite were used, which should have yielded approximately 7.9 × 10⁴ kcal of heat input for the system. Specific volumes of the H_2O estimated from the system's geometry permit the use of steam tables to determine internal energies. The results of these calculations are given in Table 12.2 and show maximum efficiencies obtainable for the measured pressures.

Controls of Explosiveness

Figure 12.10 is a plot of thermodynamic efficiency versus water-melt ratio for four thermite experiments. It shows that efficiency increases with mass ratio. The relationship is nearly linear for the limited range of ratios tested. Experiment 78-2 (Table 12.1) plots below the line, probably because of the lack of confinement. The isobaric lines show the expected slope of the efficiency curve (at $T_{vapor} = 500°C$) for each mass ratio if efficiency is not pressure dependent. The fact that the experimental efficiency increases more quickly than simple isobaric extrapolations indicates that the heat-exchange mechanism itself becomes more efficient with increasing mass ratios.

Confinement is obtained in three ways: physical (lithostatic or hydrostatic), acoustic, and inertial. These mechanisms allow pressure to build up in the system and thus influence the heat-transfer efficiency. Physical confinement maintains the pressure until the burst limit is exceeded. Acoustic confinement is short lived but results in large localized dynamic pressures until unloading occurs. Unloading takes place after the pressure wave has traveled to a free surface and is reflected back as an expansion wave. In this way, large dynamic pressures may exist for a time, the length of which depends on the sound velocity and the dimensions of the system. Inertial confinement arises through the momentum of the melt and water. In our experiments the melt drops into the water. This impact provides the initial perturbation necessary for mixing.

Explosive heat transfer results in maximum pressure rise over a few milliseconds (Sandia Laboratories, 1975). More recent studies (Reid, 1976; Buxton *et al.*, 1980) indicate even shorter time scales, on the order of microseconds. These values depend, of course, on the system volume. Conductive heat transfer, which is primarily a function of surface area, dominates in this rapid process. The experiments indicate that the size of ejected melt fragments decreases with increasing explosive energy (efficiency). But the melt surface area increases with decreasing grain size; hence, fragmentation of the melt and its mixing with water are required for explosive interactions. Vaporization energy is partitioned into ejection and fragmentation modes. The fragmentation process increases the melt surface area, ensures intimate contact with water, and promotes high thermal-energy transfer. Mechanisms explaining the fragmentation process are based on the creation of an initial, localized, pressure perturbation because of the expansion and collapse of a vapor film at the melt-water interface (Corradini, 1980; Wohletz, 1983). If the magnitude of the perturbation is sufficient in size to fragment a small part of the melt and cause mixing of the two fluids, the process is repeated, thereby fragmenting greater portions of the melt. Mixing, which results in increased surface area, promotes successive perturbation pulses (on a milli- or microsecond time scale). The magnitude of each pressure perturbation determines whether the next one will be larger. If the magnitude increases, the system grows in energy transfer exponentially. In this way each pressure perturbation promotes larger volumes of mixing, finer fragmentation, and increased heat-transfer rates. The process is limited when pressure pulses explosively disrupt the whole system. This self-sustained mixing has been called autocatalytic or dynamic mixing (Colgate and Sigurgeirsson, 1973) and can increase the heat-transfer rate by at least three orders of magnitude over normal boiling (Witte *et al.*, 1970).

Temperature calculations based on equilibrium saturation

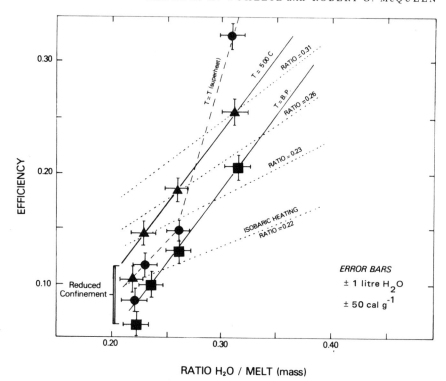

FIGURE 12.10 Plot of thermodynamic efficiency versus water-melt ratio. The solid lines connecting squares and triangles show efficiency curves for vaporization at the boiling point (BP) and for the added energy of heating vapor to 500°C, respectively. The dashed curve connecting dots shows efficiency of superheating. The dotted lines show isobaric extrapolations for efficiencies calculated at T = 500°C.

curve volumes are less than the minimum measured temperature of 500°C. This discrepancy and the molten nature of the ejecta indicate a possibility that thermal equilibrium is not reached at the explosion center. For nonequilibrium conditions water may be in the metastable state of superheat. Reid (1976) showed that superheating is to be expected, especially when a cold fluid is heated above its boiling point by a hot fluid. In the metastable state of superheating, where liquid water exists at temperatures well above its boiling point, the chemical potential for vaporization is less than that needed to overcome the confining pressure of the surrounding liquid. Many investigators (see Sandia Laboratories, 1975; Reid, 1976) have suggested that the limit of superheating, T_{sl} (the spontaneous nu-

cleation temperature), is the temperature that water must reach for explosive vaporization.

Calculated efficiencies based on the limit of superheating are given in Table 12.2. For Experiment 77-1 (above $P_{crit.}$), T_{sl} is extrapolated along the line given by Reid (1976) to the point of intersection with the maximum isobar. Figure 12.10 is a plot of calculated efficiencies of superheating compared with nonsuperheating systems. Again, experiment 78-2 plots below the line. The curve demonstrates the added efficiency of superheated systems in which maximum efficiency may occur at mass ratios near 0.35 to 0.70. A tentative conclusion on the effect of confinement is that confinement appears to increase the efficiency of equilibrium systems. For nonequilibrium systems,

TABLE 12.2 Experimental Efficiency Calculations

Experiment	Mass H_2O (kg)	Mass Thermite (kg)	Ratio	Burst P (bars)[a]	Vapor T (°C)[a]	Superheat T (°C)[b]	Efficiency[c] (1)	(2)	(3)
77-1	28	90	0.31	350	426	710	0.20	0.25	0.32
77-2	21	90	0.23	1/2(210)	368	370			
				1/2(70)	284	344	0.10	0.14	0.11
78-1	22	90	0.26	3/4(210)	368	370			
				1/4(140)	335	359	0.12	0.18	0.14
78-2	20	90	0.22	28	228	336	0.06	0.10	0.08

[a]T and P are time averaged for experiments 77-2 and 78-1.

[b]Superheat, T, is calculated from Reid (1976).

[c]Efficiency is (1) for vaporization at the boiling point, (2) for the additional enthalpy of heating the vapor to 500°C, and (3) for the enthalpy of superheating.

confinement appears to promote superheating and increase the efficiency of those systems (Reid, 1976).

Applications to Geologic Systems

Previous studies (Sandia Laboratories, 1975) conclude that the most explosive interactions occur when $R < 1.0$, which is in agreement with our preliminary data. Figure 12.11 shows a schematic diagram of explosive efficiency versus water-melt ratio. Divisions for Strombolian, Surtseyan, and submarine activity are postulated from experiments and field studies (Wohletz, 1979; Wohletz and Sheridan, 1983). For basalt with a lower melt enthalpy, maximum yield would be near $R = 0.1$ to 0.3. The sharp slope increase in efficiency near $R = 0.1$ is due to the onset of superheating and explosive fragmentation. The gradual efficiency decrease at R near 1.0 is due to the quenching by excess volumes of water, which hinders development of high pressures and superheating and, hence, results in lower heat-transfer rates.

A limitation in application of this model to natural hydromagmatic systems involves the contact geometry of the melt with water. As mentioned earlier, contact geometry is a control of explosivity. The geometry affects the initial surface area of melt-water contact and the tendency of the melt to fragment during contact. This effect has been qualitatively observed in the various experimental designs. In natural systems, magma rises into a surface water body or a water-saturated country rock, as pictured in Figure 12.1. Other less-common situations of contact are discussed by Wohletz and Sheridan (1983). The case of a surface body of water is most closely modeled by our present experiments; however, the second case, that of water-saturated country rock, involves some complications. Sufficient mixing ratios are achieved generally where groundwater is concentrated along a fault. Magma rising along that fault forms a tabular shape that may expose enough surface area with the structurally trapped water to initiate small vapor explosions. If the country rock is incompetent, such as alluvium or sand, the small initial explosions excavate the fault zone, thereby increasing the magma-water contact area. Field studies (Wohletz and Sheridan, 1983) show that eruptions beginning this way start out as low-energy ejections of dominantly fragmented country rock and subordinate juvenile material. As the zone of mixing increases in size, eruptions progressively increase in intensity as a result of greater fluxes of water contacting the rising magma. By this process, eruptions evolve from low-energy Strombolian to high-energy Surtseyan explosions. This process also operates in reverse—as was the case for Surtsey. Initial contacts of magma and water on the ocean floor result in high water-melt ratios. As pillow lavas build up around the vent and attain levels near the surface of the sea, ratios decrease and Surtseyan activity results. Eventually, water no longer had access to the vent, and Strombolian activity with passive lava emplacement characterized final eruptions.

CONCLUSIONS

Interaction of a basalt-like melt with water varies in explosivity from passive quenching of the melt to highly energetic fragmentations. A spectrum of explosive phenomena may be experimentally produced and compared with different kinds of hydromagmatic activity, including Strombolian, Surtseyan, and submarine volcanic activity. Furthermore, experiments indicate that the nature of activity and its explosiveness is primarily controlled by the water-to-melt mass ratio and confining pressure. Varying contact geometries also control the nature of melt-water interaction and ejection mode.

Small amounts (less than 10 mass percent) of water contacting

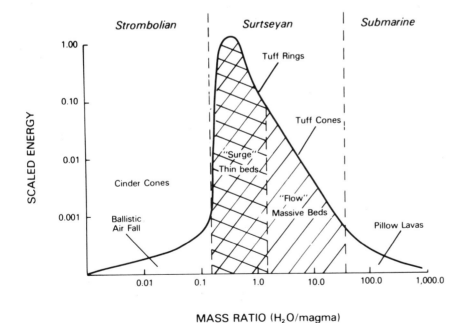

FIGURE 12.11 Plot of explosive energy versus water-melt ratio for volcanic systems. Energy is scaled to maximum yield. The sharp rise in the curve marks the onset of dynamic mixing and superheating. Surge eruptions are pulsating, high-energy, superheated steam explosions of dominantly fine ash. Flow eruptions are a more steady ejection of coarse tephra in saturated steam.

basalt produce Strombolian eruptions of dominantly centimeter-size fragments. Mass ratios of water and melt near 0.3 result in highly explosive Surtseyan blasts of millimeter and micron-sized material. Abundant amounts of water (greater than ratios of 2 or 3) generally result in passive chilling and the formation of pillow basalts.

Future experiments can be designed to model a variety of melt compositions, extending our results to more silicic compositions. Field studies (Sheridan and Wohletz, 1981; Sheridan *et al.*, 1981) show that hydromagmatic activity produces predictable eruption and ejecta transport phenomena depending on the amount of water present at the vent. Transport of pyroclastic material in water vapor is a function of the ejection rate, vapor-to-pyroclast ratio, and physical state of water in the ejecta cloud (saturated-condensing steam, superheated-expanding steam, and liquid water). These parameters can be experimentally varied, which provides a method for conducting similitude studies of ejecta emplacement. Studies could, for example, model the extent of blast or surge transport; emplacement of hot ejecta on ice, snow, or a standing body of water; topographic effects on ejecta flows; and development of geothermal systems in and around a volcanic vent.

This feasibility study has demonstrated the usefulness of these volcano simulation experiments. It is now apparent that the experiments can be documented better. In particular, our study has led to a new design (Wohletz and Sheridan, 1982), shown in Figure 12.12.

These new experiments are aimed at refining calculations for the efficiency curve and extending them over greater ranges of *R*. The new design allows quick restoration of the system after experimentation and use of smaller amounts (about 10 kg) of thermite. The vent is directed downward and the devise acts as a rocket that weighs about 250 kg. The mechanical

FIGURE 12.13   Experiment 80-5, series II, showing vertical lift as a measure of explosive energy.

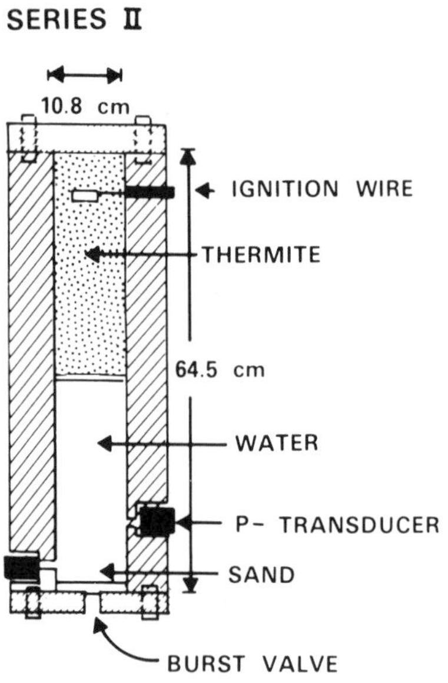

energy is calculated as a function of the acceleration and vertical lift measured using high-speed cinematography (see Figure 12.13).

ACKNOWLEDGMENTS

This work was initiated following the enthusiastic recommendations of Thomas McGetchin. Technical support and materials were provided by the shock-wave physics group, M-6, at Los Alamos National Laboratory. Michael Sheridan helped with the applications to volcanic problems. NASA provided partial support through grants NSG-7642 and NAGW-245.

SERIES II

10.8 cm

← IGNITION WIRE

← THERMITE

64.5 cm

← WATER

← P– TRANSDUCER

← SAND

← BURST VALVE

FIGURE 12.12   Series II experimental design for efficiency studies.

REFERENCES

Board, S. J., C. L. Farmer, and D. H. Poole (1974). Fragmentation in thermal explosions, *Int. J. Heat Mass Transfer* 17, 331-339.
Board, S. J., R. W. Hall, and R. S. Hall (1975). Detonation of fuel coolant explosions, *Nature* 254, 319-321.

Buxton, L. D., and W. B. Benedict (1979). Steam explosion efficiency studies, Sandia Laboratories, *SAND79-1399*, 62 pp.

Buxton, L. D., W. B. Benedict, and M. L. Corradini (1980). Steam explosion efficiency studies, part II, corium experiments, Sandia Laboratories, *SAND80-1324*, 36 pp.

Chouet, T. R., N. M. Hamicevicz, and T. R. McGetchin (1974). Photoballistics of volcanic jet activity at Stromboli, Italy, *J. Geophys. Res. 79*, 4961-4976.

Colgate, S. A., and Sigurgeirsson (1973). Dynamic mixing of water and lava, *Nature 244*, 552-555.

Corradini, M. L. (1980). Analysis and modelling of steam explosion experiments, Sandia Laboratories, *SAND80-2131*, 114 pp.

Dullforce, T. A., O. J. Buchanan, and R. S. Peckover (1976). Self-triggering of small-scale fuel-coolant interactions; I. Experiments, *J. Phys. Oil. Appl. Phys. 9*, 1295-1303.

Fisher, R. V., and A. C. Waters (1970). Base surge bed forms in maar volcanoes, *Am. J. Sci. 268*, 157-180.

Frazzetta, G., L. LaVolpe, and M. Sheridan (1983). Evolution of the Fossa cone, Vulcano, *J. Volcanol. Geotherm. Res. 17*, 329-360.

Fröhlich, A., A. Müller, and G. Unger (1976). Experiments with water and hot melts of lead, *J. Non-Equilib. Thermodyn. 1*, 91-103.

Heiken, G. H. (1972). Morphology and petrography of volcanic ashes, *Geol. Soc. Am. Bull. 83*, 1961-1988.

Jagger, T. A. (1949). *Steam Blast Volcanic Eruptions*, Hawaii Volcano Observ., 4th Spec. Rep., 137 pp.

Kieffer, S. W. (1977). Sound speed in liquid-gas mixtures: Water-air and water-steam, *J. Geophys. Res. 82*, 2895-2904.

Lipsett, S. G. (1966). Explosions from molten materials in water, *Fire Tech. (May)*, 118-126.

Livshits, L. D., and L. G. Bolkhoritinov (1977). Weak shock waves in the eruption column, *Nature 267*, 420-421.

McQueen, R. G., and K. H. Wohletz (in preparation). Experimental hydromagmatic volcanism.

Mercali, G., and O. Silvestri (1891). Le eruzioni dell'isola di volcano, incominciate il 3 Agosto 1888 e terminate il 22 Marzo 1890: Relazione Scientifica, 1891, *Ann. Uff. Cent. Meteor. Geodin. 10, No. 4*, 1-213.

Moore, J. G. (1967). Base surge in recent volcanic eruptions, *Bull. Volcanol. 30*, 337-363.

Nairn, I. A. (1976). Atmospheric shock waves and condensation clouds from Ngauruhoe explosive eruptions, *Nature 259*, 190-192.

Nairn, I. A., and S. Self (1978). Explosive eruptions and pyroclastic avalanches from Ngauruhoe in February 1975, *J. Volcanol. Geotherm. Res. 3*, 39-60.

Nelson, L. S., and P. M. Duda (1981). Steam explosion experiments with single drops of CO_2 laser-melted iron oxide, *Trans. Am. Nucl. Soc. 38*, 453-454.

Peckover, R. S., D. J. Buchanan, and D. E. Ashby (1973). Fuel-coolant interactions in submarine volcanism, *Nature 245*, 307-308.

Reid, R. C. (1976). Superheated liquids, *Am. Sci. 64*, 146-156.

Sandia Laboratories (1975). Core-meltdown experimental review, Sandia Laboratories, *SAND74-0382*, 472 pp.

Self, S., L. Wilson, and I. A. Nairn (1979). Vulcanian eruption mechanisms, *Nature 277*, 440-443.

Sheridan, M. F., and K. H. Wohletz (1981). Hydrovolcanic explosions, the systematics of water-tephra equilibration, *Science 212*, 1387-1389.

Sheridan, M. F., F. Barberi, M. Rosi, and R. Santacroce (1981). A model for plinian eruptions of Vesuvius, *Nature 289*, 282-285.

Thorarinsson, S. (1966). *Surtsey: The New Island in the North Atlantic*, Almenna Bokafelagid, Reykjavik, Iceland, 47 pp.

Walker, G. P. L., and R. Croasdale (1971). Characteristics of some basaltic pyroclastics, *Bull. Volcanol. 35*, 305-317.

Waters, A. C., and R. V. Fisher (1971). Base surges and their deposits, Capelinhos and Taal volcanoes, *J. Geophys. Res. 76*, 5496-5614.

Witte, L. C., J. E. Cox, and J. E. Bouvier (1970). The vapor explosion, *J. Metals 22*, 39-44.

Wohletz, K. H. (1979). Evolution of tuff cones and tuff rings, *Geol. Soc. Am. Abstr. Progr. 11*, 543.

Wohletz, K. H. (1983). Mechanisms of hydrovolcanic pyroclast formation: Size, shape, and experimental studies, *J. Volcanol. Geotherm. Res. 17*, 31-63.

Wohletz, K. H., and D. H. Krinsley (1983). Scanning electron microscope analysis of basaltic hydromagmatic ash, in *Scanning Electron Microscopy in Geology*, B. Whorley and D. Krinsley, eds., Geo Abstracts, Inc., Norwich, England.

Wohletz, K. H., and M. F. Sheridan (1981). Rampart crater ejecta: Experiments and analysis of melt-water interactions, *NASA Tech. Memo. 82325*, 134-135.

Wohletz, K. H., and M. F. Sheridan (1982). Melt-water interactions, series II, experimental design, *NASA Tech. Memo. 84211*, 169-171.

Wohletz, K. H., and M. F. Sheridan (1983). Hydrovolcanic explosions. II: Evolution of tuff rings and tuff cones, *Am. J. Sci. 283*, 385-413.

Volcanologists, Journalists, and the Concerned Local Public: A Tale of Two Crises in the Eastern Caribbean

13

RICHARD S. FISKE
Smithsonian Institution

INTRODUCTION

Explosive volcanism has long been a topic of great interest to scientists, as witnessed in part by the variety of reports contained in this volume. Even more obvious, however, is the fact that explosive volcanism has always been of vital concern to the people and governments immediately affected by such activity. When a potentially explosive volcano enters a period of crisis, interested scientists and a concerned, if not terrified, local community are brought into what euphemistically might be called a challenging relationship.

This paper is concerned with the challenging relationships that developed among scientists, civil authorities, and journalists during two volcanic crises that took place in the eastern Caribbean—on Guadeloupe in 1976 and on St. Vincent in 1979 (see Figures 13.1 and 13.2). Despite the close geographical proximity of these two islands and the similar geologic framework in which the erupting volcanoes are situated, the relationships between scientists and various aspects of the "real world" were markedly different. Brief consideration of these relationships can be instructive, not only to understand what happened in the eastern Caribbean during the 1970s but also to improve scientist/real-world relationships during the volcanic crises that are certain to occur elsewhere in the years ahead.

GUADELOUPE, 1976

When La Soufrière volcano on the eastern Caribbean island of Guadeloupe thundered to life on July 8, 1976, one of the most interesting chapters in modern volcanology began to unfold. By mid-August 1976, wire services and newspapers around the

world were carrying daily accounts of the worsening situation, as an apparent volcanic disaster approached. One French official flatly announced, "We have begun what we think is the countdown. The volcano cannot turn back." Fear spread rapidly from Guadeloupe, and people living as far away as Puerto Rico and Central America began to gird themselves for the onslaught of tidal waves that they thought would be triggered by a large and disastrous eruption. The anxious island, and indeed the world, awaited the coming paroxysm, which never occurred.

The first indication that La Soufrière was returning to life was provided by an increase in the number of tiny earthquakes that originated beneath the volcano. During a normal month, 1 to 10 earthquakes might have been recorded, but in July 1975 the number of events jumped suspiciously to 30. After returning to normal levels for several months, the earthquake count in November 1975 surged to 209. In March 1976, 607 local earthquakes were recorded, a number of them large enough to be felt by local inhabitants. The seismicity remained at abnormally high levels through the spring and early summer of 1976, but at about 9 A.M. on July 8, 1976, the situation took a dramatic turn for the worse.

The volcano, which up to that point had not actually erupted, suddenly began to belch forth steam and ash. Billowing clouds of sooty grey ash boiled from fissures at the summit of the volcano, and the prevailing winds carried this material to the west, showering it on the city of Basse Terre, the capital of Guadeloupe. The roaring emission of steam and ash was accompanied by continuous shaking of the ground, which added to the consternation of nearby residents. As the ash clouds became more dense, the mid-day Sun was almost completely obscured, and an eerie pall settled over the leeward flank of the volcano. Terrified residents, having almost to feel their way through the murky air, jumped into cars and trucks and some-

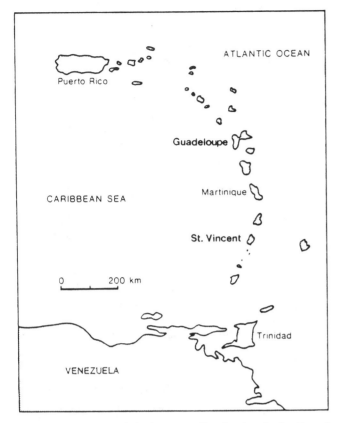

FIGURE 13.1 Map of the lesser Antilles showing the location of Guadeloupe, Martinique, and St. Vincent.

how made their way down the side of the volcano. Many of those who clung to the outside of these vehicles were completely covered by the sticky gray ash, and they were said to resemble living statues when they arrived at the foot of the volcano—unhurt but thoroughly terrified.

The tempo of activity at La Soufrière continued to increase. Steam issued from the summit of the volcano continously, and larger quantities of gray ash erupted on several occasions. The number of local earthquakes continued to rise—1220 events in July, almost 6000 in August. Increasing numbers of these quakes jolted the ground and were easily felt by people living on the flanks of the volcano.

On August 12, 1976, scientists studying La Soufrière issued a statement that unquestionably influenced the course of events on Guadeloupe. They announced that the ash, which was being erupted from the volcano in ever increasing amounts, now contained a significant component of fresh volcanic glass. The clear implication was that new magma had risen to high levels within the volcano and that tiny bits of this material were being carried upward to the surface along with the plumes of steam that were continuously boiling into the sky. Because this glass had not been identified in the ash prior to August 12, its appearance seemed to indicate that the volcano was about to enter a very dangerous phase of activity.

In this atmosphere of grim expectation and on the advice of one group of French scientists, the *prefet* (governor) of Guadeloupe, Jean Claude Arroseau, ordered the immediate evacuation of all 72,000 people living on the slopes of the volcano (see Figure 13.3). This order triggered a series of events that embroiled scientists and civil authorities in arguments and confrontations that grew more bitter as the evacuation continued for almost 4 months. Not known at the time, however, was a fact that would have colored, if not altogether altered, the course of events in those days of increasing apprehension. The identification of the fresh volcanic glass in the erupted ash—which seemed to indicate that the volcano was about to erupt violently—was a mistake. For all intents and purposes, fresh volcanic glass was not contained in the ash, and there was in fact no clear evidence that magma was about to erupt to the surface. The authorities on Guadeloupe, of course, did not realize this at the time, and so the volcano was thought to be far more dangerous than it actually was.

The evacuation was carried out with considerable efficiency. Within 2 days about one quarter of the island's population, 72,000 people, were ordered from the flanks of the volcano to other parts of the island, far from the areas of danger. Besides imposing an obvious hardship on the displaced people, the evacuation resulted in severe political and economic problems. Massive amounts of financial aid were required from France in order to provide shelter and food for the evacuees, and tax revenues collected by the government of Guadeloupe were sharply reduced because of the drop in the island's productivity. But perhaps almost as bad was the virtual cessation of tourism on Guadeloupe. The word that La Soufrière was about to erupt had been spread far and wide in both North America and Europe, and tourists responded by staying away.

The stillness of the evacuated cities and towns surrounding the volcano contrasted dramatically with the bustle of scientific activity on the slopes of the volcano. Most of the scientists present were French, grouped into two main teams, but other geologists and geophysicists from Trinidad and the United States also arrived. At times, 20 to 25 scientists were present, many of whom had never worked on the slopes of an active volcano. A volcano observatory was established deep inside the dungeon-like chambers of the seventeenth-century Fort St. Charles, on the southwest flank of the volcano, and a highly sophisticated volcano watch was put into operation. The network of seismometers, installed to monitor local seismicity, was expanded. Recording magnetometers were deployed and were beginning to yield intriguing information on internal changes of the volcano.

Volcanic gases were collected and analyzed. Precise leveling surveys (the so-called dry tilt technique) were initiated to detect the possible inflation of the volcano that often precedes eruptions. But despite the diversity of this scientific program, it is important to note that most of these monitoring techniques were not deployed on La Soufrière until July or August 1976—about a year after the initial buildup had begun. Thus, because there were inadequate baseline observations against which telltale changes could be compared, there was always an imperfect understanding of the physical changes that may have taken place within the volcano during the early stages of its activity. Had these early changes been monitored, it is possible that

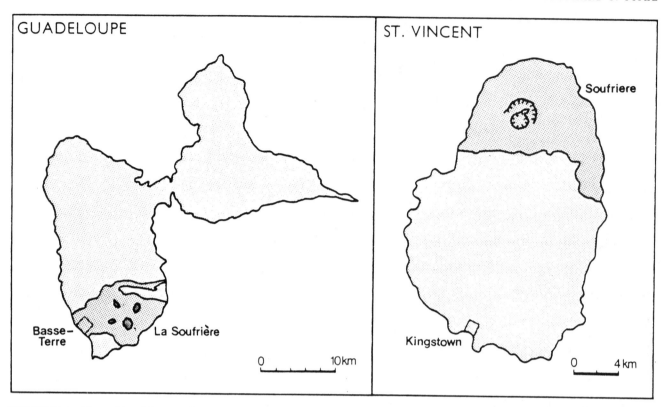

FIGURE 13.2 Maps of Guadeloupe and St. Vincent. La Soufrière and Soufriere volcanoes are shown in the denser stipple pattern. The capital cities of Basse-Terre and Kingstown also are shown.

the actual hazard posed by the volcano might have been far more accurately assessed.

In some ways, August 30, 1976, might be regarded as the climax of the eruption. At that time, just 2 weeks after the evacuation, several hundred earthquakes were being recorded every day, and a plume of gritty ash-laden stream drifted almost constantly from the top of the volcano. On the morning of August 30, 12 scientists climbed to the summit to inspect the activity. Shortly after 10 A.M., just after the party had stepped back from the gaping fissure at the top of the volcano, steam and ash began to explode to the surface, traveling at such high velocity that jagged blocks up to 1 m in diameter were carried high into the air. This debris fell back to Earth, narrowly missing the scientists as they scrambled to safety. Four members of the group were struck glancing blows by smaller fragments, and, had it not been for the hard hats strapped firmly to their heads, their injuries would have been far more serious.

This explosive emission of steam and debris was certainly impressive to those who had the misfortune to view it at close quarters, but from a volcanological point of view it represented a rather trivial outburst. However, for an hour or two after the event, and especially during the period when the extent of injuries sustained by the summit party was still uncertain, journalists sent out the news that the volcano had entered a serious episode of eruptive activity. In North America, due to garbling of the story after it had been transmitted from Guadeloupe, the American public was informed in front-page headlines that

La Soufrière had at last delivered up the long-awaited eruption. Misled and enticed by the grossly exaggerated stories, a Miami-based team of newsmen scrambled into a chartered jet and raced to Guadeloupe—only to be told, during the course of a filmed interview with me at 1 A.M., that the day's eruption, while providing momentary excitement, actually constituted a trivial volcanic event and in no way was to be confused with the expected "big bang."

From that time onward, North American journalists behaved as if they had somehow been jilted by La Soufrière. Frustrated by the absence of a catastrophe and no loss of life and property, they relegated the story to the inside pages of newspapers and then dropped it altogether. From a North American point of view, the ongoing crisis simply ceased to exist.

In Guadeloupe, and in metropolitian France, this was hardly the case. About 72,000 citizens were still in a state of evacuation, Guadeloupe's economy was in serious trouble, and the volcano was still belching forth steam and ash. It was in this highly charged atmosphere that disagreements between the two main teams of French scientists intensified and spilled into the public arena, via newspaper and newsmagazine articles, television debates, and at least one full-length book.

One team supported the evacuation and maintained that the volcano indeed posed a continuing hazard. The other team maintained from the outset that the volcano posed little danger and that there would be adequate advance warning if the situation took a turn for the worse. In this atmosphere of scientific

FIGURE 13.3 Deserted street in Basse-Terre, Guadeloupe, during the evacuation of 1976. The smoldering summit of La Soufrière is seen in the background.

controversy, civil authorities did not know whom to believe, and the evacuation rapidly became a nightmare.

In my judgment there is no question that the situation was exacerbated by the fact that journalists seemed to have unrestricted access to the disagreeing scientific teams during this period of crisis. Even though an evacuation was supposedly under way, hardly a day passed without television or radio crews appearing at the scientific headquarters at Fort St. Charles. Scientists were interviewed singly or in groups, and the journalists eagerly sought out differences in opinion regarding the state of the volcano and the hazard that it posed. In many instances, more media attention was focused on the disagreements among the scientists than on what was actually happening at the volcano.

As the weeks dragged by, into September and October 1976, there was increasing agitation to allow the 72,000 people to return to their homes. To complicate matters, it was becoming clear by mid-October that the tempo of the eruption was beginning to slow; the number of recorded earthquakes was di-

minishing, and the volume of ash reaching the surface was noticeably reduced. But the volcano was still showing signs of activity, and some of France's most respected scientists were locked in fierce public disagreement over the handling of the crisis. In this difficult moment, what could possibly be the rationale for sounding an "all clear" and ending the evacuation?

In what must be considered an unusual, if not unprecedented, move, authorities in Paris decided to seek the advice of foreign scientists, and an ad hoc *Comité Scientifique International sur La Soufrière* of six non-French scientists was convened in Paris in November 1976. Two Americans were named to the *Comité*, then MIT Professor Frank Press (who served as chairman) and myself. Others included Franco Barbari and Paulo Gasparini from Italy, Gudmundur Sigvaldason from Iceland, and Shigeo Aramaki from Japan. The *Comité* heard statements from various French scientists who had been involved with the La Soufrière situation, and at the end of 3 days prepared a report stating that the volcano, while still requiring close monitoring, appeared to pose less of a hazard at that time. Within hours of receipt of this report, Overseas Territories Minister Oliver Stirn announced an end to the evacuation.

As people returned to their homes and everyday life gradually returned to normal on Guadeloupe, the volcano appeared to oblige by becoming even more quiet. The number of local earthquakes and the abundance of steam emissions continued to diminish, and by early 1977 it was over. The crisis had come and gone; the disaster never occurred.

ST. VINCENT, 1979

It came as a considerable surprise that just 3 yr later, in April 1979, the Soufriere Volcano of St. Vincent burst into activity to initiate another volcano crisis in the eastern Caribbean. In contrast to La Soufrière, Guadeloupe, there was absolutely no doubt that the Soufriere Volcano of St. Vincent posed a hazard to the surrounding populus. From April 13 to 26 a series of powerful vertical explosions blasted through the lake and island that formerly occupied the crater of the volcano and sent clouds of debris to heights as high as 20 km (see Figure 13.4). Small pyroclastic flows and mud flows were triggered by this activity, and they poured down several of the valleys that head on the volcano. By the end of April 1979 the explosive phase of the eruption ended and a new phase began, characterized by the quiet emission of a dome of basaltic andesite on the floor of the summit crater. This dome continued to show measurable growth until October 1979.

In contrast to Guadeloupe, where relatively few baseline geophysical measurements had been made prior to 1976, a number of such preeruption measurements had been made on St. Vincent. During the 1970s a seismic network was established in the lesser Antilles by the staff of the Seismic Research Unit of the University of the West Indies, and seismic signals from St. Vincent and other islands in the lesser Antilles were routinely telemetered to Trinidad for real-time monitoring and interpretation. In 1977, 25 months prior to the eruption, 2 dry tilt stations were established on the eastern flank of Soufriere, St. Vincent, and these stations provided the first indications

FIGURE 13.4 Eruption cloud rises from Soufriere, St. Vincent, on April 22, 1979.

FIGURE 13.5 Dry tilt measurements being made on St. Vincent in March 1977, more than 2 yr before the 1979 eruption. Data obtained from this station helped confirm that the volcano was inflating.

that the volcano was beginning to inflate (see Figure 13.5). Routine temperature measurements were made in the crater lake of the volcano during the 1970s, and these gave telltale indications before April 1979 that the volcano was behaving anomalously.

When the volcano exploded to life in the early morning hours of April 12, the 22,000 people living on the slopes of the volcano did not have to wait for an order to evacuate. Remembering the story of the disastrous 1902 eruption of Soufriere, St. Vincent, when more than 1500 islanders were killed, people living on the slopes of the volcano removed themselves quickly and efficiently from the areas of danger (see Figure 13.6). The government of St. Vincent enforced a continuation of this evacuation throughout the explosive phase of the eruption, while the volcano obviously posed a hazard.

Only 5 to 7 scientists were working on St. Vincent during the explosive phase of the eruption, and these people constituted a single group or team. Most came from the scientific staff of the Seismic Research Unit of the University of the West Indies, although the team was augmented occasionally by sci-

entists from the United States and England. This team supplied a single stream of information to the government of St. Vincent, via telephone conversations and written communications. This information was straightforward in content, represented only one concensus of events at the volcano, and offered a minimum of speculation as to what might happen.

Especially noteworthy was the way in which civil authorities on St. Vincent dealt with the journalists who rushed to the island when the eruption began. Stated simply, the journalists were absolutely prohibited from entering the danger zone. Because the hurriedly established volcano observatory was located only 9 km from the volcano, well within the area of evacuation, the journalists were not able to visit the observatory or to interview scientists in the field. The members of the press corps received information about the volcano from government officials in the capital city of Kingstown, who passed along informaton that had been supplied to them by the scientific team. To my knowledge, only one journalist insisted on being allowed to enter the evacuated zone. He had indicated that, if

FIGURE 13.6 Residents fleeing areas of danger near Soufriere, St. Vincent. The populace responded spontaneously to the hazards posed by the volcano.

such permission was not granted, he would write stories describing the censorship of information on St. Vincent rather than the events that were taking place at the volcano. I am not aware of the details of what transpired between this journalist and the St. Vincent government, but we later learned that he was deported summarily.

DISCUSSION

There were, of course, many differences between the 1975-1977 crisis at La Soufrière, Guadeloupe, and the 1979 crisis at Soufriere, St. Vincent. Among these, however, the following points (see Table 13.1) seem especially pertinent when considering the relationships between scientists, journalists, and the public.

There is little doubt that the volcanic crisis was far more challenging on Guadeloupe than it was on St. Vincent. If more scientific information had been available on Guadeloupe, in the form of a geologic and geophysical data base, the anomalies observed there in 1976 could have been assessed in a better way. In addition, the relatively slow buildup of activity at Guadeloupe and the lingering threat of violence allowed all too much time for apprehensions and expectations to grow. There was nothing inherently disadvantageous in having many more scientists on the scene in Guadeloupe and having the scientists grouped into two main teams. Nor was there anything wrong with the fact that the scientific teams on Guadeloupe had differing interpretations as to what was likely to happen at La Soufrière. The main problem was that these teams were not communicating effectively with each other, and they were espousing markedly different opinions directly to competing journalists. Because the journalists seemed to have total and unrestricted access to all levels of the local scientific hierarchy, disjointed and often conflicting media coverage resulted. In the cacophony that surrounded the entire Guadeloupe affair, the fact was largely overlooked that the citizens of Guadeloupe, and especially the 72,000 people who had been ordered from their homes during the evacuation, were being presented with a very confusing account of the ongoing situation—an account that was laced with invective between some of the scientists involved. This made a very bad situation much worse than it really had to be.

Considering these aspects of the Guadeloupe and St. Vincent experiences, the communication between scientists, journalists, and the public during a future volcanic crisis—not simply an erupting volcano but also one that poses a demonstrable threat to nearby populations—could be improved if the following actions were taken:

1. Gather some amount of geologic and geophysical information about threatening volcanoes *before* a crisis begins, so that scientists can have a basis for assessing abnormal patterns of volcanic behavior. Of course, it would not be feasible to maintain a program of continuous monitoring at all volcanoes of this type, but modest programs involving the intermittent gathering of baseline data, such as patterns of local seismicity, ground deformation, or thermal emissions, can yield enormous dividends, especially at potentially dangerous volcanoes located near population centers. Realizing in many cases that the "past

TABLE 13.1 Differences Between Hazard Response During the 1975-1977 La Soufrière (Guadeloupe) and the 1979 Soufriere (St. Vincent) Volcanic Crises

GUADELOUPE	ST. VINCENT
Less complete data base	More complete data base
Slow buildup of activity	Rapid buildup of activity
Threatened violence	Demonstrated violence
20 to 25 scientists	5 to 7 scientists
Several scientific teams	One scientific team
Differing interpretations	One interpretation
Poor communication between working scientists	Good communication between working scientists
Unlimited access for journalists to working scientists	No access for journalists to working scientists

is prologue," there is also great advantage in knowing earlier patterns of eruptive behavior by means of preeruption geologic studies as a guide to what is likely to happen in the future.

2. Once a crisis evolves, have an experienced chief scientist who, while not suppressing scientific disagreements, would attempt to coordinate the activities of the scientists involved into a single group effort, to increase communication between the scientists, and to help ensure that a single and complete stream of information is made available to civil authorities and journalists. Crises in developed parts of the world will often attract teams of workers from more than one organization, and it should be the responsibility of the chief scientist to stimulate the activities of the various teams and to make them feel part of the group effort. A volcano in crisis needs the attention of many different specialists working together and communicating freely. A certain amount of redundancy in the scientific arena is not necessarily undesirable, because scientists will be gathering large amounts of information in a short period of time. Competition between scientific teams for media attention, however, is simply irresponsible. Such competition can lead to biased and incomplete news accounts, which almost certainly will heighten the anxieties of an already apprehensive population.

3. Working scientists—those who gather the data that become part of the body of scientific understanding—should realize that their responsibilities often extend far beyond their individual scientific activities, especially with regard to their interactions with journalists. In general, individual scientists should acknowledge that they do not necessarily have a balanced view of all of the scientific investigations under way at a particular crisis. Thus, when interviewed by a journalist seeking an angle or exclusive aspect of a story, both the scientist and the journalist should realize that undue emphasis on the part of the story that happens to be the specialty of that particular scientist can lead to unbalanced, and in many cases misleading, reporting. Added to this is the all-too-human temptation of the scientist to pontificate far beyond the confines of his or her true expertise—in response to the somehow intox-icating influence of having the undivided attention of a journalist. I have found myself doing this on more than one occasion and have been shocked at the results in the following day's newspaper.

4. Journalists must realize that they should not feel entitled to monitor scientists as they communicate with each other during times of crisis. Scientists need to discuss among themselves a multitude of speculative possibilities, and the most efficient way to do this is in their own jargon, without concern for misinterpretation by nonspecialists. The mere presence of a journalist at such discussions limits free communication between scientists, which is vital during a volcanic crisis.

5. If the crisis is of such proportions that numerous journalists are attracted to the scene, an information officer (ideally drawn from the scientific ranks) should be appointed. Such a person should work closely with the chief scientist to ensure that a single and complete stream of information is made available. This person should not suppress scientific disagreements that might exist between working scientists but should express them freely in terms of the limits of overall scientific understanding. It is common for scientists to disagree, and such disagreements should be reported publicly in a balanced and nonpersonalized way.

In conclusion, it is worth emphasizing that the two examples of volcanic crises cited here, Guadeloupe in 1976 and St. Vincent in 1979, clearly represent extremes with regard to scientist/journalist interactions. Few people would endorse either the open-door policy of Guadeloupe or the closed-door policy of St. Vincent. What we are left with, therefore, is the search for the most appropriate middle ground. In times of a genuine volcano crisis, we must seek to maximize the flow of information from scientists to the public in a way such that journalists (who must play an essential role) can enhance the process. It is up to us to provide this flow of information in a balanced, forthright, and timely way. If there is any lesson to be learned by the scientific community from the volcanic crises in the eastern Caribbean during the 1970s it is that the challenge posed by the scientist/journalist relationship is mostly ours to solve.